D1321186

An Introduction to Biotechnology

An Introduction to Biotechnology

The science, technology and medical applications

W. T. Godbey
Tulane University
New Orleans, Louisiana

AMSTERDAM • BOSTON • HEIDELBERG • LONDON
NEW YORK • OXFORD • PARIS • SAN DIEGO
SAN FRANCISCO • SINGAPORE • SYDNEY • TOKYO
Academic Press is an imprint of Elsevier

Academic Press is an imprint of Elsevier
32 Jamestown Road, London NW1 7BY, UK
525 B Street, Suite 1800, San Diego, CA 92101-4495, USA
225 Wyman Street, Waltham, MA 02451, USA
The Boulevard, Langford Lane, Kidlington, Oxford OX5 1GB, UK

British Library Cataloguing in Publication Data
A catalogue record for this book is available from the British Library

Library of Congress Cataloging-in-Publication Data
A catalog record for this book is available from the Library of Congress

ISBN: 978-1-907568-28-2

For information on all Academic Press publications
visit our website at elsevierdirect.com

Typeset by SPi Global, India

14   15   16   17   10 9 8 7 6 5 4 3 2 1

Working together
to grow libraries in
developing countries

www.elsevier.com • www.bookaid.org

Contents

List of Figures

List of Tables

Preface

This book was written for the student with little to no biology background. The goal of the book is to introduce the student to the world of biotechnology but in a way that runs deeper than a mere survey. There are a myriad of biotechnologies in the world today, and the number continues to grow. I happen to find the world of biotechnology a very exciting world, and I want to share that excitement with you, the reader. However, to fully appreciate just how cool some of these technologies are, one must understand some of the science underlying the glamour.

This is not intended to be a comprehensive book on all biotechnologies. One could spend the better part of a decade becoming an expert in only one area. Likewise, while I state that one must understand the science behind a technology to fully appreciate it—and healthy portion of that science is covered in this book—the text is not intended to be a complete, rigorous reference book for the biological or chemical sciences. Instead, it was my aim to produce a book that serves as a mix of both basic science and biotechnological applications so that you, the reader, might become energized about part or all of the field on a level that is deep enough to allow you to continue further pursuits of it with the knowledge that a solid foundation has been laid. If you still have a passion for one of the subjects contained in this book after studying it, that passion is probably quite real. I want you to get enough foundation so that you can appreciate what is out there in the world of biotechnology, so that you will understand new developments in greater depth than what you might hear in the news, and so that you will not be fooled by unsubstantiated claims you might read on the Internet.

The book is divided into three units. In the first, basic science is covered to introduce the reader to the cell, how it behaves, and what it is made of. For instance, if you want to design a drug that you want to enter the cell, you must know and understand the barrier that separates the inside of the cell from the exterior environment. If you want that drug to affect how the cell is behaving, then you should understand how a cell functions so that you can pick a cellular target upon which your drug can act. Perhaps, you want the cell to produce a product that you can isolate and sell, such as recombinant insulin, ethanol, or a novel protein that you have designed. To be successful in this endeavor, you must understand how the cell would go about producing the product, in addition to knowing exactly what your product would be. While we do not cover every possible product that could be produced by a cell, we do cover some of the building blocks such as amino acids/proteins and nucleic acids/DNA/RNA.

Godbey, 978-1-907568-28-2

The second unit in the book is aimed at the biotechnological application of scientific principles in the laboratory. In the first unit, we will see how things work, but in the second unit, we will see how we can employ such knowledge to design, produce, and analyze products on a laboratory scale. These applications have allowed for the identification of pathogens so that appropriate antibiotic selections can be made, the production of cell cultures for the study of investigational drugs or engineered tissue constructs, and the amplification of DNA or RNA for the engineering and construction of entire genes.

The third unit of the book presents biotechnologies "in the real world." This should not imply that the laboratory is not the real world! Certainly, the laboratory makes up the primary world of many biotechnologists. The unit title refers to technologies that are used for practical purposes to aid nonscientists. Examples include recombinant proteins that are available to millions of patients, plants that have been engineered to produce food that has been made available to people around the world, and regenerative medicine that may someday allow patients to receive organs that have been grown from their own cells. From fighting crime to removing fingernail polish to powering automobiles without the use of fossil fuels, biotechnology is being applied in wide applications that affect the lives of millions of people every day.

When I first undertook the writing of this book, I was under the impression that biotechnology was a new and cutting-edge field. While the cutting-edge part is correct, upon citing important references for the book, I came to realize that the information is not necessarily what is commonly referred to as "new." Sure, there are famous experiments from the 1950s that we learn in school, but it was a little surprising when I found myself citing a paper from 1948 and then heavily relying upon a referenced paper from 1852. However, when I came across texts that predated the Bible, my outlook on biotechnology changed from it being a new field to it being a timeless field. Progress in and applications of biotechnology perhaps occur at a greater rate these days, but it is entirely incorrect to think that the only "real" science has occurred in recent times. We may utilize more sophisticated equipment today, but the science that took place in, for example, G. G. Stokes' laboratory in the mid-1800s was just as real (and the mathematics just as complex) as what is performed today.

Finally, this book was written primarily for the student, not the teacher. The style is informal throughout, which will undoubtedly irritate some professors. Not to worry, they simply will not adopt the book for their course. However, the text is written in a style that is easy to follow, perhaps a little light-hearted in places, with numerous figures to reinforce the material that is being presented. The material is solid, however. The student and teacher can both feel confident that, just because the material may be entertaining and easy to understand, the topics covered in the book are based on decades of personal experience, education, and consultation with experts in the individual topics.

About the Author

Prof. W. T. Godbey has been working with biotechnology as a bioengineer for over 17 years. He has performed research at the Texas Medical Center while a PhD student at Rice University and then at Children's Hospital, Boston, as a postdoctoral research associate at Harvard Medical School. He is currently a professor at Tulane University in New Orleans, Louisiana.

Godbey has started two businesses, one being a biotechnology company, and is the primary inventor on a patented technology for a targeted cancer therapeutic. He has worked directly with most of the technologies presented in this book, including gene therapy, tissue engineering, regenerative medicine, stem cells, fermentation, biofuels, and genetically modified organisms. His current research interests include the use of gene therapy for carcinoma treatment, controlled release applications for efficient gene delivery, and the use of gene delivery for cellular engineering.

Chapter 1

Membranes

The biotechnologist will inevitably have to deal with cells at some point. Whether an extracellular matrix is being isolated to serve as a tissue engineering scaffold, a gene is being delivered to treat a genetic disease, or a biofuel is being produced as a type of renewable energy, cells will be involved during the research and development process (if not the final application of the technology itself). A basic understanding of the cell is therefore critical to the development of biotechnologies.

We will begin our survey of the cell with (one of) the exterior layer(s) of the cell—the plasma membrane. At the cellular level, this is the first point of contact for technologies such as immunotherapy, gene delivery, and patterned cell attachment. When designing a particle that is to be taken up by a cell, a thorough knowledge of the plasma membrane is advisable.

1.1 MEMBRANE LIPIDS

Both eukaryotic and prokaryotic cells are surrounded by a plasma membrane (Figure 1.1). The plasma membrane is not the outermost layer of a prokaryotic cell, but this point will be addressed in detail later in this chapter. The plasma membrane is made up of lipids, carbohydrates, and proteins, and various combinations thereof, but the primary constituent is the phospholipid. There are many different types of phospholipids, but the ones used in the plasma membrane all follow some basic principles.

Membrane phospholipids contain a hydrophilic head and two hydrophobic tails, and as such are said to be amphipathic. Phospholipids can be considered as three constituents linked to a glycerol backbone via reactions with each of the three hydroxyls. The structure of glycerol is given in Figure 1.2.

Of course, there are exceptions, but the following rules of thumb help to describe the structure of a phospholipid built off of a glycerol backbone:

1. The first hydroxyl will have been used to react with the C-terminus of a fatty acid to form an ester linkage. The fatty acid is typically 14-24 carbons long, with an even number of carbons. The fatty acid will be saturated.

An Introduction to Biotechnology. http://dx.doi.org/10.1016/B978-1-907568-28-2.00001-0
Copyright © W T Godbey, 2014.

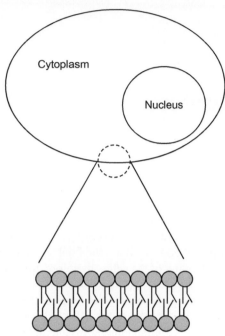

FIGURE 1.1   The eukaryotic plasma membrane.

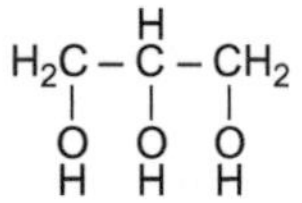

$$H_2C - \overset{\overset{\displaystyle H}{|}}{C} - CH_2$$
$$\underset{\underset{\displaystyle H}{|}}{O} \quad \underset{\underset{\displaystyle H}{|}}{O} \quad \underset{\underset{\displaystyle H}{|}}{O}$$

FIGURE 1.2   The structure of glycerol, the backbone of the phospholipid structure.

2. The second hydroxyl will also be attached to a fatty acid with an even number (14-24) of carbons in an ester linkage. The fatty acid differs from the first one in that it is typically unsaturated. The unsaturated fatty acid will be kinked, which results in less dense packing in the membrane and therefore a lower freezing temperature.

3. The third hydroxyl will be linked to a phosphate to form an ester bond. The arm of the molecule that contains the phosphate group will constitute the hydrophilic portion of the phospholipid. If this arm has only a phosphate group, the lipid is phosphatidic acid (Figure 1.3). However, additional moieties can be attached to the phosphoryl group to give rise to the range of phospholipids found in the plasma membrane.

Three of the most common phospholipids of the plasma membrane are phosphatidylethanolamine, phosphatidylserine, and phosphatidylcholine. Figure 1.4 gives general structures of these phospholipids, but it can also be seen that there is not one single formula for a given phospholipid. The compositions of the two

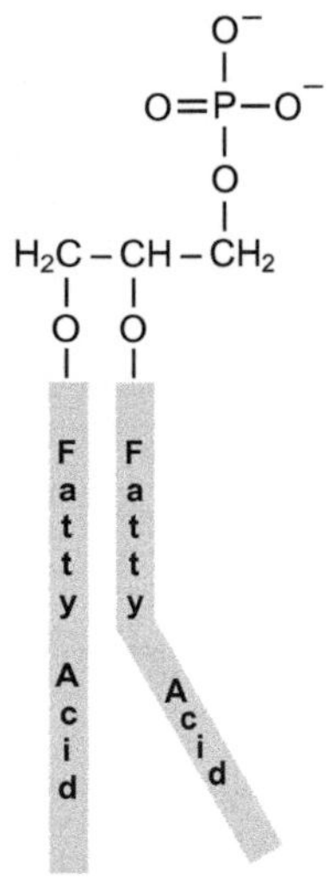

$$
\begin{array}{c}
O^- \\
| \\
O=P-O^- \\
| \\
O \\
| \\
H_2C-CH-CH_2 \\
| \quad | \\
O \quad O \\
| \quad |
\end{array}
$$

F a t t y	F a t t y
A c i d	A c i d

FIGURE 1.3   Phosphatidic acid.

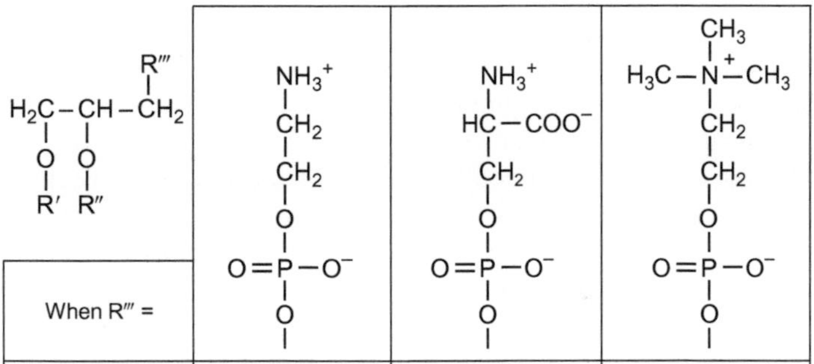

$\begin{array}{c} R''' \\ | \\ H_2C-CH-CH_2 \\ | \quad | \\ O \quad O \\ | \quad | \\ R' \quad R'' \end{array}$	$\begin{array}{c} NH_3{}^+ \\ | \\ CH_2 \\ | \\ CH_2 \\ | \\ O \\ | \\ O=P-O^- \\ | \\ O \\ | \end{array}$	$\begin{array}{c} NH_3{}^+ \\ | \\ HC-COO^- \\ | \\ CH_2 \\ | \\ O \\ | \\ O=P-O^- \\ | \\ O \\ | \end{array}$	$\begin{array}{c} CH_3 \\ | \\ H_3C-N^+-CH_3 \\ | \\ CH_2 \\ | \\ CH_2 \\ | \\ O \\ | \\ O=P-O^- \\ | \\ O \\ | \end{array}$
When R''' =			
Name	Phosphatidyl-ethanolamine	Phosphatidyl-serine	Phosphatidyl-choline
Abbreviation	PE	PS	PC
Overall charge	Neutral	Negative	Neutral
Found in which layer of PM	Inside	Inside	Outside

FIGURE 1.4   Structures, charges, and general locations of three of the most common phospholipids in the plasma membrane.

fatty acids attached to the glycerol backbone are not fixed, which means that there can be several different phosphatidylcholines, for example.

There are two additional types of lipids that are common members of the plasma membrane that bear mention. The first is sphingomyelin, a phospholipid that has a choline head group but differs from phosphatidylcholine in that instead of a glycerol backbone holding two fatty acid tails, the hydrophobic

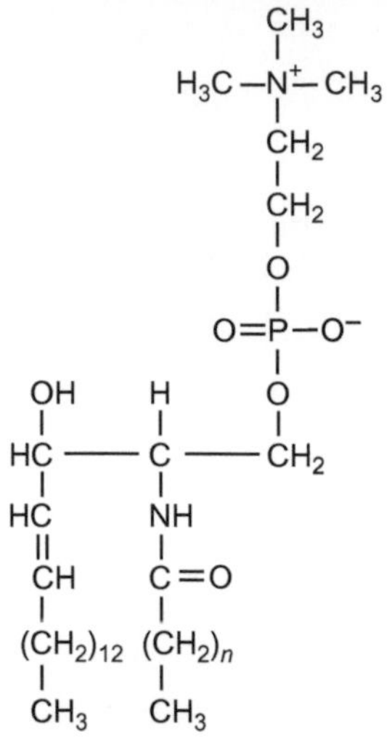

FIGURE 1.5   Sphingomyelin.

portion of the molecule is ceramide (Figure 1.5). As the name implies, sphingomyelin is found in the myelin sheath that surrounds the axons of certain nerve cells, but it also makes up a significant percentage of the plasma membranes of cells such as liver and red blood cells.

Glycolipids also bear mention because of their prevalence in certain plasma membranes. Regarding the plasma membranes of animal cells, they are found only on the exterior surface, which implies that they may have some function in cell-cell interactions, cellular identification, or cell signaling. Although the glycolipid composition of the plasma membrane differs greatly between cells of different species or even between tissues in the same animal, it is believed that all cells contain at least some glycolipids in the exterior face of their plasma membranes. This is a feature that has yet to be fully capitalized upon in fields such as stem cell engineering or tissue engineering, although the medical science and immunology disciplines have recognized the glycolipid signatures of various microbes and have developed treatments accordingly.

The plasma membrane is asymmetrical due to the preferential locations of the above phospholipids to one side of the membrane or the other. As a general rule, the phospholipids with a terminal amino group tend to be on the inner (cytoplasmic) face of the plasma membrane (PE and PS), while those with a choline in their head group (PC and sphingomyelin) are typically found on the external face. Glycolipids are effectively always on the external face (see Figure 1.6a). These guidelines do not dictate the absolute positions of membrane constituents, however—the plasma membrane is a fluid structure. Molecules on the outer face can generally move about like people on the second floor of a crowded mall, and molecules on the inner face can move around like people on the first floor. Some

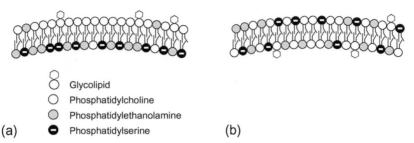

Glycolipid
Phosphatidylcholine
Phosphatidylethanolamine
(a)   Phosphatidylserine   (b)

FIGURE 1.6   (a) A representative distribution of membrane phospholipids. (b) During apoptosis, scramblase randomizes phospholipid distributions. The redistribution of PS can be detected by other cells, such as macrophages.

lipids will stay together (in lipid rafts) much like how a family might stay together in the mall, while other lipids can change the side of the membrane they are on like somebody taking the escalator. Such translocation can be seen in specific circumstances, such as when a membrane is being actively formed by the addition of new lipids, as takes place in the endoplasmic reticulum. Enzymes known as phospholipid translocators cause the phenomenon, which is known as flip-flop. Another example of phospholipid flipping occurs during the process of programmed cell death known as apoptosis. During the early stages of this complex process, an enzyme known as scramblase randomizes phospholipid locations relative to which side of the membrane they usually reside (Figure 1.6b). One result is that PS will be flipped to the exterior face of the plasma membrane. The sudden appearance of these negatively charged phospholipids on the exterior portion of the cell is thought to serve as a signal to macrophages to engulf and degrade the apoptotic cells.

1.2 CHOLESTEROL

Cholesterol is another constituent of the plasma membrane. The structure of cholesterol is given in Figure 1.7. Although it is not a phospholipid, cholesterol has a similar structure in that it has a hydrophilic portion (the OH group) and a hydrophobic region (the rest of the molecule). Cholesterol fits between adjacent phospholipids in the plasma membrane, with the hydroxyl aligning itself with polar head groups and the rest of the molecule fitting in among the fatty acid tails (Figure 1.8). Cholesterol can be found in high abundance in animal cells, sometimes at concentrations of one cholesterol molecule for every phospholipid molecule. While many people are aware that high cholesterol content in the blood is considered a bad thing, cholesterol is nevertheless a necessary component of the cell membrane of every cell because of the several functions it serves. First, the four steroid rings are relatively rigid, which serves to decrease the fluidity of the membrane and increase its stability. With the hydroxyl group being positioned as it is, the rings will be firmly placed in the area of the glycerol backbones and first several carbons of the fatty acid tails of adjacent phospholipids. This effectively immobilizes those carbons of the adjacent

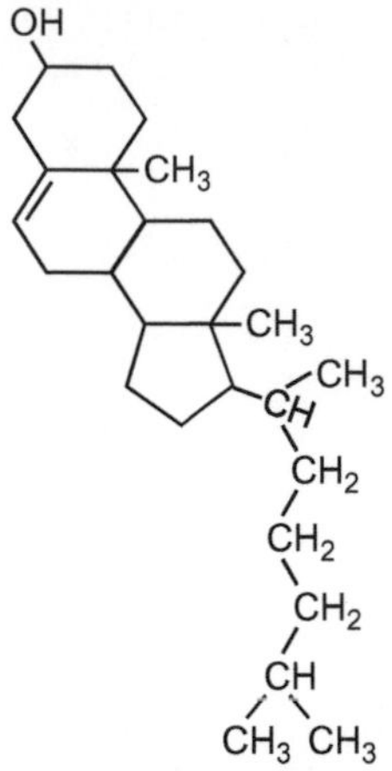

OH

CH₃

CH₃

CH₃

CH

CH₂

CH₂

CH₂

CH

CH₃ CH₃

FIGURE 1.7   The structure of cholesterol.

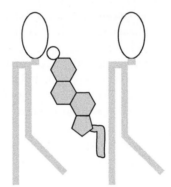

FIGURE 1.8   Cholesterol fits between adjacent phospholipids.

phospholipids and renders the membrane less fluid. Second, while making the membrane less fluid, cholesterol also serves to make the membrane more flexible. While the lower-numbered carbons of phospholipid fatty acid tails are held in relative rigidity, the higher-numbered carbons (the furthest from the glycerol backbone) are free to gyrate. With cholesterol serving as a spacer, there is less chance for the carbons at the end of the tails of one phospholipid to associate with the tails of adjacent phospholipids. As a result, fatty acid tails are prevented from coming together to form ordered, crystalline structures, as will happen during phase transitions at lower temperatures (freezing).

1.3   MEMBRANE PROTEINS

A discussion of the plasma membrane would not be complete without a presentation of membrane proteins. However, membrane proteins cannot be adequately described without an introduction to proteins in general. Entire careers have been devoted to proteins, and many wonderful books exist devoted to the subjects of protein structure and function. While an exhaustive introduction to proteins is not appropriate for an introduction to biotechnology, a rudimentary knowledge of proteins is indeed required when learning about the structure and function of the cell. We will begin our survey in the next chapter with the building block of the protein: the amino acid.

QUESTIONS

1. What does cholesterol do for the cell membrane?
2. A phospholipid head has a nitrogen group attached to four carbons. Can it be found on the inside or outside layer of the plasma membrane, or both? Explain your answer.
3. Describe the usual structure of a phospholipid built off of a glycerol backbone.
4. What is the most common molecule found in the plasma membrane?
5. Why do phospholipids in solution form micelles?
6. Which of the above phospholipids might be a common constituent in the cell membrane? For the ones that are not expected to be common, give a reason why not.

7. Why is the charge on phosphatidylserine useful?

Chapter 2

Proteins

2.1 AMINO ACIDS

Amino acids are so named because they contain both an amino group and a carboxylic acid. The structure of an amino acid at neutral pH (pH = 7.0) is given in Figure 2.1. Disregarding the side group R for a moment, the figure shows that the molecule contains both a positive charge and a negative charge, making it a *zwitterion*. A zwitterion is a molecule that has both positive and negative regions of charge. In the common amino acids, the carbon that connects the amino and carboxylate groups will be chiral if the R group is anything except a hydrogen atom. Chirality is used to describe the property of certain molecules to rotate polarized light, with L-forms rotating polarized light to the left (counterclockwise) and D-forms rotating it to the right. Examples of L- and D-forms of an amino acid are given in Figure 2.2. Note that the two forms are mirror images of one another. While chirality may be thought of as interesting only to the organic chemist, it is important to the biotechnologist as well. In the case of amino acids, it is the L-form that is used in constructing proteins such as enzymes, receptors, and signaling peptides. D-form amino acids are sometimes used, with varying success, to produce biologically inert molecules.

While all amino acids have the basic structure shown in Figure 2.1, the identity of the R group defines each specific molecule. There are 20 amino acids commonly found in nature to form proteins, each with a defined R group. These amino acids are shown in Figure 2.3.

After examining Figure 2.3 and the accompanying figure caption, a couple of points may need further clarification. First, notice that aspartate and asparagine are grouped together, as are glutamate and glutamine. The difference between each pair of side groups is that the negatively charged oxygen has been replaced by an amine group (NH_2), which is hydrophilic. These amino acids are paired together because (1) aspartate and asparagine are very similar chemically, as are glutamate and glutamine, and (2) they are all hydrophilic. Appearing on the second page of the figure are the positively charged amino acids, which are hydrophilic, followed by the aromatics, which tend to be hydrophobic. A case can be made that tyrosine can be considered hydrophilic because of the hydroxyl group on the aromatic ring. This is the reason that the one-letter code

An Introduction to Biotechnology. http://dx.doi.org/10.1016/B978-1-907568-28-2.00002-2

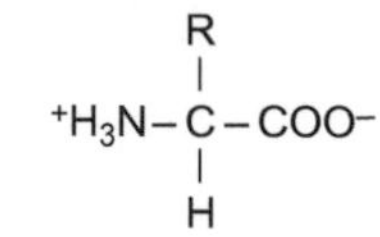

$$\text{}^+H_3N - \underset{\underset{H}{|}}{\overset{\overset{R}{|}}{C}} - COO^-$$

FIGURE 2.1   The structure of an amino acid at neutral pH.

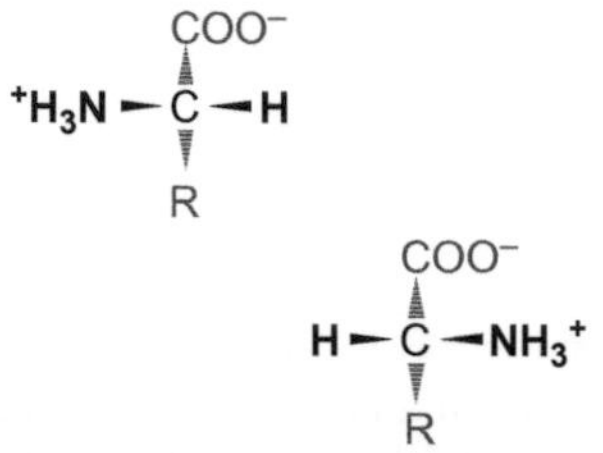

$$\text{}^+H_3N - \underset{R}{\overset{COO^-}{C}} - H$$

$$H - \underset{R}{\overset{COO^-}{C}} - NH_3{}^+$$

FIGURE 2.2   L-(top) and D-forms of the same amino acid. Heavier fonts indicate atoms coming toward the reader out of the plane of the page, while lighter fonts indicate atoms oriented away from the reader below the plane of the page.

"Y" is not circled for tyrosine. Note that the other hydroxyl-containing amino acids are also hydrophilic.

Cysteine is a special amino acid in that its side group contains a sulf-hydryl group. In a protein, oftentimes, cysteines will occur in pairs. The paired cysteines do not have to be next to each other (or even close to each other) in terms of the order of amino acids in a polypeptide. However, when cysteines pair in three dimensions as a part of protein folding, they will be in close proximity and form a disulfide bond. The pair of bonded cysteines will be known as a cystine (note the spellings). The disulfide bonds can be broken via reduction, which means that acidic environments or proton do-nors can convert cystines back into a pair of cysteines. As we will see, this is a property one must contend with in the laboratory, especially in terms of identifying the primary sequence of a protein.

Even though biotechnology students might at first be overwhelmed with the prospect of learning the structures of all 20 common amino acids, they should nevertheless strive to acquaint themselves with the structures and functions of these molecules. At the very least, they should have a working knowledge of the different classes of amino acids as they delve into different biotechnical applications. For instance, if one were interested in looking at specific amino acids for their propensity to bind with DNA or perhaps serve as a delivery ve-hicle for siRNA, negatively charged side groups (D and E) should be avoided because they would be repelled by the negative charges in the DNA or RNA molecules. Similarly, using a polypeptide made up entirely of nonpolar and un-charged amino acids (G, A, V, L, I, P, F, W, and M) for the purpose of a signaling

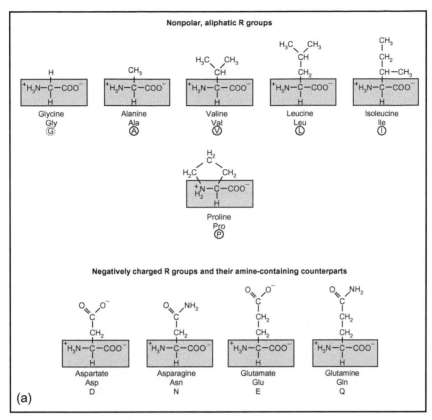

FIGURE 2.3 Structures, names, three-letter abbreviations, and one-letter codes of the common 20 amino acids used to form proteins. These formulas reflect the prevailing ionization states at pH=7.0. The shaded portions of the structure are the entities common to each amino acid and would form the backbone in a polypeptide; the R group of each amino acid is unshaded. Because the pK_a of the histidine R group is relatively close to 7.0, both ionization states are shown (with a hydrogen and positive charge in parentheses). The unionized form of the histidine R group will predominate at pH=7.0. Circles around single-letter codes indicate nonpolar side groups. The circle around G is dashed because, although the small hydrogen side group is most easily classified as nonpolar, it contributes very little to hydrophobic interactions. Also note that C is considered polar here, although the side group is commonly bound to another C via a disulfide linkage, which is strongly nonpolar.

(Continued)

molecule to be delivered via the blood would not be advisable because the molecules would be hydrophobic and therefore not soluble in the aqueous environment of the blood. Any sort of protein engineering should be accompanied by the knowledge of which amino acids are small, large, inflexible, charged, polar, hydrophobic, or cross-linkable.

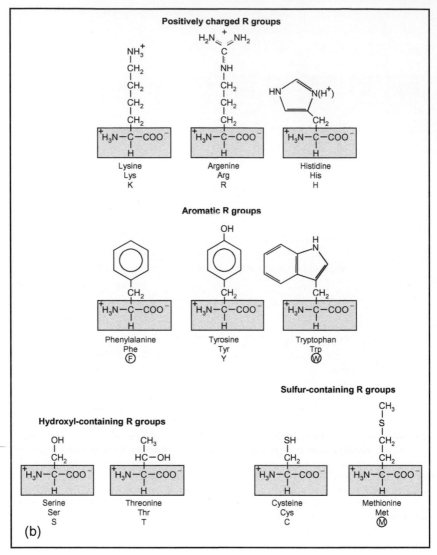

FIGURE 2.3, cont'd

2.1.1 pK_a

Consider a pH scale and the side group for aspartate ($-CH_2-COO^-$). Some will often refer to this amino acid as aspartic acid, which has the side group $-CH_2-COOH$. The naming convention has the "acid" form as the form with its complete set of protons. When a proton is lost and the species carries a negative charge, we refer to this as the "ate" form, hence "aspartate," as opposed to "aspartic acid."

Not all acids ionize at the same pH. A way to characterize different acids with respect to the pH at which they will tend to lose a proton is known as the $p\mathbf{K_a}$ **value**. Without deriving it, the $p\mathrm{K_a}$ value is related to the dissociation constant. The definition of $p\mathrm{K_a}$ is the pH at which an ionizable species is 50% ionized. Considering the side group for aspartate/aspartic acid, the $p\mathrm{K_a}$ is 3.65. This implies that if one has a beaker of aspartate/aspartic acid in an aqueous, buffered solution held at pH 3.65, half of these amino acid molecules will be aspartate and half will be aspartic acid. What this tells us is that for aspartate/aspartic acid at physiological pH (7.2 inside the cell), the pH is a good distance from the $p\mathrm{K_a}$ value on the number line, so virtually all of these molecules will be in the form of aspartate. As a general rule, a "good distance" in these cases generally refers to being at least one pH unit away from the $p\mathrm{K_a}$. The $p\mathrm{K_a}$ value of aspartic acid being 3.65 does not imply that at pH = 3.66, all of these molecules will be in the form of aspartate; one would expect that slightly more than 50% would be. As the pH is increased further from the $p\mathrm{K_a}$ value, a greater percentage of the side groups would be in the "ate" form over the acid form. In the body, changes in pH by 0.2 units can have profound effects on protein folding.

When deciding on the ionization state of a species, a handy rule to keep in mind is that at pH values below the $p\mathrm{K_a}$ for that species, there will be a relative abundance of protons relative to the species, and it will carry as many protons as possible. At pH values above the $p\mathrm{K_a}$ for the species, there will be a relative shortage of protons relative to the species, so it will tend to release protons into the solution. In the case of carboxylic acids, low pH will mean that the species is in its acid (–COOH) form. At pH = 1, the amino acid Asp would be in the form of aspartic acid. The same is true for Glu, for which the side group has a $p\mathrm{K_a}$ of 4.25, so the amino acid would be in the form of glutamic acid.

The same rules apply to species that ionize to carry a positive charge, such as the side group for lysine ($-CH_2 - CH_2 - CH_2 - CH_2 - NH_3^+$). In this case, the side group has an ionizable amine, with a $p\mathrm{K_a}$ value of 10.53. This means that at pH = 10.53, half of these amines will be $-NH_2$ and half of them will be $-NH_3^+$. At physiological pH, since it is a good distance below the $p\mathrm{K_a}$ value for the Lys side group, the amine will carry as many protons as possible and be in the form $-NH_3^+$.

Carrying the general ionization rule one step further, at "low" pH, where there is a relative excess of protons and the ionizable groups will carry as many protons as possible, the ionizable species will carry the highest charge that they can. For an amino acid such as aspartate, low pH means that the side chain will carry a charge of 0 instead of -1. For an amino acid such as lysine, low pH will cause the side chain to carry a charge of $+1$ instead of 0.

Using the $p\mathrm{K_a}$ values for every ionizable species in the molecule, one can determine the *isoelectric* point. The isoelectric point is the pH at which the predominant net charge of a species of a molecule in a solution is zero. It's a term typically applied to proteins and polypeptides.

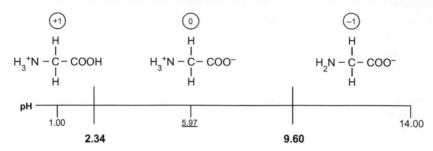

FIGURE 2.4   Ionization states of glycine at different pH values. Each ionizable species has a distinct pK_a value, shown on the number line in bold. The isoelectric point for glycine, which is the average of the two pK_a values, is underlined. Circled numbers above the structures indicate the predominant net charge of the molecule in that pH range.

Consider glycine, which is the amino acid for which the side group is a hydrogen. At physiological pH, it can be described with the formula $H_3{}^+N-CH_2-COO^-$. Note that the N-terminus and the C-terminus can each be ionized, which implies that each terminus has its own pK_a value. In fact, this is true and the pK_a values are 9.60 and 2.34 for the N- and C-termini, respectively. From these two values, the ionization state of the entire molecule can be written for any given pH value (Figure 2.4). At a low pH such as 1.0, each of the termini will carry as many positive charges as possible, so the structure of glycine will be $H_3{}^+N-CH_2-COOH$ and the overall charge of the molecule will be +1. If the pH were to be gradually increased, after the lowest pK_a value is crossed, then the corresponding molecular group will change to the ionization state carrying a charge one lower than before. For glycine, the lowest pK_a value (2.34) corresponds to the C-terminus, which will lose a proton to yield $H_3{}^+N-CH_2-COO^-$. The overall charge of the molecule is now 0 (neutral), and the molecule will tend to retain the same state until the pH reaches the next pK_a value (9.60). At this point, the N-terminus will lose a proton and the molecule will have the structure $H_2N-CH_2-COO^-$, with a net charge of −1. To determine the isoelectric point, simply take the average of the pK_a values that surround the pH range where the charge of the molecule is zero. For glycine, we would take the average of 2.34 and 9.60 to obtain an isoelectric point of 5.97.

As another example, consider lysine. For this amino acid, in addition to N- and C-termini, there is a side group that can be ionized. The pK_a value for the N-terminus is 8.95, while it is 2.18 for the C-terminus. (The pK_a value for any ionizable group depends upon the structure of the entire molecule, which explains why the N- and C-termini have pK_a values that are different from those of glycine.) The side group of lysine has a pK_a value of 10.53. The ionization states of this molecule with respect to pH are shown in Figure 2.5. Note that, once again, the molecule will hold as many protons as possible at low pH. When the pH is gradually raised, as each pK_a value is crossed, one proton will be lost from the species that corresponds with the pK_a value just crossed. Likewise,

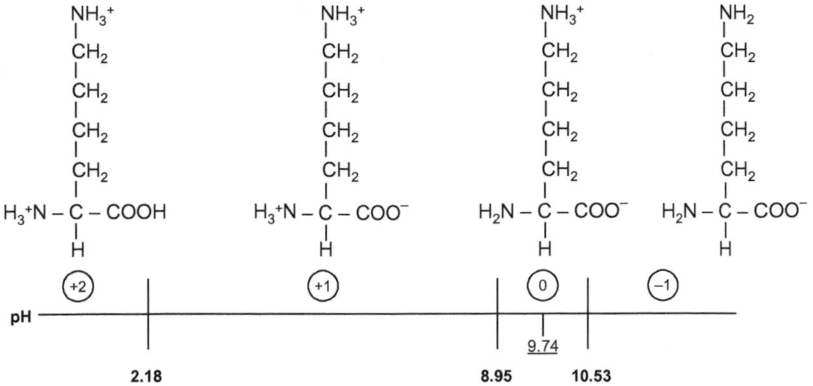

FIGURE 2.5   Ionization states of lysine at different pH values. To determine the isoelectric point (pI), take the average of the two pK_a values (shown in bold) that bound the pH range where the net charge$=0$. Circled numbers indicate the predominant net charge of the molecule in the given pH range. In this case, pI$=9.74$ (underlined).

as pH is raised, the overall charge of the molecule will go down by 1 as each pK_a value is crossed, coinciding with the loss of a proton. Even though there are more than two pK_a values for lysine, determining the isoelectric point is straightforward as long as one is aware of the different ionization states of the molecule for each pH. Applying the same rule that was used for glycine, the isoelectric point is the average of the two pK_a values that surround the pH region for which the net charge of the molecule is neutral. For lysine, the isoelectric point is $(8.95 + 10.53)/2 = 9.74$.

Once you figure out what the overall charge of the molecule is at a very low pH, every time you cross a pK_a value, it reduces the overall charge by 1. Using this rule, it should be fairly straightforward to determine the isoelectric point of a polypeptide with hundreds of ionizable side groups. Figure out how many pK_a's must be crossed to obtain an overall charge of zero, and then, take the average of the two bounding pK_a values. To find the overall charge of the molecule at low pH, it's simply a matter of counting the number of groups that ionize to a positive value. For lysine, there are two groups that ionize to a positive value, meaning that the overall charge of the molecule at pH 1 is equal to $+2$, which means that two pK_a values must be crossed to bring the molecule to a neutral state, which means that the isoelectric point is going to be the average of the second and third pK_a values (as ordered by increasing value on the number line).

2.2   PROTEIN STRUCTURE

2.2.1   Primary Structure

Amino acids can be polymerized to form a polypeptide, which gets its name from the peptide bonds formed by the loss of water during polymerization

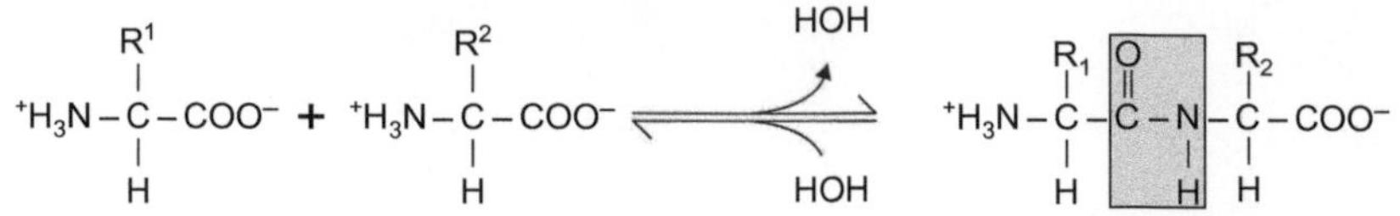

$$^+H_3N - \underset{\underset{H}{|}}{\overset{\overset{R^1}{|}}{C}} - COO^- \ + \ ^+H_3N - \underset{\underset{H}{|}}{\overset{\overset{R^2}{|}}{C}} - COO^- \ \underset{HOH}{\overset{HOH}{\rightleftharpoons}} \ ^+H_3N - \underset{\underset{H}{|}}{\overset{\overset{R_1}{|}}{C}} - \overset{\overset{O}{\|}}{C} - \underset{\underset{H}{|}}{N} - \underset{\underset{H}{|}}{\overset{\overset{R_2}{|}}{C}} - COO^-$$

FIGURE 2.6   The polymerization of two amino acids results in the loss of a water molecule and the formation of a peptide bond, shown by the shaded box.

(Figure 2.6). The peptide bond is rigid and planar but can be broken by a water molecule (hydrolyzed) in the reverse direction of the reaction shown. When forming peptide bonds in this way, notice that there will still be a primary amine on one side of the polymer backbone and a carboxyl group on the other, termed the N-terminus and C-terminus, respectively. It is customary to draw a polypeptide in the N → C direction.

Rather than draw a polypeptide, one can imply its structure through proper naming. Using the same convention of beginning with the N-terminus, a pentapeptide made of phenylalanine, histidine, lysine, isoleucine, and threonine would be called phenylalanylhistidinyllysinylisoleucinylthreonine. What a great way to win bets over who knows the longest word! However, naming polypeptide sequences in this manner is not very practical. One way to write the ordered sequence of amino acids, or primary sequence, of a polypeptide more clearly is to use the three-letter codes: Phe-His-Lys-Ile-Thr. A convention used for longer sequences is to use the one-letter codes, still adhering to the N-terminus always appearing first. One would write the pentapeptide in this example as FHKIT.

2.2.2   Secondary Structure

The secondary structure of a protein describes the spatial arrangement of the atoms in the protein backbone. The repeating pattern of α-carbon peptide bond can exist in a disorganized array (called a random coil) or in a distinctly well-defined manner, with the angles of the two planar peptide bonds attached to each α-carbon repeating in a regular fashion. The most common secondary structures are α-helices, β-pleated sheets, and β-turns. The α-helix is a right-handed helix with amino acid residues spaced at 3.6 residues per turn and a rise of 0.54 nm per turn. (A "right-handed helix" is wound such that if one were to curl the fingers of the right hand around the backbone of the helix with the fingers pointing in the N → C direction, the thumb would point along the helical axis in the direction of N → C helix progression.) Figure 2.7 illustrates approximate amino acid placement and rise for an α-helix. Notice that amino acid 0 is not directly below the third or fourth subsequent residue. If the coil were a number line, amino acid 0 would be beneath 3.6, which is where "3.6 residues per turn" comes from. Because of this, amino acid 0 can form hydrogen bonds with the third or fourth subsequent residue, as is shown in the second panel of the figure. The side groups of each amino acid would point out from the helix (not shown).

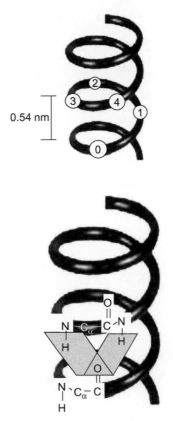

0.54 nm

FIGURE 2.7   Placement and interactions of amino acids in an α-helix. *Top*: The helix has 3.6 residues/turn and a pitch of 0.54 nm/turn. *Bottom*: The structure is made stable by internal hydrogen bonds: a peptide oxygen can interact with a peptide hydrogen three or four residues away.

The β-pleated sheet is composed of multiple, relatively straight segments within one polypeptide, with the strands lining up adjacently (Figure 2.8). It is the zigzag nature of the N–C_α–C backbone that gives the sheet its pleats. The sheet is stabilized by hydrogen bonding between peptide oxygens and amide hydrogens on a neighboring segment. The straight segments can be parallel, meaning that the N → C directions of all β-sheet segments point the same way or antiparallel, meaning that the segments alternate in direction. Antiparallel β-sheets have hydrogen bonds perpendicular to strand directions, while parallel sheets have hydrogen bonds occurring at alternating angles. Both forms of the β-pleated sheet have amino acid side groups occurring parallel and on the same side of the sheet. In the figure, take note of the N → C directions, hydrogen bonding, and alignment of the side groups for both parallel and antiparallel configurations.

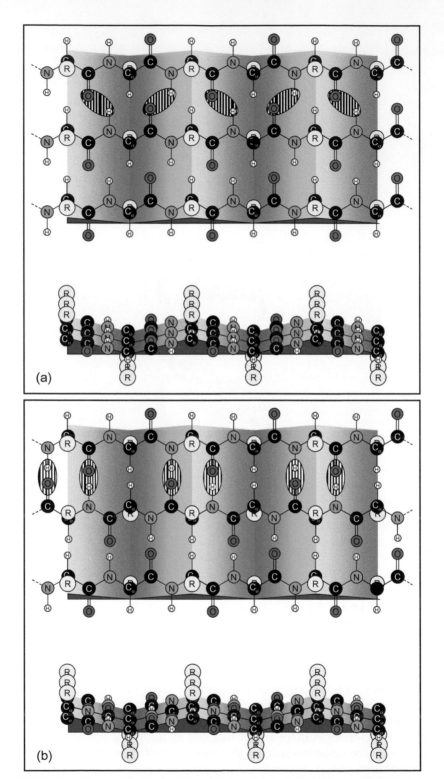

FIGURE 2.8 Placement and interactions of amino acids in β-pleated sheets. Hydrogen bonds between the first two strands are circled. (a) parallel β-sheets. (b) antiparallel β-sheets.

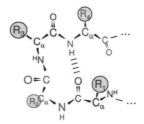

FIGURE 2.9 Four amino acids make up a β-turn. Amino acids #1 and #4 lend stability to the turn through a hydrogen bond, which is shown. The majority of β-turns are type 1, which means that they have a proline residue in position #2. Another common turn, type 2, is defined by residue #3, which is a glycine.

A β-turn (also known as a β-bend) is a structure that allows for sharp turns in the middle of a polypeptide. One way in which β-turns appear in proteins is during the formation of β-sheets. A β-turn is made up of four amino acid residues. Residues 1 and 4 often interact as illustrated in Figure 2.9. The amino acids glycine and proline are often present in β-turns. Glycine works well in a beta turn because of its small side group (–H) that is sterically favorable for the compact structure, rendering the residue relatively flexible. (Keeping this in mind, one might also correctly predict that alanine would be a relatively common amino acid in β-turns.) Proline works well because of some interesting properties, which merit deeper discussion.

Proline is a unique amino acid in that its side group is attached to its backbone atoms in two places. The fact that the side group attaches to both the chiral carbon and the amino nitrogen of this amino acid renders proline a rigid amino acid. As such, it is not often found in α-helices. Prolines are found in β-turns, though. There are two different conformations for proline (Figure 2.10). In the *trans* conformation, the polypeptide backbone will be somewhat straight. In the *cis* conformation, the backbone makes a sharp turn. Because the β-turn itself is a sharp turn, the *cis* conformation is overwhelmingly favored for prolines involved in β-turns.

In summary, for β-turns, amino acids 1 and 4 interact with each other, and glycine and proline are commonly present—glycine because it is small and proline because it is rigid and can naturally form this sharp turn.

2.2.2.1 Supersecondary Structure

Supersecondary structure is simply a combination of secondary structures. For instance, consider a piece of paper. If the piece of paper were rolled into a tube, it would still be a piece of paper, but it would have been modified into a more complex structure. The same can be done to a β-sheet, and in this case, the resulting structure would be a β-barrel. Another form of a supersecondary structure is the helix-loop-helix, where two α-helices are connected by a β-turn. Any combination of secondary structures can be used to produce a supersecondary structure.

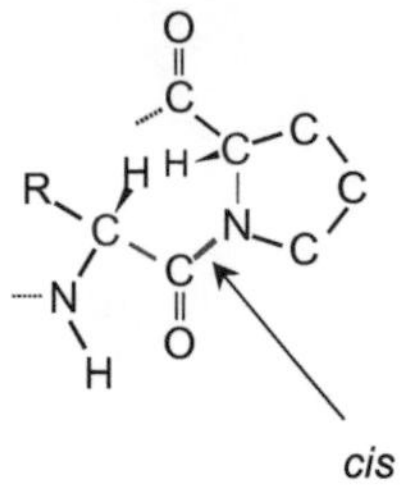

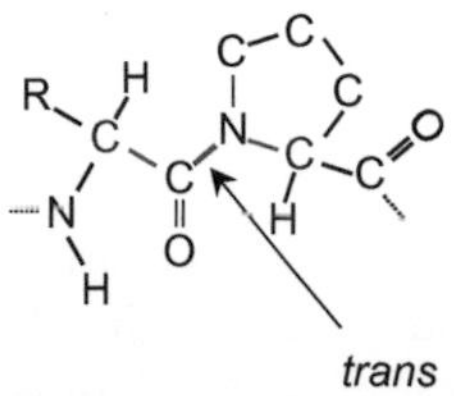

FIGURE 2.10 Proline in *cis* and *trans* conformations. Alpha carbons are shown in blue, arrows point to peptide bonds. It is the arrangement of the three bonds in red (the positions of the two alpha carbons relative to the peptide bond) that determine the conformation.

2.2.3 Tertiary Structure

Tertiary structure is the next level of complexity in protein folding. Tertiary structure is the three-dimensional structure of a protein. While individual amino acids in the primary sequence can interact with one another to form secondary structures such as helices and sheets and individual amino acids from distant parts of the primary sequence can intermingle via charge-charge, hydrophobic, disulfide, or other interactions, the formation of these bonds and interactions will serve to change the shape of the overall protein. The folding that we end up with for a given polypeptide is the tertiary structure.

2.2.4 Quaternary Structure

Quaternary structure is the interaction of two or more folded polypeptides. Many proteins require the assembly of several polypeptide subunits before they become active. If the final protein is made of two subunits, the protein is said to be a dimer. If three subunits must come together, the protein is said to be a trimer, four subunits make up a tetramer, etc. If the subunits are identical, the prefix "homo" is used, as in "homodimer." If the subunits are different, we use "hetero," as in "heterodimer."

Hemoglobin is the protein responsible for carrying oxygen in the blood. It is made up of four polypeptides: two α- and two β-subunits. One α-subunit and one β-subunit will come together to form a heterodimer, and two of these heterodimers will interact to form one hemoglobin molecule. Hemoglobin can therefore be thought of as a dimer of dimers, which come together to give the final protein its quaternary structure.

2.3 THE HYDROPHOBIC EFFECT

It was presented earlier that some individual amino acids can carry charges, while others are not charged. At physiological pH, the negatively charged amino acids are not charged. At physiological pH, the negatively charged amino acids are aspartate and glutamate, while the positively charged amino acids are lysine, arginine, and histidine. It is the interaction of amino acid side groups that plays the pivotal role in protein folding. Hydrophobic amino acids, including alanine, valine, leucine, isoleucine, and phenylalanine, tend to fold toward the interior of a protein because of the *hydrophobic effect*. Oil and water do not mix, or, in other words, hydrophobic and hydrophilic molecules do not mix well. Proteins within the cell must exist in the hydrophilic environment of the cytoplasm, even though hydrophobic amino acids within these proteins do not mix well in the aqueous environment of the cell.

The hydrophobic effect is the primary driving force behind protein folding. It is directly related to the thermodynamic property of a system known as the Gibbs free energy. Interactions and reactions occur in ways that tend to minimize the Gibbs free energy, and processes that have a negative value for the Gibbs free energy changes are said to be *spontaneous*. Changes in free energy are given by the formula

$$\Delta G = \Delta H - T\Delta S,$$

where Δ denotes change, G the Gibbs free energy, H the enthalpy, T the temperature (in Kelvin), and S the entropy (disorder).

Note that increases in entropy (disorder) contribute to negative free energies. This implies that increasing the order in a system is not favored. To understand the hydrophobic effect and its relation to the Gibbs free energy, let us consider an example where we have hydrophobic molecules (such as methane) in an aqueous solution (Figure 2.11a). In the two shaded beakers, the one on the left

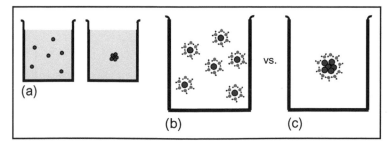

FIGURE 2.11   (a) Hydrophobic molecules in an aqueous solution might disperse throughout the solution or they might coalesce. (b) When the hydrophobic molecules are dispersed, water molecules (represented by) will form a cage around each one. (c) When the hydrophobic molecules coalesce, a cage of water molecules will still form around them. Although this cage will be larger than the cages in (b), the total number of water molecules in the cage will be less than the sum of all of the water molecules in all of the cages in (b). This will leave a greater total number of molecules in a disordered state, which will increase entropy and decrease the Gibbs free energy. The formation in (c) is therefore favored.

appears to have more entropy because the methane molecules are dispersed throughout the beaker with a great disorder. However, you probably know from your own experience that if we placed some oil into a beaker of water, the oil would coalesce more like the figure on the right (as it floated to the top). How can this possibly be, since this scenario appears to have less entropy? To resolve the issue, one must consider the entropy of the entire system, not just the hydrophobic molecules. Water molecules will form a cage around a single hydrophobic molecule, and in forming the cage, they form a very ordered structure. If additional hydrophobic molecules were added and kept separate from each other, separate cages would form around each of the added molecules (Figure 2.11b). This is different from the situation where the hydrophobic molecules come together and become surrounded by a larger, yet singular, cage (Figure 2.11c). The larger cage will incorporate more water molecules than an individual small cage, but the larger cage will involve an overall lesser number of water molecules versus forming many separate cages around each individual hydrophobic molecule. Water molecules that are not involved with cages are free to move virtually anywhere, which contributes to the overall entropy of the system.

The above discussion illustrates the hydrophobic effect, and the hydrophobic effect explains why hydrophobic amino acids tend to fold to the inside of proteins. As a corollary, since hydrophobic amino acids tend to be on the inside of a folded protein, then the exterior portion of the protein would tend to be composed of charged and polar amino acids.

Protein folding is one reason why the charge of individual amino acids in the primary sequence is important. Another reason is molecular recognition.

Consider the protein shown in Figure 2.12. In this example, as is the case for many proteins, there is an indentation that might appear as a hollowed out spot.

FIGURE 2.12   A hypothetical protein, with an active site in a cleft as indicated by the arrow.

This hollowed out portion could serve one or two functions. First, there could be two or more polar or charged amino acids that interact with one another to help hold the shape of the cavity and thereby preserve the shape (the *conformation*) of the protein. Second, the region could serve as a point of interaction with another molecule. Consider that the protein we are talking about is a receptor. The molecule for which the protein is a receptor (the *ligand*) may have positive, negative, or polar regions that line up with negative, positive, or polar regions of the receptor. The receptor and ligand will fit some degree of precision—the greater the amount of precision, the greater the *specificity* of the receptor for its ligand. A similar example is that of *enzymes*, which are proteins that catalyze chemical reactions. An enzyme will interact with its *substrate*(s) via the *active site* of the enzyme. There are over 4000 enzymes within a human cell, and each enzyme catalyzes a specific chemical reaction or type of reaction. Even the polymerization of amino acids into proteins (Figure 2.6) is catalyzed by enzymes made of proteins.

2.4   A RETURN TO MEMBRANES

The early material of this book is focused on cells and the basic entities that a bioactive agent will first encounter as it attempts to gain entry into a cell. As already mentioned, cells are surrounded by a plasma membrane, the membrane is a lipid bilayer, and the bilayer is primarily made up of phospholipids. It was also mentioned that mixed in among these phospholipids are other molecules such as cholesterol and transmembrane proteins. Now that proteins have been more thoroughly introduced, let us now discuss them in the context of the plasma membrane.

Transmembrane proteins can be embedded partially in the membrane, cross the entire membrane once, or cross it multiple times (Figure 2.13). While the proteins can be anchored in membranes by, for example, fatty acids, when a protein spans the membrane, the portion of the protein that interacts with the lipid tails is typically in the form of an α-helix.

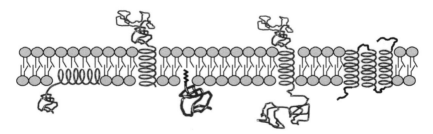

FIGURE 2.13   Membrane proteins can be exposed on one or both sides of the membrane, can cross the membrane once or several times, or can be anchored in the membrane via amino acids or other molecules such as fatty acids.

The number of times that a transmembrane protein spans the lipid bilayer gives rise to the terms *single-pass* and *multipass* transmembrane proteins. One important class of multipass transmembrane proteins is the seven-pass trans- membrane proteins. These are often receptors, and they are used to transmit a signal from outside of the cell to the inside of the cell, thus initiating a chemical cascade inside the cell without transporting the signaling molecule across the membrane. Other transmembrane proteins make up *channels* that allow certain molecules to cross the plasma membrane into and out of the cell, as is the case with sodium channels. Some membrane proteins that do not completely span both layers of the plasma membrane are anchored within the hydrophobic portion of the lipid bilayer. Remember that the middle of the lipid bilayer contains fatty acid tails, so the anchoring moieties will also be hydrophobic, as is the case with common anchoring groups such as fatty acids and prenyl groups.

2.4.1   Protein Movement Within the Plasma Membrane

As mentioned in Section 1.1, components of the plasma membrane can be lik- ened to people walking around in a crowded mall. Individual components, be they phospholipids, transmembrane proteins, lipid rafts, etc. are able to move around in the membrane. While two individual phosphatidylcholine molecules might be next to each other in one minute, they are able to move around and may become separated in the next minute. Following is an example to illustrate this point.

Consider a mouse cell, which has the same phospholipids that were dis- cussed for the general cell. The cell also has transmembrane proteins, but some are unique to mouse cells. At the same time, let us also consider a human cell, which will have the same phospholipids and will also have trans- membrane proteins, many of which are unique to the human cell. These two cells can be fused together. Specific proteins can be labeled using antibodies that carry a fluorescent *tag*. These tags are molecules that will glow with a specific color when exposed to the right light conditions. (Fluorescence will be discussed in a later chapter.) For this example, let us suppose that certain mouse proteins are labeled with antibodies that carry a green tag and certain human proteins are labeled using a red tag. If the hybrid cell were to be ex- amined immediately after the fusion event, the green and red labels would be separate, with green labels on one side of the cell and red labels on the other. However, after only 40 minutes, the cell would display a pattern where the labels appeared to be randomly dispersed (Figure 2.14). This experiment is a demonstration that membranes are dynamic, with components such as trans- membrane proteins being able to travel about as claimed in the membrane mall example.

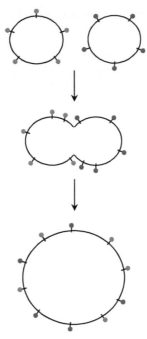

FIGURE 2.14   Proteins are able to migrate in the plasma membrane. Individual cells with labeled membrane proteins have been fused, with the redistribution of labeled molecules occurring in a matter of minutes.

Aside: The Creation of Fusion Cells

The way that cells are fused together is a two-step process. First, the cells must be touching each other, and then, an electrical current is delivered, which will cause a perturbation of the phospholipid bilayer. After the current is removed, pores and other results of the membrane disruption will resolve, but individual phospholipids may flip into the phospholipid bilayer of the other cell in the process. The result will be that where once there were two distinct phospholipid bilayers, there will now be one. To get the cells to touch each other, one might use poly(ethylene glycol). The electrical pulse will be on the order of 1-20 kV/cm, delivered for ~1-20 μs ($1\text{-}20 \times 10^{-6}$ s). When such an electrical pulse is delivered to cells, pores will spontaneously open up in the plasma membrane. (The delivery of a current to produce pores is also a technique used for gene delivery known as electroporation, which will be discussed later.) When two cells are touching during the membrane perturbation, a result can be the creation of fusion cell.

2.4.2   Restriction of Protein Movement Within the Plasma Membrane

Although it was just demonstrated that individual members of the plasma membrane are able to migrate around the entire cell, this is not always the case. There are several instances where multiple membrane components must be held together as a unit. For some cells, such as those comprising the gut epithelium, those of the gastric lining, or sperm cells, some proteins are displayed in only one region of the plasma membrane because of the spatial location required for certain functions in these cells. Consider the gut epithelium and the gastric lining, where the absorption of nutrients occurs. There are proteins that serve as receptors/transporters for certain nutrients (such as glucose) that the body would need to pull out of the stomach or out of the intestine for transport into the body. If these proteins were displayed on both sides of the cell, then between meals, the body might pull glucose out of the bloodstream and dump it back into the alimentary tract, which would not be a good thing. There are different ways that such membrane proteins can be sequestered on one area of the plasma membrane: aggregation, tethering, and blocking via intercellular junctions (Figure 2.15).

Aggregation is where proteins of the cell membrane stick together. Tethering can occur by attaching the protein to an extracellular protein of another cell or some other extracellular structure. Similarly, the proteins could be tethered to an internal structure such as the cytoskeleton. Consider cuboidal cells of the alimentary tract. These cells have to stick together very tightly, and they do so via intercellular junctions. Intercellular junctions, while holding two cells together, will also serve to prevent the movement of membrane proteins from one side of the cell to the other.

So now, we know that our membranes contain phospholipids, cholesterol, and proteins. We have special proteins that can act as identifiers of self and others that serve as receptors. There are other components of the plasma membrane as well, such as channels, which we will get to in the next chapter. Before we branch off into a discussion of how cells use membrane proteins for the transport of molecules into and out of the cytoplasm, let us discuss how proteins can be isolated for further study. The discussion might be of interest because the procedure involves molecules very similar to the phospholipids we have already studied.

2.4.3   Protein Isolation Often Involves Detergents

To isolate transmembrane proteins, the membranes must be dissolved. The dissolution can be accomplished with surface-active agents ("*surfactants*") such as detergents. Detergents have structures similar to fatty acids and soaps in that there are a polar head group and a hydrophobic tail.

One might wonder why detergents are used instead of soaps to disrupt plasma membranes to isolate transmembrane proteins. To answer this question, let's begin with the difference in chemical structure between the two classes of molecules. While both are surfactants, they have different polar head groups.

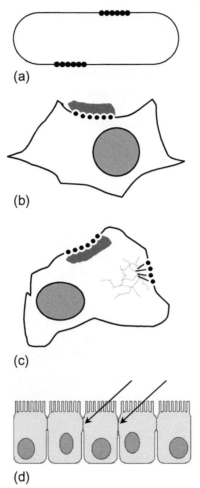

FIGURE 2.15 Not all constituents are free to diffuse throughout the membrane. (a) Some membrane proteins are held together via aggregation (b). (c) Still, other proteins are held in place via tethering to macromolecules outside (b) or inside (c) the cell. In many instances, proteins are tethered to the cytoskeleton (c). (d) Tight junctions (noted by arrows) can prevent proteins in the outer layer of the plasma membrane from diffusing to the opposite face of the cell and vice versa.

Soaps are sodium salts of fatty acids. One common soap is sodium stearate, an 18-carbon molecule made of a long hydrocarbon tail that terminates in a carboxylic acid (Figure 2.16, left side). For comparison, consider the detergent sodium lauryl sulfate, which has a long hydrocarbon tail that terminates with a sulfate group (Figure 2.16, right). The different head groups make a difference in practicality. Soaps, by virtue of the carboxylate head groups, will form precipitates in the presence of calcium or magnesium, two ions found in cells. If the soaps precipitate out of solution, they will not be available to form the structures needed to isolate membrane proteins. Detergents, on the other hand, will not precipitate out of solution in the presence of these ions, so they are more effective for protein isolation.

Aside: Don't Pour Hand Soap into the Dish Washer!

Soaps and detergents are found in the home as well. The calcium that is present in hard water is responsible for some of the soap scum that accumulates in bathtubs and sinks. When washing clothes, calcium acts to inactivate soaps by removing them from the cleaning solution via precipitation. Such precipitates can harm appliances and are the main reason why dish washers and washing machines require detergents instead of soaps as cleaning agents.

The amphipathic structure of surfactants allows them to form micelles in aqueous solutions. Consider what happens when clothes are washed. One might put water into a bucket, throw a little detergent into the water, and then throw a dirty, greasy shirt in the bucket. The shirt is agitated in the water/detergent mixture, rinsed, and finally, a sparkling clean shirt is pulled out of the mix. Where did the oil go? The oil interacts with the detergents to form micelles. As shown above, detergents have a structure that is similar to the structure of the phospholipids that are a part of the plasma membrane, but typically, the detergent has only one fatty acid tail. The hydrophobic portion does not have to be a long hydrocarbon tail, but the detergent does have to contain a polar head

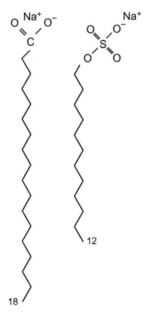

FIGURE 2.16   The structures of sodium stearate, a soap (left), and sodium lauryl sulfate, a detergent (right).

group and a hydrophobic portion. If several detergent molecules are added to water, they will spontaneously form structures known as micelles (Figure 2.17). In this scenario, the hydrophobic portion will face the interior of the micelle. Hydrophobic particles, such as skin oil, could also be carried inside the micelle as *cargo* molecules. That is what happens in the washing machine. The same principle is used for certain types of drug delivery.

Hydrophobic drugs are often delivered using surfactants to form micelles, with the drugs being carried as cargo. Hydrophilic head groups will make up the exterior of the micelle, making it free to interact with water such as is found in blood or in the cell cytoplasm. The head groups may be charged or uncharged and polar, because both are hydrophilic.

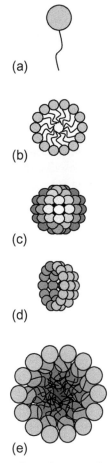

(a)

(b)

(c)

(d)

(e)

FIGURE 2.17   (a) General structure of a surfactant, such as a detergent. (b) Amphipathic molecules can come together to form a micelle. (c) Three-dimensional view of a spherical micelle from the outside, (d) cut open to show it is hollow and (e) turned to reveal hydrophobic interior that can carry cargo molecules.

Example 2.1

Let's consider a cancer cell as being a good cell gone bad. If the cancer cell were a foreign cell, our immune systems would sense it as foreign and destroy it almost immediately. But a cancer cell is a derivative of one of your own cells, so it expresses all the normal proteins that your cells typically express. In fact, cells of the immune system might bump into a cancer cell and examine it, but if the cancer cell is displaying a signal that identifies it as one of your own cells, it will not be destroyed. That is great if you are the cancer cell, but if you are a person, it is not so great.

So, now let's say that you're a researcher who is trying to come up with a biotechnology that will recognize cancer cells based on the proteins that they are displaying. As a researcher, you might wish to isolate these proteins for further study. The way that such proteins are isolated is by disrupting the phospholipid bilayers of the isolated cells. Similar to the example of throwing a greasy shirt into a bucket with some detergent, in this case, you are in essence going to throw some detergent onto the cells. The detergent will disrupt the plasma membrane and allow you to isolate the embedded proteins, so that you can identify (through more testing) one of the unique proteins displayed by this class of cancer.

What detergent might you use for the above example? One common detergent for laboratory applications is sodium dodecyl sulfate, which is the chemical name for sodium lauryl sulfate, the detergent shown in Figure 2.16. The term "dodecyl" means 12 and is used to describe the hydrophobic tail, which has 12 carbons in it. The sulfate portion is an SO_4^-. Because of the charged head group, the detergent is classified as an *ionic detergent*.

Triton X-100 is another detergent used for cellular disruption and, being a detergent, has a polar head group and a hydrophobic tail group (Figure 2.18). However, Triton X-100 does not carry a charge, so it is known as a *nonionic*

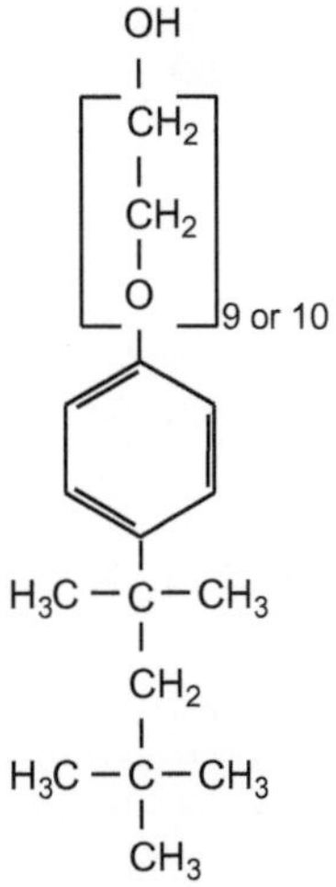

FIGURE 2.18   Triton X-100, a nonionic detergent.

detergent. Nonionic detergents are often used in protein research because they typically do not *denature* (cause the unfolding of) proteins. They can surround membrane proteins that are embedded within membranes without changing their native conformations, so the proteins will retain their functions and folding patterns for molecular recognition.

To isolate individual membrane proteins according to size, one would use sodium dodecyl sulfate poly(acrylamide) gel electrophoresis (SDS-PAGE). The SDS molecules will interact with the plasma membrane to form micelles. They will also interact with membrane proteins. Recall that transmembrane proteins typically have at least one hydrophobic section, which we know because the transmembrane protein must either cross or be anchored in the hydrophobic region of the plasma membrane. Since this protein has a hydrophobic section, the hydrophobic tail of SDS can interact with that section of the protein as it also disrupts the phospholipid bilayer. At the same time, interactions between the hydrophobic tails of several SDS molecules and the protein will serve to denature the protein. The detergent is mixed with the cells and energy is added—the mixture can be vortexed, heated, or sonicated—and in doing so, micelles will be formed with the phospholipids and the detergent, and the membrane proteins will be unfolded.

Aside: Counterions

In the laboratory, *sodium* dodecyl sulfate is used, as opposed to dodecyl sulfuric acid (which would have a hydrogen in place of the sodium). The problem with the acid is that when it is placed in water, it would give up the hydrogen and thereby lower the pH of the surrounding solution. So, rather than having the hydrogen associated with the dodecyl sulfate, a sodium counterion is used, thus forming the sodium salt. Sodium dodecyl sulfate salt is another name for the molecule. The common name in the laboratory, however, is SDS. When you look at the chemical names of many pharmaceuticals, you will see the counterion listed either at the beginning or at the end of the name. Examples include atorvastatin calcium (Lipitor®), alendronate sodium (Fosamax®), and doxacurium chloride (Nuromax®).

Since the SDS is an ionic detergent, the now-unfolded proteins can be separated with the aid of an electrical field. The negative charges of the SDS molecules will be pulled toward the anode, carrying the unfolded proteins with them. Note that it does not matter if a given protein is large or small; the concentration of SDS per unit length of protein is going to be roughly the same regardless of the protein size. Larger proteins will complex with more SDS molecules, but the number of SDS molecules per unit length will remain roughly the same for large and small proteins alike. This implies that large and small proteins will be pulled with the same relative amount of force toward the anode. Separation is achieved because the differences in physical size of the two unfolded proteins limit the rate at which each can traverse the twists and turns of the poly(acrylamide) gel. This is the principle behind protein separation via SDS-PAGE.

QUESTIONS

1. Identify the level of structure (1°, 2°, 3°, or 4°) for the following:
   ____ β-barrel
   ____ Interactions between polypeptides
   ____ Sequence of amino acids
   ____ 3-D structure of polypeptide
   ____ α-helix
   ____ Heterodimer
   ____ Helix-loop-helix
   ____ Asp-leu-tyr

2.
   (a) Which is the more likely in a β-turn, *cis* or *trans* proline? Why?
   (b) Which is the more likely in a membrane phospholipid, a *cis* or *trans* double bond in an unsaturated fatty acid? Why?

3. Could histidine act in place of proline in a beta-bend structure? Why or why not?
4. Identify the bonds/interactions that lend to the stability of secondary structures and which portion of an amino acid is responsible for these interactions.
5. What is the difference between a polypeptide and a protein?
6. Why are glycine and proline usually involved in a β-turn?
7. Why is the charge of a given protein not important when SDS-PAGE is used to separate proteins by size?
8. Why do we use detergents instead of soaps to dissolve membranes?
9. Why don't we use Triton X-100 for gel electrophoresis?
10. Compare and contrast ionic and nonionic detergents.
11. Does it make a difference for the SDS-PAGE analysis if a protein is denatured before beginning the procedure? Why or why not?
12. What are the general structures of soaps and detergents? How do the structural differences between the two affect their functionality?
13. If a small, neutral protein and a large, negatively charged protein were to be separated via an SDS-PAGE, which protein would be expected to move faster through the gel?
14. Below is the result of an SDS-PAGE experiment:

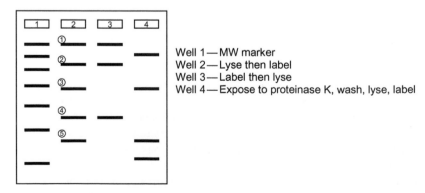

Well 1—MW marker
Well 2—Lyse then label
Well 3—Label then lyse
Well 4—Expose to proteinase K, wash, lyse, label

(a) Do the bands in lane 2 represent intracellular (I), extracellular (E), or transmembrane (T) proteins?

(b) To which of the numbered bands in lane 2 does each band in lane 4 correspond?

(c) From the following list, which is most likely to be the classification of protein ①? Explain your answer.

Adhesion molecule

Signaling protein

Nuclear protein

Phospholipid

Glycolytic enzyme

15. Of the three amino acids with positively charged R groups, which one would be least likely to aid in transporting a negatively charged molecule across a cell membrane? Explain your answer.

RELATED READING

Alberts, B., Johnson, A., Lewis, J., Raff, M., Roberts, K., Walter, P., 2007. Molecular Biology of the Cell, fifth ed. Garland Science, Taylor & Francis Group. New York, NY.

Benz, R., Beckers, F., Zimmermann, U., 1979. Reversible electrical breakdown of lipid bilayer membranes: a charge-pulse relaxation study. J. Membr. Biol. 48, 181–204.

Neumann, E., Rosenheck, K., 1972. Permeability changes induced by electric impulses in vesicular membranes. J. Membr. Biol. 10, 279–290.

Sugar, I.P., Förster, W., Neumann, E., 1987. Model of cell electrofusion. Membrane electroporation, pore coalescence and percolation. Biophys. Chem. 26, 321–335.

Teissie, J., Tsong, T.Y., 1981. Electric field induced transient pores in phospholipid bilayer vesicles. Biochemistry 20, 1548–1554.

Zimmermann, U., Pilwat, G., Riemann, F., 1974. Dielectric breakdown of cell membranes. Biophys. J. 14, 881–899.

Chapter 3

Cellular Transport

3.1 MEMBRANE TRANSPORTERS

Although we do not have an adequate definition of life, we do know that part
of a cell's being considered alive has to do with the maintenance of a different
environment inside versus outside the plasma membrane. As we have already
learned, the plasma membrane serves as an excellent barrier for most molecules. Some molecules can cross more freely than others, as has been shown
experimentally with synthetic membranes (Figure 3.1). The hydrophobic tails
of phospholipids provide a phase that does not mix well with charged molecules
such as ions, so the phospholipid bilayer is very effective at separating the ionic
content of the cytoplasm from that of the extracellular environment. Conversely,
although students occasionally consider oxygen or carbon dioxide to be polar
because of the high electronegativity of oxygen atoms, these molecules are in
fact nonpolar because the dipole moments are symmetric. Nonpolar molecules
have a greater ability to cross the plasma membrane.

While it might seem apparent that certain molecules can freely get into and
out of the cell, the question remains as to how a cell can acquire polar molecules
or ions, especially at concentrations that are different from what is in the extracellular environment. The answer is that the cell contains transporters, which
are membrane proteins that are specific for the molecules they carry. There are
different classes of transporters based upon the direction of molecule travel
(Figure 3.2). Transporters that carry a single molecule in only one direction (the
direction can be into or out of the cell) are called *uniporters*. Transporters that
carry two (or more) molecules across the cell membrane at the same time and in
the same direction are termed *symporters*. Transporters that carry two (or more)
molecules across the cell membrane simultaneously, but in opposite directions,
are termed *antiporters*. If energy in the form of ATP is directly required for the
transporter to carry molecules across the membrane, the carrier is said to be an
active transporter. If no ATP hydrolysis is required, the carrier is known as a
passive transporter.

Channels are another means by which molecules can enter a cell. Ion channels are proteins that span the plasma membrane, allowing for diffusion of specific ions when the channels are open. Some channels are always open (*leak
channels*), and some are opened in response to a stimulus (*gated channels*, such

An Introduction to Biotechnology. http://dx.doi.org/10.1016/B978-1-907568-28-2.00003-4
Copyright © W T Godbey, 2014.

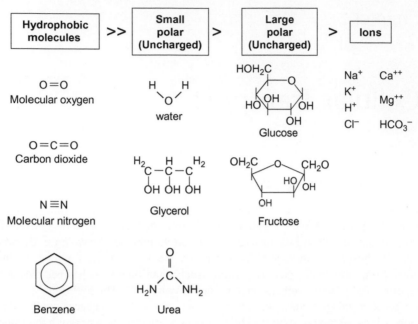

FIGURE 3.1 Relative abilities of types of molecules to cross a synthetic lipid bilayer membrane, with examples.

as voltage-gated channels). Ion channels gain their specificity because of their size. Consider for a moment a K^+ channel. As a K^+ moves through the channel, it will interact with specific amino acids in the channel, commonly via their electronegative carbonyl oxygens. The space is so atomically precise that the water molecules that used to surround the ion will be stripped off, being replaced spatially by the amino acids in the channel's selectivity filter as the ion is positioned within the channel. From the periodic table, we can see that a potassium ion is smaller than a rubidium ion, and it is larger than a sodium ion.

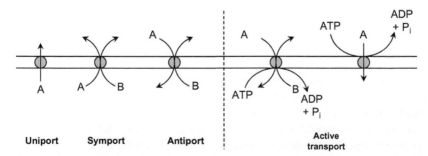

FIGURE 3.2 Examples of transport across a cell membrane. Note that uniport and antiport are shown twice, as passive and active transporters.

The channel is sized so that the larger Rb^+ cannot fit. On the other hand, although a Na^+ can fit through the opening of the channel, this ion is too small to interact with the amino acids in the selectivity filter in an energetically favorable way, so the water molecules surrounding the ion will not be stripped off, and the Na^+ will not be able to interact properly with the amino acids of the channel, so it will not be able to proceed through the channel. This is why ion channels have great specificity.

We will now look at several examples of transporters, starting with the sodium/glucose symporter. As you read the descriptions, remind yourself of what types of molecules are being discussed. For the sodium/glucose symporter, it is not only a symporter but also a passive transporter and a transmembrane protein.

3.1.1   The Sodium/Glucose Symporter

Figure 3.3 is a sketch of the sodium/glucose symporter. Initially, sodium ions and glucose molecules that will be transported are on the outside of the cell—the *extracellular* side—but are free to diffuse into the active sites on the symporter. Outside of the cell, there is a relatively high sodium concentration as compared with the inside of the cell. (The reason for this difference will be covered later.) As a result, there will be numerous sodium ions that could bind to the sodium binding sites in this transporter. Once sodium is bound, there will be a conformation change in the transporter to create active glucose binding sites in the same transporter. Once glucose binds to the newly available acceptor sites, there will be a second conformation change in the transporter protein, which will effectively result in a closing of the external side of the symporter and an opening on the *cytoplasmic* side. At the same time, there will be a conformation change to the sodium binding sites to cause a release of the sodium ions. Upon their release, there will be another conformation change, this time causing a release of the glucose molecules. Once the glucose has been released, there is the final conformation change that leaves the transporter in its original state.

A second ion-driven symporter is lactose permease, which works in a similar manner to the sodium/glucose symporter. In this case, the ion is H^+.

3.1.2   Transporters That Control pH

3.1.2.1   Examples of Passive Transport to Control pH

$$HOH + CO_2 \Leftrightarrow H_2CO_3 \Leftrightarrow HCO_3^- + H^+$$

The above chemical reaction shows that carbon dioxide plus water can combine to yield carbonic acid (H_2CO_3). In a solution at physiological pH, carbonic acid will dissociate from its acid form into bicarbonate plus a proton. This dissociation is important for a couple reasons. First, carbon dioxide is nonpolar because of its symmetry. Being nonpolar, it can freely cross the plasma membrane. It can get out of the cell and into the blood, which is an aqueous solution. This means that carbon dioxide produced within the cell can end up in the blood as

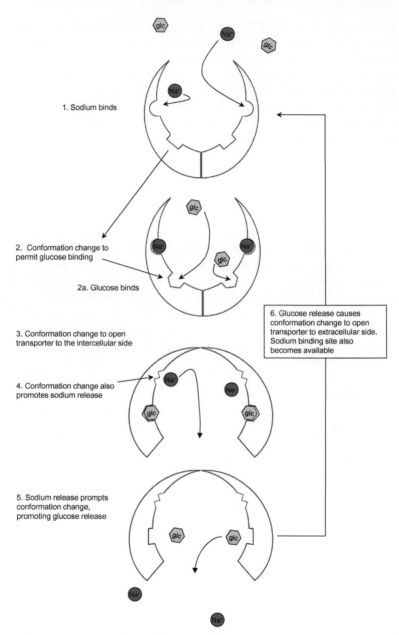

1. Sodium binds

2. Conformation change to permit glucose binding

2a. Glucose binds

3. Conformation change to open transporter to the intercellular side

4. Conformation change also promotes sodium release

5. Sodium release prompts conformation change, promoting glucose release

6. Glucose release causes conformation change to open transporter to extracellular side. Sodium binding site also becomes available

FIGURE 3.3   The mechanism of the sodium/glucose symporter.

bicarbonate plus a proton. This does two things: in addition to being transported as CO_2, it gives the body a second route for transporting the waste carbons and oxygen through the blood. Second, carbonic acid dissociation into a bicarbonate ion plus a proton is involved in altering the pH of the blood.

Suppose that you were to hold your breath and run up the stairs, which would create a buildup of carbon dioxide in the body. (Carbon dioxide is a product of aerobic metabolism, the citric acid cycle in particular. Even though you are holding your breath in this example, the initial exertion of running up the stairs will utilize aerobic metabolism.) The carbon dioxide created in your muscle cells will diffuse into the blood and dissociate. Normally, as your cells respire and you are breathing normally, when the bicarbonate and H^+ reach the lungs, they will encounter a lower partial pressure of CO_2 due to the freshly inspired air, which will drive the above reaction to the left, and carbon dioxide will be reformed from the bicarbonate and proton. This carbon dioxide is normally removed from the body upon exhalation. However, if you do not exhale, you will have a buildup of HCO_3^- and H^+ in the blood, meaning that you will have a lower blood pH, which is called *acidosis* (in this specific case, it is *respiratory acidosis*).

Using a similar logic, you can intentionally hyperventilate as you read this, which will reduce the partial pressure of carbon dioxide in your lungs even lower than normal. This will pull the above reaction further to the left because the concentrations of the reactants on the left and right sides of the reaction will be held constant at equilibrium. In other words, the ratio of CO_2 to HCO_3^- and H^+ is constant at equilibrium. So, if you get rid of CO_2 by hyperventilating, this reaction is going to go to the left to reestablish that constant ratio. As a result, you will be removing protons from the blood and blood pH is going to increase. The term for this is *respiratory alkalosis*. So, when somebody hyperventilates and passes out, they're going to have a higher blood pH. The pH change is not so drastic that it reaches values of 10, or even 7.8, because that would lead to protein denaturation (which would be catastrophic), but it will rise above the normal blood pH of 7.4.

Because pH helps to determine fundamentally important things such as protein folding, it is important that we examine ways that cells regulate their own pH. There are several transporters that are used for this purpose. There is a difference in pH between the inside and the outside of the human cell, with both environments being slightly alkaline (pH = 7.2 and 7.4, respectively). Figure 3.4 shows three common transporters used by cells to regulate pH.

The first transporter shown in the figure is the sodium/proton antiporter. The net effect of this transporter is the removal of one proton from the cell. This transporter is useful when the pH is too low inside the cell, which is the same as the cytoplasm being too acidic, which is the same as having too many protons inside the cytoplasm. Notice that there is no change in charge: the electrical gradient from the outside to the inside the cell is preserved as one positive charge comes in and one positive charge comes out. Keep in mind, however, that only protons contribute to pH.

The second transporter shown in Figure 3.4 is the sodium-driven chloride/bicarbonate exchanger. This antiporter transports a sodium ion into the cell along with a bicarbonate ion, while at the same time, a proton and a chloride ion

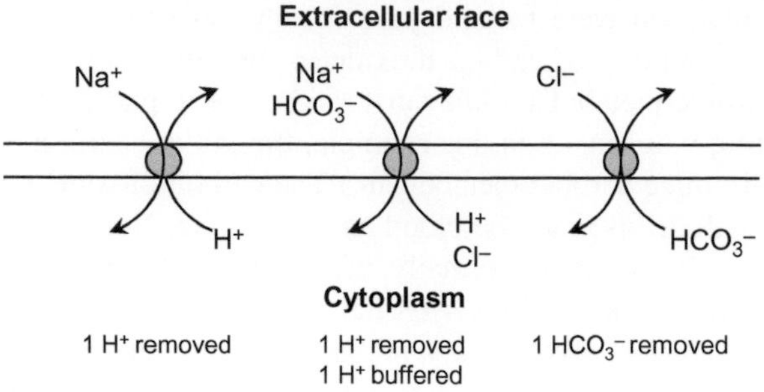

Extracellular face

| 1 H⁺ removed | 1 H⁺ removed
1 H⁺ buffered | 1 HCO₃⁻ removed |

FIGURE 3.4   Three passive antiporters that are used by cells to control cytosolic pH. Left: Sodium/proton antiporter. Middle: Sodium-driven chloride/bicarbonate exchanger. Right: Sodium-independent chloride/bicarbonate exchanger.

are removed from the cytoplasm. Notice again that the electrical gradient is not affected by this transporter, but the pH inside the cell is. The net effect on pH is that one proton is removed from the cytoplasm and an additional proton can now be buffered via $HCO_3^- + H^+ \Leftrightarrow H_2CO_3$ in the cytoplasm. This could be considered a more efficient buffering mechanism versus the previous transporter because, for every transport event, two cytoplasmic protons will effectively be removed.

A variation of the above theme is the sodium-*independent* chloride/bicarbonate exchanger (Figure 3.4). As with the previous exchanger, a chloride ion and a bicarbonate ion are being exchanged. The net result is the loss of one buffering molecule from the cytoplasm. Note that cytoplasmic pH is being affected without the transport of any protons: one buffering molecule is lost, which effectively results in one extra (unbuffered) proton in the cytoplasm, so the pH will be decreased ever so slightly. This transporter is used when cytoplasmic pH is too high (meaning that the concentration of H^+ must be increased).

3.1.2.2   Examples of Active Transport to Control pH: The Proton ATPases

Proton ATPases are used by cells to transport protons against the electrochemical gradient by harnessing the energy of ATP hydrolysis. Because ATP is hydrolyzed, the type of transport is referred to as active transport. One use of a proton ATPase is to acidify a cellular compartment, such as a lysosome. (The function of this organelle is covered below.) As shown in Figure 3.5a, a proton will be transported into the vesicle at the expense of one ATP. This particular transporter is known as a V-ATPase. (The "V" stands for "vesicular.")

There is another type of proton ATPase, found on the inner mitochondrial membrane, which may appear to be set up to pump protons out of the mitochondrion at the expense of an ATP similar to V-ATPase. Known as F-ATPases, these membrane-bound transporters typically function in the reverse direction (Figure 3.5b). There is a relatively strong proton gradient just outside of the

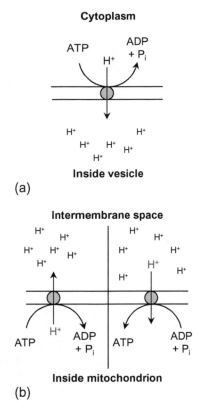

FIGURE 3.5 Actions of two proton ATPases. (a) V-ATPases drive protons against their gradient to acidify vesicles such as lysosomes. (b) F-ATPases are set up similarly to V-ATPases (left), except that the proton gradient is used to drive the production of ATP by mitochondria (right).

inner mitochondrial membrane in what is known as the intermembrane space of the mitochondrion. The proton gradient is high enough that it drives protons back into the mitochondrion, and the ATPase reaction is reversed, meaning that the energy from the proton gradient is harnessed by this transporter to form ATP. This is the last step of the electron transport chain, which is used by the mitochondrion to harness energy from certain energy-rich molecules via redox reactions, to ultimately reduce oxygen to water plus energy in the form of ATP.

There are good reasons why a biotechnologist should be aware of how cells maintain or alter pH levels. First, if one is going to work with cells, it is important to understand how they function and how various drugs might affect the ability of a cell to maintain its own pH. Second, biotechnologists interested in drug delivery or gene therapy should be aware that intracellular vesicles with low pH known as *lysosomes* present a significant barrier to successful drug or gene delivery. There will be more on this specific point in a moment, but first, let us examine what lysosomes are.

3.1.2.3 Lysosomes

Lysosomes are cytoplasmic organelles that degrade materials that are transported into the cell. If a cell ingests something, that something may very well be broken down inside a lysosome. If that something is a protein, there are proteases inside the lysosome that will accomplish the task. (One can think of proteases as "protein-ases" or "enzymes that act upon proteins.") If that something is a lipid, lysosomes contain molecules known as lipases ("lipid-ases") that break down lipids. There also are nucleases ("*nucle*ic acid-*ases*") inside of lysosomes that break down DNA and RNA. The degradative enzymes inside lysosomes require a low pH to be active. The low pH is required for proper folding of these molecules, which happen to be proteins. In this case, "low pH" means in the range of 4.5-5.5. Compare that to the normal cytoplasmic pH of 7.2 and you can get a feel for the disparity. This disproportion serves a purpose. Sometimes, vesicles burst. If a lysosome were to burst and release proteases, lipases, and nucleases into the cytoplasm, then the cell would be destroyed if these enzymes remained active. However, these proteins will be folded incorrectly at pH 7.2, and therefore, they will be inactive. A lysosome is relatively small compared to the size of the entire cell (Figure 3.6), so the bursting of a single lysosome will have virtually no effect upon cytoplasmic pH. One can therefore consider the requirement of low pH as a failsafe mechanism for the cell.

The reason that lysosomes might be of interest to the biotechnologist is that they are important in the areas of drug delivery and gene delivery. Consider gene delivery for a moment. Nonviral gene delivery complexes typically enter cells via endocytosis and will be subjected to the harsh interiors of late endosomes and especially lysosomes. There has been a valid concern in the field of gene therapy about how to cause the gene delivery complexes to escape from

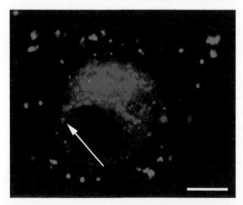

FIGURE 3.6 Image of a cell undergoing gene delivery using green-labeled gene delivery complexes. Lysosomes have been labeled with a red dye. The green spots delineate the cell exterior in this micrograph. Note the size of one lysosome (denoted by arrow) relative to the size of the cell. Scale bar = 10 μm.

or get around late endosomes and lysosomes. All of this relates directly back to how cells can regulate their pH.

The proton pump that is responsible for acidifying the (endo)lysosome is the proton V-ATPase. The antimalaria drug chloroquine is *lysosomotropic*, meaning that it acts upon lysosomes. It acts by shutting down proton V-ATPases. This drug was once thought to hold value for the field of gene therapy, if only in a laboratory setting. By preventing the acidification of (endo)lysosomes, chloroquine prevents the activation of the associated degradative enzymes and therefore prevents the destruction of gene delivery complexes in the (endo)lysosome.

3.1.3   Another Active Transporter: The Sodium/Potassium ATPase

The regulation of pH is not the only reason to transport molecules into and out of the cell. The active transporter shown in Figure 3.7 is perhaps the most famous of the active transporters and is known as the sodium/potassium exchanger. This antiporter belongs to a class of transporters known as the P-ATPases, which bind an inorganic phosphate as part of their mechanism (hence the "P" in the name). By using ATP hydrolysis for energy, the sodium/potassium exchanger is able to pump three sodium ions out of the cell while at the same time pumping two potassium ions into the cell. A couple of things are in play here. (1) The cell is establishing an *electrical gradient* (or voltage gradient) by pumping more positive charges to the outside than what are being transported to the inside. For every ATP used, there will be a net change of +1 toward the outside of the cell. (Do not confuse a change in net charge with a change in pH. Only protons are considered in pH.) (2) The cell is building a *chemical gradient*, because for every ATP used, there will be three sodium ions pumped to the outside for every two potassium ions that are transported back in. The cell uses the established sodium gradient to drive other transporters.

Considering the two points at the same time, we say that the cell is establishing an *electrochemical gradient*.

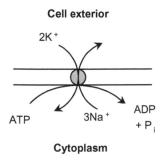

Cell exterior

$2K^+$

Cytoplasm

ATP $3Na^+$ ADP $+ P_i$

FIGURE 3.7   The sodium/potassium exchanger. This transporter establishes an electrochemical gradient for the cell by pumping three sodium ions out of the cell while moving two potassium ions into the cell. Energy to drive the transporter is supplied by the hydrolysis of ATP.

3.1.4   Transporters can be Coupled: The Sodium-Driven Calcium Exchanger

The sodium-driven calcium exchanger will pump three sodium ions into the cell while removing one calcium ion (Figure 3.8). Ca^{2+} concentration outside the cell is about $1\,mM$ ($10^{-3}\,M$), while inside the cell, it is about $10^{-7}\,M$. The very low concentration of calcium ions inside the cell means that even a small influx of Ca^{2+} will make a marked difference on Ca^{2+} concentration in the cytoplasm, permitting the cell to use Ca^{2+} as a signaling molecule. Therefore, the cell needs a way of obtaining and maintaining a low cytoplasmic Ca^{2+} concentration. There is a straightforward active transporter that pumps Ca^{2+} out of the cell at the expense of one ATP, but a perhaps more interesting Ca^{2+} transporter is the sodium-driven calcium exchanger. This exchanger pumps calcium outside the cell against its concentration gradient without direct use of ATP. With the gradient entailing about 10,000 times more calcium ions outside the cell than inside, energy is required to push the calcium out. The energy for this exchanger comes from the sodium chemical gradient outside of the cell that was established by the sodium/potassium exchanger just described. Because of the sodium gradient, there is a force that drives reentry of sodium ions into the cytoplasm. This

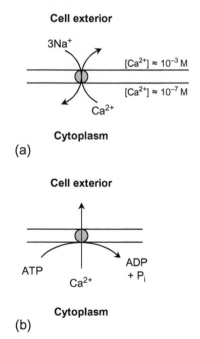

Cell exterior

$3Na^+$

$[Ca^{2+}] \approx 10^{-3}\,M$

$[Ca^{2+}] \approx 10^{-7}\,M$

Ca^{2+}

Cytoplasm

(a)

Cell exterior

ATP

ADP + P_i

Ca^{2+}

Cytoplasm

(b)

FIGURE 3.8   (a) The sodium-driven calcium exchanger. Using the Na^+ gradient that was established via the Na^+-K^+ ATPase, Ca^{2+} is transported up a steep concentration gradient. This is an example of coupled transporters. (b) Ca^{2+} can also be transported out of the cell by directly using the energy from ATP hydrolysis.

force is harnessed to drive the export of calcium ions against their gradient. In chemistry, when the energy from one reaction is used to drive another reaction, we refer to the reactions as being *coupled*. As shown by this example, transporters can also be coupled.

3.1.5   ABC Transporters

The ABC transporters are a class of active transporters that are used by both prokaryotes and eukaryotes. While prokaryotes use these transporters to import hydrophilic molecules, both prokaryotes and eukaryotes use them to export molecules such as lipids, steroids, and toxins. ABC transporters are fairly complex proteins that span the plasma membrane multiple times. The transporter consists of two integral membrane proteins that span the membrane six times each; two peripheral membrane proteins that bind and hydrolyze ATP (the *ATP-binding* cassette, which gives the transporter class its name); and a substrate-binding protein.

Within the ABC transporter class of proteins, some of the exporters are responsible for pumping toxins or other hydrophobic molecules out of a cell. Cancer cells often utilize ABC transporters to pump drugs out before they are negatively affected. In this case, from the perspective of the cancer cell, the drugs are toxins that must be removed. The drugs are hydrophobic so they can cross the plasma membrane to gain access to the cell interior. However, the cell may be able to recognize the drug via the substrate-binding region of certain ABC transporters and is then able to transport the drug back out using the energy gleaned from ATP hydrolysis. Such transporters confer *multidrug resistance* to the cancer cell, and the specific ABC transporter in this case is known as multidrug resistance protein.

As another example of ABC transporter use in nature, consider that four species in the genus *Plasmodium* are able to produce the disease malaria. Malaria is often treated with the drug chloroquine, which works as a lysosomotropic agent (discussed in the section on lysosomes elsewhere in this chapter). However, some strains of *Plasmodium* have ABC transporters that recognize and remove chloroquine from the cytoplasms of these microorganisms. The result is that chloroquine is ineffective at treating malaria cases when they are caused by one of these strains of drug-resistant microbes.

3.1.6   Hydrophilic Molecule Transport and Electrochemical Gradients

A general point to keep in mind is that while hydrophobic molecules diffuse across the plasma membrane, charged molecules are largely prevented from doing so. But the cell *must* get charged, hence polar molecules across the plasma membrane if it is to survive. One example of a polar molecule required by the cell is glucose, which is used for energy. (Certain cells, such as those of the brain, use glucose almost exclusively as their primary energy source.) An example of the need for charged molecules to cross cellular membranes is that of proton transport to regulate the pH inside the cell or across the membranes of cellular organelles such as the mitochondria for ATP production. In addition, keep in mind that while sodium is not used directly to drive certain reactions, the sodium ion gradient that is established by the sodium/potassium ATPase is often used to drive the transport of other molecules across the plasma membrane. An example of this is, once again, glucose. Recall the sodium/glucose cotransporter discussed earlier. The transporter gets its energy from the electrochemical gradient of sodium ions, generated in large part by the sodium/potassium ATPase, which hydrolyzes ATP to establish the gradient.

Ion channels were mentioned earlier. Recall the specificity of ion channels, such as a potassium channel, but realize that channel specificity alone does not dictate the direction of ion movement; electrochemical gradients play a defining role in determining the direction of movement. Consider for a moment the chemical gradient for potassium ions: using the transporters discussed so far, which way would K^+ be pushed? Into or out of the cell? The sodium/potassium ATPase will pump K^+ inside the cell, establishing a chemical gradient pointing toward the outside of the cell (down the chemical gradient) for potassium. However, this antiporter also pumps more positive charges out of the cell than into the cell and thus establishes an electrical gradient. Considering potassium ions specifically, although the chemical gradient is pushing them toward the outside of the cell, the electrical gradient pushes back in the other direction. For sodium ions, the chemical and electrical gradients push in the same direction to drive sodium ions back into the cell. That is how the cell can harness the transport of sodium back into the cell to drive things like the sodium-driven calcium pump. But for potassium leak channels, the situation is different. While the chemical gradient is forcing potassium ions out of the cell, the electrical gradient serves to force them in. Both must be considered together, as an electrochemical gradient. When the chemical gradient balances the electrical gradient across the plasma membrane, the cell is said to be at its *resting membrane potential*. Note that, although in balance with the chemical gradient, the electrical gradient (resting membrane potential) is not equal to zero, lest the cell be dead.

3.1.6.1   The Nernst Equation

The Nernst equation is used to relate the electrical and chemical gradients to each other to determine at what membrane potential the gradients are in balance. While the chemical and electrical gradients can be considered separate entities, this equation shows how they can work with or against each other. The equation is given by

$$V = \frac{RT}{z\Im} \ln\left(\frac{C_{\mathrm{o}}}{C_{\mathrm{i}}}\right),$$

where R is the gas constant (=8.315 J/mol K = 2 cal/mol K), T is the temperature (in K), z is the charge of the ion in question, $\Im$ is Faraday's constant

($= 96{,}480\,J/V/mol = 23{,}0000\,cal/V/mol$), C_o is the concentration outside the cell, and C_i is the concentration inside the cell.

For K$^+$, suppose that the concentration inside the cells is 139 mM and the concentration outside the cell is 4 mM. At normal body temperature (37 °C), the equilibrium potential for K$^+$ is given by

$$V_{K^+} = \frac{(2)(310)}{(1)(2.3\times10^4)}\ln\left(\frac{4}{139}\right) = -0.0956 = -95.6 \ \ mV.$$

This value represents the membrane potential required to keep K$^+$ from flowing out of the cell. You might think of this in terms of excess positive charges inside pushing K$^+$ out of the cell, or it might be more straightforward to think of the equilibrium potential as indicating the amount of negative charge (perhaps from Cl$^-$, phosphatidyl serine, or negatively charged proteins) needed to pull on K$^+$ to keep these ions in the cytoplasm. Notice that if the extracellular concentration of K$^+$ were increased to 139 mM, which is equal to the intracellular concentration, the ion would be at chemical equilibrium and no negative charges would be needed to keep K$^+$ in the cell, a situation that is also reflected by an equilibrium potential $=0$ (verify this with the above equation).

3.2 VESICULAR TRANSPORTERS: ENDOCYTOSIS

3.2.1 Phagocytosis

Endocytosis is a process through which cells take up exogenous material: macromolecules, particulates, or even other cells. *Phagocytosis* is a special form of endocytosis, where large items such as bacterial cells are taken up into phagosomes. Certain cells are known as professional phagocytes; their main function is to phagocytose particles, debris, or bacterial cells within tissues. Examples of professional phagocytes include macrophages, neutrophils, and dendritic cells. When one of these cells phagocytoses a foreign cell or particle, it bumps into the foreign body and recognizes something that marks it as a target for phagocytosis.

Phagocytosis is a triggered event. Four different triggers are

1. recognition of antibodies,
2. recognition of a complement,
3. recognition of apoptotic cells,
4. recognition of oligosaccharides or glycoproteins.

Recognition of antibodies. Let's say that your body is a host to a bacterium, perhaps group A *Streptococcus* (the microorganism responsible for strep throat), and that you were going to rely on your body to get rid of it. Your body is going to secrete immunoglobulins, specific protein molecules that are used by the immune system. Immunoglobulins such as the common immunoglobulin G

(IgG) have the general structure shown in Figure 3.9. The sketched IgG molecule can be thought of as having two regions: an antigen-binding region (F_{ab}) and a constant region (F_c). These two regions can be separated by enzymatic digestion followed by fractionation, hence the "F" in the region names. F_c is short for "fraction crystallizable": to make a crystal, one must have a pure substance. The F_c region is crystallizable because it is constant for all members of this family of immunoglobulins. The F_c region is constant over all members of a species.

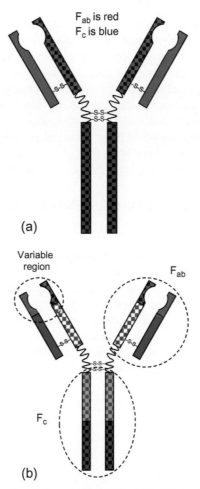

FIGURE 3.9 Schematics of an IgG molecule. In each panel, the two heavy chains are denoted by checkered patterns and the two light chains are solid. The chains are held together by disulfide bonds (−S−S−). (a) Note how the heavy and light chains interact to create the antigen-binding region. (b) Each heavy chain has one variable region (red, checkered) and three constant regions (all with blue checks). The light chains also have a variable (red, solid) region and a constant region (blue, solid). Different fractions and regions are shown by dotted lines.

Within the antigen-binding region are a constant portion and a highly variable portion that gives the antibody its specificity. The variable region for a given IgG is specific to whatever *antigen* is being displayed by the cell. IgG antibodies that are secreted by the immune system are always bumping around and "looking" for things to stick to. If the F_v region is able to bind to an antigen that is displayed by a bacterial cell, the antibody will stick to the bacterial cell. Eventually, the bacterial cell will be coated with antibodies. Professional phagocytes, such as macrophages, are always wandering around "looking" for something to engulf. Macrophages will recognize the exposed F_c regions of the bound antibodies and begin to adhere to them, gradually enveloping the entire antibody-coated bacterium (Figure 3.10).

Recognition of complement. The complement cascade is a complex series of molecular events that results in the permeabilization of a microbial membrane. In this molecular cascade, molecules of the microbial membrane will be coated by antibodies, which can activate the cascade member "C1," which then serves to activate other members of the cascade, and so on. Alternatively, liposaccharides and polysaccharides on the surface can active the complement cascade. (The cascade, with its two pathways, is beyond the scope of this book and will not be presented in great detail.) Chemoattractants will be released, causing professional phagocytes such as macrophages to come toward the area. In addition to recruiting phagocytes, the cascade can result in the drilling of holes through the plasma membranes of targeted microorganisms. These holes will allow a free flow of ions, causing the loss of any chemical or electrical gradients, which will lead to the death of the microbe.

Recognition of apoptotic cells. Apoptosis is also known as programmed cell death. When cells undergo irreparable damage, they may commence a complex series of events involving a molecular cascade. Apoptosis culminates in the cell destroying its own cytoskeleton and genome, packaging the degraded material into apoptotic bodies that will be engulfed and carried off by phagosomes. While the details of the apoptosis cascade will not be covered here, one important aspect that is relevant to the biotechnologist is the change in the plasma membrane associated with the early events in the cascade. Most notably, phosphatidylserine, which typically resides on the inner face of the plasma membrane, will be flipped to the outside of the cell. This greatly changes the charge of the outside of the cell, signaling that the cell is undergoing apoptosis and marking it for destruction by professional phagocytes.

Recognition of oligosaccharides or glycoproteins. Recall that the sugars displayed by cells on their extracellular face are used for cell-cell communication. Bacteria display different patterns of sugars than do mammalian cells. These patterns can be recognized by professional phagocytes to help keep the body free of microbial invaders. Cells such as dendritic cells will phagocytose and degrade the microbes and then attach portions of the degraded cells (such as membrane liposaccharides) to *major histocompatibility complexes* (MHCs) for display to other immune cells: helper T and cytotoxic T cells. Think of the

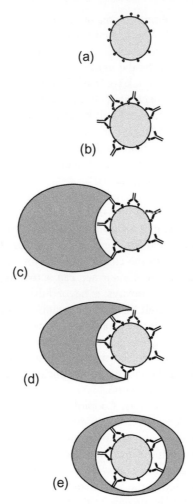

FIGURE 3.10   A macrophage endocytosing a foreign cell. (a) The cell has antigens such as glyco-proteins on its exterior. (b) These particles are recognized as foreign through the binding of antibodies such as IgG. (c–e) The macrophage binds to the F_c regions of the antibodies, gradually engulfing the foreign cell.

dendritic cell's act of display like a warrior advertising a victory by making a flag out of parts of his victim (or the victim's clothing). The flag portion comes from the vanquished microbe, but the flag pole is an MHC. The MHC causes helper T or cytotoxic T cells to take note of the antigen displayed on the dendritic cell and on other cells they will interact with in the future (like B cells, infected macrophages, or virus-infected cells). Interestingly, the flag pole turns out to be an important marker of self. If the wrong MHC is displayed, which is what generally happens if a foreign tissue is transplanted into the body, the presenting cell is seen as foreign and will be destroyed by the host's immune cells. This is what happens when implants such as donated organs or tissue-engineered constructs are rejected by the host.

3.2.2 Pinocytosis

Endocytosis is the uptake of molecules from the outside of the cell by vesicle formation. Phagocytosis typically has to do with the endocytosis of particles, while *pinocytosis* is typically associated with the uptake of solutes. Mechanistically, pinocytosis is associated with colchicine, while phagocytosis is associated with cytochalasin B. More generally, phagocytosis has been termed "cellular eating" and is associated with larger particles, while pinocytosis has been termed "cellular drinking" and is associated with smaller particles, generally <100 nm in diameter. There is no definitive particle size for distinguishing between phagocytosis and pinocytosis. For example, Percoll particles with 30 nm diameters have been shown to enter rat macrophages via both pinocytosis and phagocytosis. (Notice that *particles* were pinocytosed, but also something under 100 nm was phagocytosed.) This example shows that a hard-and-fast rule of size cutoff or architecture to define phagocytosis and pinocytosis is not appropriate (at least not in rats). However, the chance of entry via phagocytosis has been shown to increase with increasing particle diameter.

Gene delivery is associated with pinocytosis. Endocytosis encompasses both phagocytosis and pinocytosis. In the body, phagocytosis is most often associated with specific cells that wander around looking for things to engulf. In the case of gene therapy, however, gene delivery complexes can be endocytosed by more than phagocytes; they can be taken up by most cell types via the numerous pinocytotic events that are continually occurring within the plasma membrane. Gene delivery complexes have sizes on the order of 100 nm, so their uptake is generally associated with pinocytosis. There are two types of pinocytotic vesicles: those utilizing *caveolin* and those utilizing *clathrin*. Caveolae, which come from a word meaning "little cavities," are a type of lipid raft that utilize the protein caveolin. Caveolin is a multipass integral membrane protein. We have now seen two specific examples of multipass transmembrane proteins in this book: first ABC transporters and now caveolin. The other type of pinocytotic vesicle, involving clathrin and clathrin-coated pits, is presented in the next section.

3.2.3 Endocytosis via Clathrin-Coated Pits

Individual clathrin molecules have an interesting shape: a *triskelion* (Figure 3.11). Inside the cell, triskelions will self-assemble to form a cage similar to a soccer ball. When bound to the plasma membrane, the three-dimensional geometry of clathrin puts force on the membrane to make it concave. Imagine several clathrin molecules floating around the cytoplasm, when one eventually sticks to the cell membrane. Then, another clathrin molecule attaches to the first triskelion,

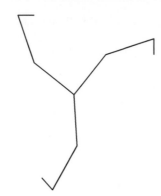

FIGURE 3.11   Clathrin molecules have the shape of a triskelion.

then another, etc. until a cage that is fairly sturdy has been formed. As the clathrin molecules continue to form this cage, they will pull in on the membrane until it is shaped like a ball (Figure 3.12).

Sometimes, clathrin molecules seem to be specific for certain receptors. The way that specificity is achieved is through the use of a molecule called *adaptin*. The cytoplasmic side of individual receptors may bind to specific adaptin molecules, which are bound to clathrin. This chain of associations can give specificity to clathrin-coated pits, meaning that specific receptors can be endocytosed into one pit. In some cases, receptors are endocytosed whether they have bound a ligand or not. In other cases, the only time that the receptor takes on a conformation that is amenable to binding adaptin is after the receptor has bound its specific ligand.

There is a constant assembly of clathrin, which helps shape the membrane into a ball, which is eventually pinched off by the protein *dynamin*.

If the total number of a specific type of receptor on (and in) the cell were held constant and the rate of formation of clathrin-coated vesicles were also constant, the cell would end up with a variable amount of endocytosed ligand if the amount of substrate outside the cell were varied. If the concentration of ligand outside the cell were increased, then there would be a greater chance that one of those ligand molecules would bind to a receptor and be internalized by a clathrin-coated pit. Of course, this is not the only way that a cell can internalize greater numbers of ligand molecules. If the extracellular concentration of ligand molecules were held constant but the number of membrane receptors for the ligand were increased, then the probability of a receptor/ligand interaction would increase, ultimately resulting in greater numbers of ligand molecules being endocytosed by the cell.

If one were to blindly reach into a tank to grab an oyster and there were only one oyster to grab, then the probability of grabbing an oyster would be relatively low. However, reaching into a tank that contains dozens of oysters will provide a greater probability of grabbing one because there are simply more of them

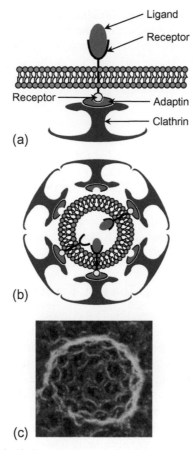

FIGURE 3.12   (a) Clathrin binds to a receptor molecule via adaptin, which gives specificity to a forming pit. Adaptin recognizes the cytoplasmic face of certain receptors. (b) A clathrin-coated pit following pinching off from the plasma membrane, a feat achieved with the protein dynamin (not shown). (c) Electron micrograph of a clathrin-coated pit. *Photo from Heuser, J., 1980. Three-dimensional visualization of coated vesicle formation in fibroblasts. J. Cell. Biol. 84: 560–583.*

around. The same principle works for cells taking up specific nutrients. If the blood contains a nutrient that is used by the cell, increasing the concentration of that nutrient will increase the probability of one such molecule binding to a receptor, which means that the cell will ultimately bring more of that nutrient into the cell. A similar parallel can be drawn from blindly reaching into the oyster tank with two hands instead of one—it will be much easier to find a yummy, yummy oyster that way. In some instances, when cells need to acquire more of a certain type of molecule, they will increase the number of receptors displayed on the plasma membrane and thereby increase the chances of pulling in the needed molecules when the receptors are endocytosed.

Sometimes, there is a conformation change after the receptor binds the ligand, which permits association with clathrin via adaptin. The clathrin-coated vesicle will form and pinch off via dynamin. Once endocytosed, the clathrin will come off of the vesicle and migrate back to the cell surface. The uncoated vesicle is then able to fuse with an *endosome*, a membrane-bound structure that is involved in processing endocytotic and pinocytotic vesicles.

3.3   RECEPTOR FATES

Receptors are proteins, which are coded for by genes. A fair amount of work goes into creating a receptor, so it makes sense that recycling receptors can be advantageous for the cell. Not all receptors are treated the same within a given cell, though. To illustrate this variation in processing, three examples of receptor fates will be presented below: one where the receptor is recycled, one where both the receptor and the ligand are recycled, and one where neither the receptor nor the ligand is recycled.

3.3.1   Receptor Recycling: The LDL Receptor

A good example of receptor recycling is the low-density lipoprotein (LDL) receptor. As we have seen, not all of the molecules in the cell are purely made of protein, nucleotides, or lipids—hybrids exist. Examples include glycolipids, glycoproteins, and lipoproteins. There exists a protein in the blood, called apoprotein B (ApoB), which can carry various forms of lipids. ("Apo" is a prefix that means "away from" and is used in biology to denote something that is incomplete.) While most people have heard of LDL, the same lipoprotein is known as ApoB (apolipoprotein B) when there are no lipids attached.

ApoB can carry certain passengers, including cholesterol. When loaded up with triglycerides, fatty acids, and cholesterol (or cholesterol esters), we call the conglomeration very-low-density lipoprotein (VLDL). When VLDL loses fatty acids and triglycerides (e.g., when it delivers them to fat cells (adipocytes) for storage), the density of the lipoprotein particle goes up and the particle may be referred to as LDL. LDL, which consists of a hydrophobic core of polyunsaturated fatty acids and esterified cholesterols surrounded by ApoB, phospholipids, and unesterified cholesterol, delivers cholesterol to cells. As we learned in Chapter 1, cells need cholesterol for fluidity and stability of the plasma membrane. Certain cells also use cholesterol as a precursor for steroid synthesis.

High-density lipoprotein (HDL) is used for a process known as "reverse cholesterol transport," where cholesterol is taken from cells back to the liver for catabolism (e.g., breakdown into bile acids that are used for digestion). HDL can also grab cholesterol from LDL and is thought to help combat atherosclerotic plaques in the vasculature via cholesterol scavenging. It might interest you to know that HDL acquires cholesterol from cells via … ABC transporters.

Aside: LDL Is *Not* Cholesterol!

The reason people might think that LDL is bad is that if one has an excess of this form of the lipoprotein, a form that is carrying cholesterol to cells, it probably means that he or she either has taken in too much cholesterol or has made too much cholesterol. If one has a large amount of LDL versus HDL, it indicates that apolipoprotein molecules in the blood have been loaded with an excess of cholesterol. (Be aware that ApoB is not the only apolipoprotein in play here.) The ratio of LDL to HDL serves as an indicator of the relative amount of cholesterol being transported in the blood and its direction.

High cholesterol levels can lead to the development of atherosclerotic plaques, which are plaques that build up in the vasculature and restrict blood flow and oxygen delivery, which can lead to unwanted events such as myocardial infarctions (heart attacks). Note that, despite the terms used in common culture, there is no "good" cholesterol nor is there any "bad" cholesterol. Cholesterol is a molecule, and there are neither good nor bad subtypes of it. Excessively high levels of cholesterol are undesirable, however, and such levels can be detected via LDL values and HDL/LDL ratios. Keep in mind that LDL is not bad cholesterol. In fact, LDL is not cholesterol at all; it is a lipoprotein.

Simply delivering a cholesterol molecule to a cell does not satisfy the cell's cholesterol needs. The cholesterol must be taken up into the cell, an act that is accomplished by LDL and LDL receptors. Study of the presence or absence of LDL receptors and their relation to blood cholesterol levels won the Nobel Prize in Physiology or Medicine for Michael Brown and Joseph Goldstein in 1985, and the work performed to this end played a major role in launching the field of receptor biology.

LDL receptors bind LDL to deliver cholesterol to cells. LDL contains cholesterol bound to fatty acid tails, as well as free cholesterol molecules. To give you an idea of the scale involved, consider that one LDL complex will contain around 1500 cholesterol or other sterol molecules bound to fatty acid tails (e.g., cholesterol esters). It will also contain about 500 free cholesterol molecules, phospholipids forming a monolayer, and a fairly large protein (~500,000 daltons) (Figure 3.13). The following illustrates the process of cholesterol uptake: suppose the cell needs more cholesterol. Perhaps you eat it or it is made by liver cells. The cholesterol is transported through the blood to the cells that need it. When a cell needs cholesterol, there will be an increase in transcription of the gene that codes for LDL receptors, leading to an increase in the amount of the associated messenger RNA, which, in turn, leads to an increase in the amount of translation of LDL receptor proteins. More translation of these proteins leads to a greater number of LDL receptors displayed on the surface of the cell. With more LDL receptors displayed on the exterior of the cell, there is a greater probability that a circulating LDL complex will bind to a receptor on the given cell, which means that there will be a greater amount of LDL entering the cell via endocytosis. This endocytosis will take place via clathrin-coated pits.

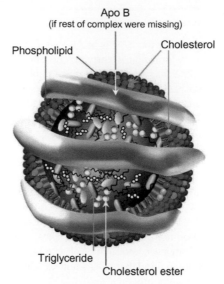

Apo B
(if rest of complex were missing)

Phospholipid

Cholesterol

Triglyceride

Cholesterol ester

FIGURE 3.13   Structure of an LDL complex. *Drawing courtesy of Natalee Buisson, Tulane University.*

Recycling of LDL receptors follows the scheme that is presented in Figure 3.14. Consider that a cell has a receptor that binds extracellular LDL. The receptors will diffuse in the membrane, with some dispersing to the area of a forming clathrin-coated pit. Endocytosis of the LDL and receptor will take place via a clathrin-coated pit. (Notice that the receptors do not necessarily have to be filled to be endocytosed in this case.) Clathrin will be shed following endocytosis, and the individual triskelions will be free to participate in additional endocytotic events. After the shedding of the clathrin coat, the vesicle fuses with an early endosome. Keep in mind that a distinguishing feature of endosomes is a relatively low pH, created by the proton ATPase that was discussed earlier. When the pH drops, the charge of some amino acids will be changed and some polypeptide folding will be altered. In this case, when the pH drops, the receptor/ligand interaction will be altered, thus allowing LDL to float free of its receptor. The early endosome will form chutes, which will contain unbound receptors. The chutes will pinch off of the endosome in a process called *budding* to yield vesicles that contain unbound LDL receptors. These vesicles will then be shuttled back to the plasma membrane where the receptors will be redisplayed on the cell exterior. Meanwhile, other vesicles that contain endocytosed ligands, including LDL, will pinch off to form late endosomes, which deliver the ligands to lysosomes via fusion to form endolysosomes. Cholesterol esters in the LDL are degraded via lysosomal enzymes to release free cholesterol.

Note: The formation of vesicles is always going to happen. It's much like the waves that wash up on the beach. Whether we are there to see it or not, waves

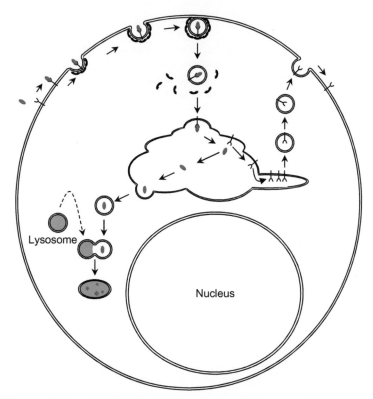

FIGURE 3.14   Endocytosis of LDL bound to its receptor, with the receptor being recycled.

will continue to wash up on the beach. Even when the beach is deserted, it is safe to assume that waves still wash up on the shore. The formation of pinocytotic vesicles by the cell is similar. Whether or not, for example, LDL is bound to LDL receptors (or is even present), clathrin-coated pits are going to form and pinch off. Vesicle formation happens all the time, whether or not receptors have been filled with ligands. The pinching off of these vesicles is going on at all times as well, as are the merging of vesicles into larger units and the formation of chutes with subsequent budding from these larger vesicles. Some of these budded vesicles may return to the plasma membrane, while others may interact with lysosomes.

Aside: Treatments for High Cholesterol

There are two common causes for a person to have high blood cholesterol levels. First, one could ingest too much cholesterol. If you wake up every morning and have an eight-egg yolk omelet, chased down with whipping cream and avocados, then you probably are going to have high blood cholesterol levels.

The good news is that such high blood cholesterol can be changed by modifying one's diet. A second cause of high blood cholesterol is due to a condition called familial hypercholesterolemia. This is a problem rooted at the genetic level and is an autosomal dominant condition that is passed down through families. Patients with this condition are unable to remove LDL from the blood due to a problem with LDL receptors (Figure 3.15). Perhaps, the receptors are unable to bind LDL. Perhaps, the receptors can bind LDL, but the cytoplasmic portions of these receptors are unable to bind clathrin. In other cases, not enough receptors are made in the first place. The first treatment for this condition is simply a modification of the diet, preferably with an increase in exercise. However, it is quite common that a dietary fix is insufficient to adequately lower blood cholesterol levels. Biotechnologists in the pharmaceutical industry have created a class of drugs known as the "statins." Examples of statin drugs include atorvastatin (with the brand name Lipitor), simvastatin (Zocor), lovastatin (Mevacor), fluvastatin (Lescol), and pravastatin (Pravachol). Cholesterol is made in the liver through a complex series of chemical reactions that are made more efficient by enzymes. The statin drugs act as competitive inhibitors to HMG CoA reductase, the enzyme

FIGURE 3.15   Problems with LDL receptors include (*top*) a defective LDL binding site, (*middle*) a defective binding site for clathrin or adaptin, or (*bottom*) too few receptors are made at all. *Photo from Heuser, J., 1980. Three-dimensional visualization of coated vesicle formation in fibroblasts. J. Cell. Biol. 84: 560–583.*

that catalyzes the rate-limiting step in the cholesterol synthesis pathway. The result is a decrease in the amount of cholesterol synthesis. A secondary, and greater, effect of the statins is that sterol regulatory element-binding protein (SREBP) will be activated in the cells. This protein is a transcription factor that causes an increase in the expression of the gene that codes for the LDL receptor. An increase in LDL receptor expression results in more LDL being taken up by the cells, thereby lowering LDL concentration in the blood.

3.3.2   Receptor and Ligand Recycling: The Transferrin Receptor

In another class of receptors, the cell will recycle both the receptor and its ligand. For example, consider transferrin and its receptor (Figure 3.16). Iron is taken up into cells via transferrin receptors, meaning that the receptors are

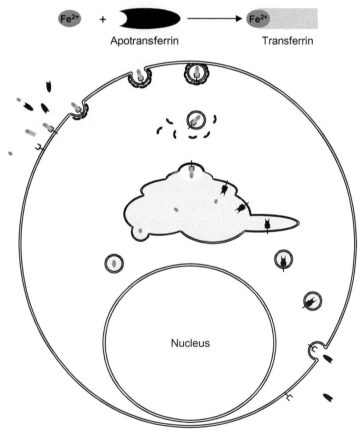

FIGURE 3.16 (Top) Apotransferrin binds Fe^{2+} to create a transferrin molecule. (Bottom) Endocytosis of transferrin bound to its receptor, with both the receptor and the (apo) ligand being recycled.

specific for transferrin. Transferrin is an iron-transporting glycoprotein found in the blood. When the protein is not bound with iron, it is called apotransferrin and will not bind with a transferrin receptor at physiological pH. Once iron is bound, however, apotransferrin will undergo a conformation change that will allow this protein (now transferrin) to bind with a transferrin receptor. The receptor/ligand complex will be endocytosed, the vesicle will undergo a drop in pH as we have already discussed, and the iron will be released from the protein. Note that the protein is still bound to its receptor. As in the previous example, chutes that contain the receptor (still bound to the protein ligand this time) will form and bud, returning the receptor/protein complex to the cell surface. The pH of the exocytotic vesicle is still low prior to fusion with the plasma membrane. Once the phospholipid bilayers of the vesicle and the plasma membrane fuse, what was the interior of the vesicle will be exposed to the extracellular environment, which has a relatively higher pH (~7.4). At higher pH, the apotransferrin goes back to its initial form, which does not fit the active site of the receptor, and it is released from the receptor. The result of this process is that both the receptor and the ligand are recycled.

3.3.3   Neither Receptor nor Ligand are Recycled: The Opioid Receptor

Let us now consider another scenario in receptor processing, which will be represented by the opioid receptor. The opioids are a class of drugs that are used to relieve pain. Members of this class of drugs include morphine, heroin, and hydrocodone (Vicodin). Opioid receptors bind their ligands *before* attaching to a forming clathrin-coated pit (unlike LDL receptors, which can be endocytosed whether or not they have a ligand bound to them). After the receptor/ligand complex is endocytosed, the complex can undergo the same pathway as LDL receptors, where the receptor undergoes recycling, or the entire complex can be directed to a lysosome for destruction. Endocytosing opioid receptors reduces their plasma membrane concentration, thereby rendering the cell less sensitive to additional stimulation by the opioid. While the cell can transport fresh receptors (although not necessarily identical, in the case of morphine receptors) to the surface, such an event requires a stimulus so replenishment is not guaranteed. The result is desensitization to the drug, sometimes a long-term condition. The result is that patients will build up a tolerance to the drug, requiring greater and greater doses to achieve the same analgesic effect.

3.3.4   Transcytosis

In addition to recycling of receptors in the fashion just described, the receptors can also participate in transcytosis, whereby the receptor and ligand are endocytosed and transported to the other side of the cell and the cargo released outside of the cell (Figure 3.17). Transcytosis is what happens when a baby

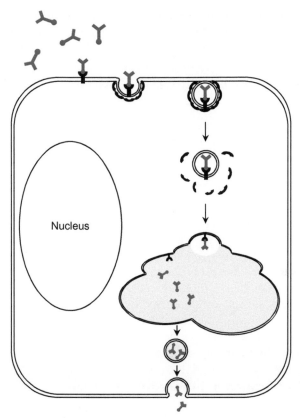

FIGURE 3.17   Transcytosis is where the endocytosed material is transported across the cell and released. In this example, antibodies (such as secretory IgA) are being transported, perhaps into an infant's body.

suckles its mother. The mother has antibodies in her breast milk, especially early on when the breast milk is termed colostrum. Colostrum is secreted by the mammary glands during the first few days following childbirth. It has a higher concentration of antibodies than breast milk, although both provide antibodies for the baby's adaptive immune system. When the baby drinks, the milk enters the acidic environment of the stomach and small intestine. The folding of the maternal antibodies (in this case, IgA) changes in the low pH, and the altered conformation is recognized by specific receptors on the lumenal side of the baby's gut. The antibodies are pulled out of the digestive tract and into the gut epithelial cells. Then, rather than destroy these antibodies, the cells relocate them to the plasma membrane on the other side of the cell where they are exocytosed, availing them to the baby's lymphatic system. The pH of the extracellular fluid on this side, the basolateral side, of the gut epithelia is near-neutral, so the antibodies regain their original conformation, which causes them to dissociate from the receptors. In other words, the antibodies are pulled from the intestinal lumen and transferred through the cell and released into the body.

3.4   LYSOSOMES ARE FOR DEGRADATION, BUT ARE THEY SAFE?

The proton ATPases are always working in the lysosomal membrane, so lysosomal pH may drop to as low as 4.5. Just as pH affected conformation for receptor/ligand complexes, pH also affects the conformation of lysosomal proteins. Specifically, the acid hydrolases of the lysosome, which include proteases, nucleases, and lipases, begin to take on an active conformation at pH values around 5-6. This dependence on acidic conditions is a safety feature for the cell. Suppose that several lysosomes rupture, perhaps for osmotic reasons. This will release proteases, nucleases, and lipases into the cytoplasm. However, because the volume of the cytoplasm is very large relative to a lysosome, and because it is buffered, the pH of the cytoplasm (7.2) will be enough to change the conformation of these hydrolases and render them inactive. Acid hydrolases are proteins and, as such, are composed of amino acids. Recall the discussion of pK_a values and how pH can alter the charge of certain amino acids. Altering the ionic state of amino acids in the primary structure of a protein can have a profound effect on the tertiary structure of the protein.

3.4.1   Identification of Intracellular Vesicles

Although all of the smaller vesicles inside the cell are interrelated, there are guidelines for defining the identities of each of the vesicles. Early endosomes have a slightly acidic pH (~6.5) and do not contain active degradative enzymes. They serve as a sorting station for freshly endocytosed vesicles. As already discussed, some vesicles will bud off of the early endosome to return to the plasma membrane, and some will be transported to late endosomes. Late endosomes have a more acidic pH (~6) and several types of *acid hydrolases*, pH-sensitive degradative enzymes. The enzymes and lower pH are the result of the late endosome serving as another sorting station, this time not only for the processing of endocytosed materials but also for the final portion of lysosome construction. While the complete process of lysosome production will not be covered here, one of the later steps in the process involves the delivery of the acid hydrolases from an organelle known as the Golgi apparatus to late endosomes. The final lysosome maturation is a gradual process, with endosomal membrane proteins being trafficked away from the developing lysosome. Lysosomes have a loose definition because they vary so much in morphology even within the same cell. Generally, however, lysosomes are characterized by a low pH (~5) and the presence of *active* acid hydrolases.

QUESTIONS

1.
- a. Given the following chart, what is the equilibrium potential for sodium, chloride, and calcium ions?

	Concentration in Cell (mM)	Concentration in the Blood (mM)
K+	139	4
Na+	12	145
Cl–	4	116
Ca²+	<0.0002	1.8

- b. Consider the ion X^z, which has an equilibrium potential of +43 mV. If it were normally more concentrated inside the cell than outside, what can you deduce about the charge of this ion?

- c. Consider the ion Y^z, which has an equilibrium potential of −57 mV. If it were normally more concentrated outside the cell than inside, what can you deduce about the charge of this ion?

2. Consider an endocytosed molecule bound to a receptor. In the early endosome, it will be released from its receptor due to a conformation change in the receptor, which is a protein.
- a. Which amino acids might contribute to the change in conformation in the early endosome because of an ionization state change?
- b. Which amino acids might contribute to a change in conformation because of an ionization state change if the protein is eventually transported to a lysosome?

3. Which would you expect to cross a synthetic membrane the easiest, a lipid or sucrose? Why?

4. Can transporters denature just like other folded proteins? Why or why not?

5. Do ions with the same charge have the same resting membrane potential? Why or why not?

6. What does the Nernst equation calculate?

7. Assuming that the membrane potential of a cell is only affected by potassium ions and given an extracellular concentration of potassium ions of 5 mM and a membrane potential of −92 mV, what is the concentration of potassium ions inside the cell?

8. Can potassium ions move through Na+ channels even if $V_{Na+}=0$? What if $V_{cell}=0$?

9. Does a chemical gradient produce a measurable force? (Consider this in the absence of an electrical gradient.)

10.
   a. What are ABC transporters? What do they do? What does "ABC" mean?
   b. Name two instances where ABC transporters are utilized. Decide whether each would be good or bad for a human patient.
11. Draw the Na^+/Ca^{++} exchanger, and describe how the energy of ATP is involved.
12. Describe how a potassium leak channel is specific for potassium ions.
13. If a lysosome were to lyse, why would it not destroy the cell?
14. What is the difference between a channel and transporter?

RELATED READING

Bie, B., Pan, Z.Z., 2007. Trafficking of central opioid receptors and descending pain inhibition. Mol. Pain. 3, 37.

Cahill, C.M., Holdridge, S.V., Morinville, A., 2007. Trafficking of delta-opioid receptors and other G-protein-coupled receptors: implications for pain and analgesia. Trends Pharmacol. Sci. 28, 23–31.

Ejendal, K.F., Hrycyna, C.A., 2002. Multidrug resistance and cancer: the role of the human ABC transporter ABCG2. Curr. Protein Pept. Sci. 3, 503–511.

Hellgren, M., Sandberg, L., Edholm, O., 2006. A comparison between two prokaryotic potassium channels (KirBac1.1 and KcsA) in a molecular dynamics (MD) simulation study. Biophys. Chem. 120, 1–9.

Heuser, J., 1980. Three-dimensional visualization of coated vesicle formation in fibroblasts. J Cell Biol. 84, 560–583.

Lever, J.E., 1984. A two sodium ion/D-glucose symport mechanism: membrane potential effects on phlorizin binding. Biochemistry 23, 4697–4702.

Mindell, J.A., 2012. Lysosomal acidification mechanisms. Annu. Rev. Physiol. 74, 69–86.

Pratten, M.K., Lloyd, J.B., 1986. Pinocytosis and phagocytosis: the effect of size of a particulate substrate on its mode of capture by rat peritoneal macrophages cultured in vitro. Biochim. Biophys. Acta. 881, 307–313.

Chapter 4

Genes: The Blueprints for Proteins

4.1 NUCLEOTIDES AND NUCLEIC ACIDS

4.1.1 The Phosphoribose Backbone

We have been learning about membranes and their composition, which includes not only phospholipids but also cholesterol, sugars, proteins, and combinations thereof. Membrane proteins serve as channels, pores, identifiers, or receptors. In this next section, we will learn about nucleic acids, because a proper look at biotechnology cannot be completed without considering the entire cell, and cells rely upon genetics for function and survival.

There exist two types of nucleic acids in the cell: RNA and DNA. The structure of RNA contains the sugar β-D-ribose (Figure 4.1a). The hydroxyl group (−OH, noted by an arrow) on carbon #1 is substituted by another molecule in RNA—we will get to that in a moment. Figure 4.1b illustrates the molecule phosphoribose. The only difference between ribose and phosphoribose is that a hydroxyl has been replaced by a phosphate on carbon #5. Phosphoribose could serve as a monomer in the formation of a poly(phosphoribose), the polymer that makes up the backbone of RNA. Polyphosphoribose is a succession of phosphoribose molecules that are attached between carbon #3 on one phosphoribose molecule and the phosphate group attached to carbon #5 on the other molecule (Figure 4.1c). Depending on the R groups, the molecule shown in Figure 4.1c might be the phosphoribose backbone for an RNA dinucleotide. If each of the ribose units was missing a hydroxyl on carbon #2, the molecule would be DNA. DNA is the abbreviation for deoxyribonucleic acid, with "deoxy" referring to the loss of the oxygen atom from carbon #2.

A closer look at the carbons of the ribose in a nucleotide reveals that carbon one holds the R group, carbon two determines whether we have DNA or RNA, carbon three is used for polymerization, carbon four is part of the ribose ring structure, and carbon five is involved with polymerization just like carbon three. Note that the phosphate group carries a negative charge. There will be one negative charge for every nucleotide in a DNA or RNA polymer (except for the one on the end with an unpolymerized phosphate, which will typically carry two negative charges).

An Introduction to Biotechnology. http://dx.doi.org/10.1016/B978-1-907568-28-2.00004-6

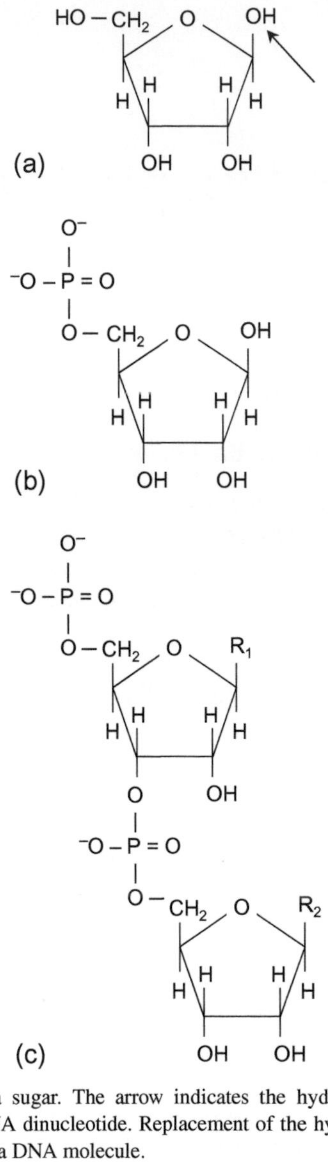

FIGURE 4.1 (a) Ribose, a sugar. The arrow indicates the hydroxyl attached to carbon #1. (b) Phosphoribose (c) An RNA dinucleotide. Replacement of the hydroxyls on carbons #2 on the ribose rings would make this a DNA molecule.

   The repeated negative charges of DNA are utilized in gene delivery by most of the synthetic delivery vectors. Most lipids used for liposomes will be cationic, polymers used for gene delivery will be polycations, and even electroporation works, in part, because of the numerous negative charges on the delivered DNA. (In electroporation, DNA flows in the same direction as the electrons in an applied current because of the negative charges on the phosphates in the polynucleotides.)

A linear DNA or RNA polymer will have two ends that are chemically distinct. On one end, there will be an exposed phosphate group because it is only attached to one (deoxy)ribose. On the other end, there will be an unpolymerized hydroxyl on the #3 carbon of the terminal (deoxy)ribose. The two distinct ends give directionality to the polynucleotide and give rise to the terms 5′ phosphate and 3′ hydroxyl or the 5′ and 3′ termini of a linear DNA or RNA molecule.

Aside: Ethers, Esters, and Phosphodiesters

An ether is two alkyl or aryl groups connected by an oxygen atom. For example, $H_3C-CH_2-O-CH_2-CH_3$ is diethyl ether. An ester involves two oxygen atoms, one that links two carbons in the same fashion as an ether and one that serves as a carbonyl oxygen on one of the two carbons attached to the linking oxygen. Another way to think of an ester is as a carboxylic acid with the hydroxyl hydrogen replaced by a hydrocarbon group.

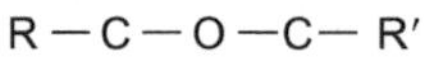

$$R-C-O-C-R'$$
Ether

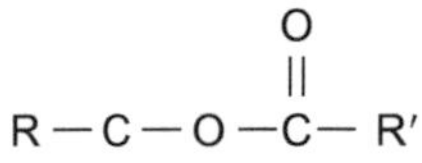

$$R-\overset{\overset{\textstyle O}{\|}}{C}-O-C-R'$$
Ester

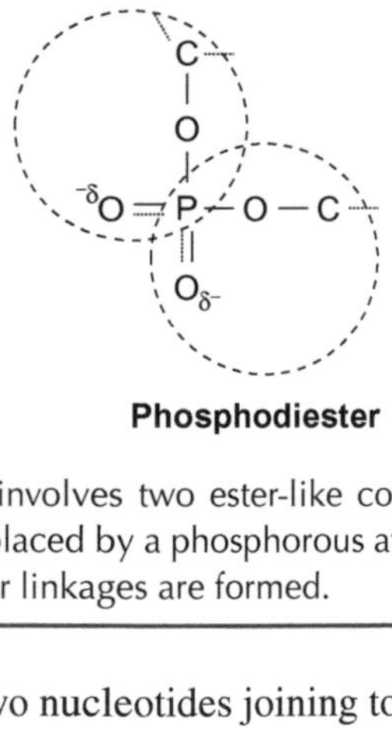

Phosphodiester

A phosphodiester bond involves two ester-like configurations (circled), with one of the carbons being replaced by a phosphorous atom. When nucleotides are polymerized, phosphodiester linkages are formed.

The polymerization of two nucleotides joining together to make a single dinucleotide requires +25 kJ of free energy per mole. The reaction could proceed in the reverse direction of what is written to yield 25 kJ of free energy per mole (the change in Gibbs free energy is −25 kJ/mol). Because the reverse direction yields free energy, this is the favored direction of the reaction. In other words, hydrolysis of polynucleotides (the breaking of polynucleotides via water) is favored. One might ask, "If this is the case, then how can we even be alive? How can our cells synthesize DNA if the reaction is energetically unfavorable?" The answer is that we do not polymerize using nucleotides with only one phosphate attached. Rather, we use nucleotides with three phosphates attached.

Let NMP stand for "nucleotide monophosphate," and let NTP stand for "nucleotide triphosphate" (Figure 4.2). Cells polymerize RNA using NTPs (and they polymerize DNA using deoxy-NTPs, or dNTPs). The triphosphate structure is a high-energy form. When the phosphates are cleaved, energy is released. The reaction $NTP \rightarrow NMP + PP_i$ yields 45.6 kJ/mol of energy ($\Delta G = -45.6$ kJ/mol). When using NTPs for nucleotide polymerization, the two reactions are coupled to require $25 + (-45.6)) = -20.6$ kJ/mol (20.6 kJ/mol will be yielded) (see Figure 4.3).

4.1.2   Nucleotide Bases, Nucleosides, and Nucleotides

The R group attached to carbon #1 of (deoxy)ribose could be a hydroxyl, although it seldom is. When a nucleotide is constructed from scratch by the cell, the R group at some point will be a pyrophosphate ($-O-PO_3^--PO_3^{2-}$). However, for nucleotides, the R group will be one of the DNA or RNA bases: guanine, adenine, cytosine, thymine, or uracil. The structures of these nitrogenous bases are given in Figure 4.4. The bases guanine, adenine, cytosine, and thymine (G, A, C, and T) are typically found in DNA. The same is true for RNA, except that thymine is replaced by uracil (U). Note that there are two types of ring structures for these bases: a bicyclic structure containing a five-member and a six-member ring and a monocyclic structure with a six-member ring. The bicyclic structure is characteristic of the bases known as *purines*, which include guanine and adenine, while the monocyclic structure is characteristic of the *pyrimidines*, which include cytosine, thymine, and uracil.

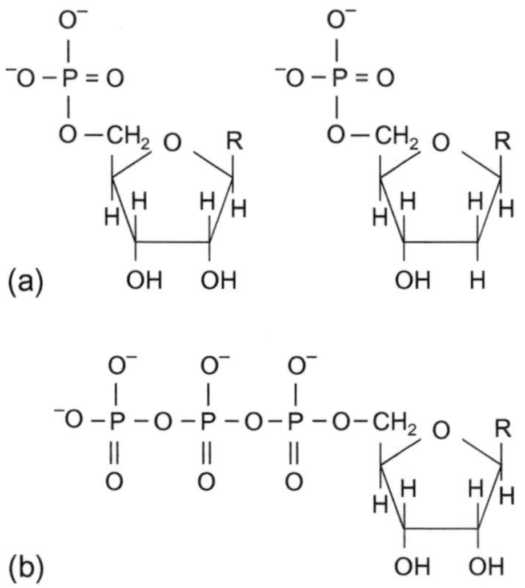

FIGURE 4.2   (a) Left: a nucleotide, also known as a nucleotide monophosphate (NMP). Right: a deoxynucleotide, or dNMP. (b) A nucleotide triphosphate (NTP).

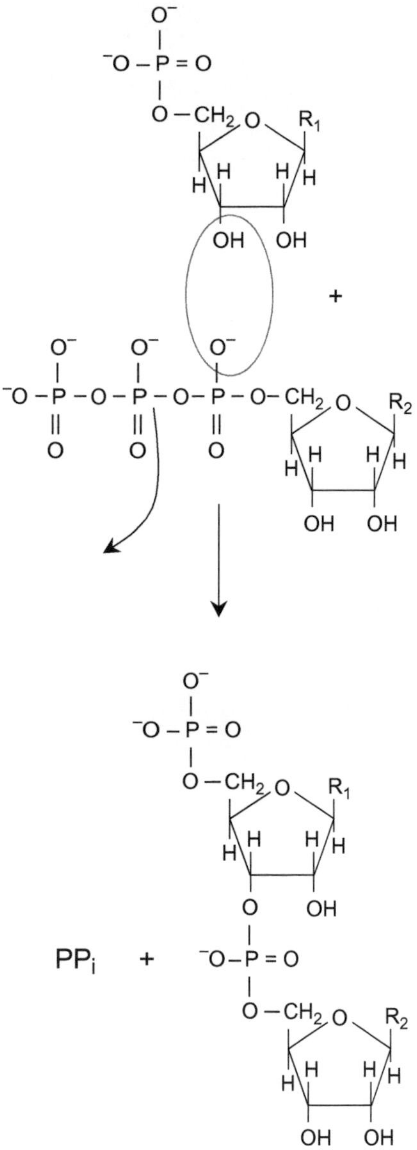

FIGURE 4.3 NTPs are used in growing a poly(nucleotide) chain. While the reaction of a 3′ hydroxyl with a 5′ phosphate requires energy (25 kJ/mol), the cleavage of the high-energy phosphate bond in the NTP to produce a pyrophosphate group (shown as PP_i) yields 45.6 kJ/mol. The two reactions are coupled to produce a net energy yield of 20.6 kJ/mol, making the polymerization reaction energetically favorable.

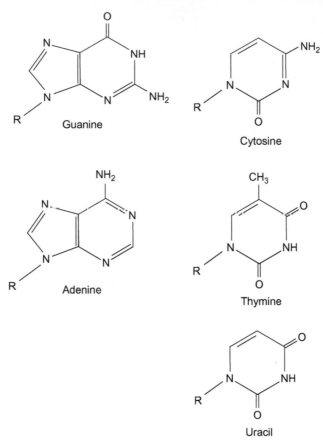

FIGURE 4.4   Structures of the nucleotide bases. Purines are in the left column, and pyrimidines are in the right column. The makeup of the R group will determine if the molecule is a base, a nucleoside, or a nucleotide (see Table 4.1).

   Again referring to Figure 4.4, note the R group attached to each of the bases. When the R group is a hydrogen atom, the molecule is a purine or pyrimidine base. If the R group is a ribose, then the molecule is a *nucleoside*. If the R group is a phosphoribose, then the molecule is a *nucleotide*. The names of the most common of these molecules are given in Table 4.1.

   It was just pointed out that DNA is polymerized using NTPs. This means that for DNA, polymerization is carried out using dGTP, dATP, dCTP, and dTTP. For RNA, UTP is also used. Note that the ATP used in RNA polymerization is the same molecule that was described previously as an energy donor for other reactions and processes, such as active transport. It is also possible for the other NTP's to serve as energy donors in specific cases.

TABLE 4.1 Names of Purine and Pyrimidine Bases, Nucleosides, and Nucleotides

Base	Nucleoside	Nucleotide
Base	Base / (deoxy)Ribose	Phosphate / (deoxy)Ribose / Base
Guanine	Guanosine	Guanylate (GMP)
	Deoxyguanosine	Deoxyguanylate (dGMP)
Adenine	Adenosine	Adenylate (AMP)
	Deoxyadenosine	Deoxyadenylate (dAMP)
Cytosine	Cytidine	Cytidylate (CMP)
	Deoxycytidine	Deoxycytidylate (dCMP)
Thymine	Deoxythymidine	Deoxythymidylate
Uracil	Uridine	Uridylate

Knowledge of the structures of the five nitrogenous bases is important for understanding DNA base pairing, as well as why certain transcription factors are specific for certain sequences. Note that there are two types of ring structures for these bases: a bicyclic structure containing a five-member and a six-member ring and a monocyclic structure with a six-member ring. The purines are defined by this bicyclic structure, while the pyrimidines are defined by the monocyclic structure. Note that, as displayed in Figure 4.4, the bottom-left nitrogen will be the atom that attaches to carbon #1 of the ribose sugar.

When a DNA sequence is written out, it is always written in the $5' \rightarrow 3'$ direction, meaning the nucleotide with the unpolymerized phosphate on the #5 carbon is on the left, and the nucleotide with the unpolymerized hydroxyl on the #3 carbon is written on the right. The structure of a trinucleotide is shown in Figure 4.5. For simplicity, however, DNA sequences are typically written out with just the one-letter abbreviations of the bases. The trinucleotide shown in the figure can therefore be written as CTG. From this nomenclature, not only the directionality of the nucleotide sequence but also the composition in terms of DNA and RNA can be determined. A sequence such as AGCCAUGC would represent an RNA sequence, because of the presence of uracil.

4.1.3   DNA Is the Genetic Material

In the early days of molecular biology, it was not readily apparent what the genetic material was made of. In fact, for many years, it was thought that proteins

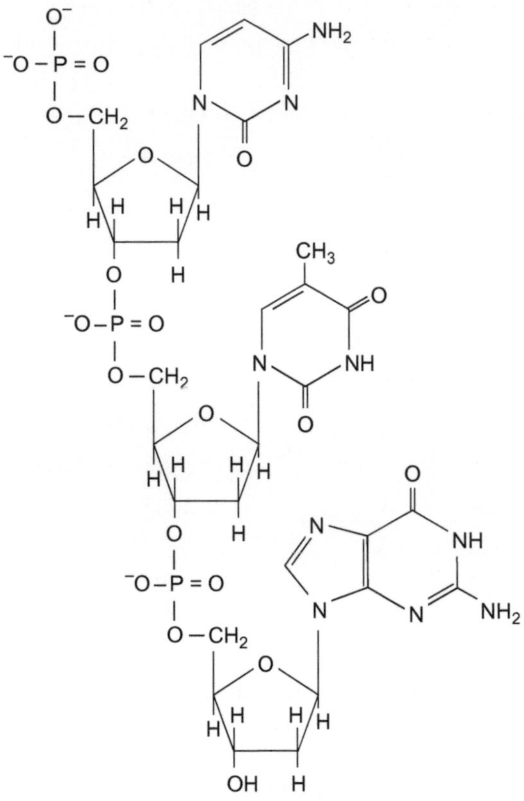

FIGURE 4.5 The trinucleotide CTG. Note that the one-letter abbreviations for DNA nucleotides can be written the same as for RNA (*e.g.*, CTG vs. dCdTdG), especially when a sequence is being depicted. Also note the terminal phosphate on the 5′ terminus of the structure and that the DNA sequence "CTG" is written with the 5′ terminus on the left.

contained the genetic blueprints for heredity. In 1952, however, Alfred Hershey and his research assistant Martha Chase published the results of the experiments that proved that it is DNA that serves as the genetic material.

   Hershey and Chase capitalized on the fact that DNA and proteins can be distinguished by the presence of distinct atoms (Figure 4.6). A phosphorus atom can be found in DNA but not in polypeptides. Conversely, sulfur atoms can be found in most proteins (by virtue of the amino acids cysteine and methionine) but not in DNA. The team capitalized on these differences through the radioactive tracers ^{32}P and ^{35}S to distinctly label DNA and proteins. Viruses, which have proteins in their coats and DNA in their payloads, were labeled with ^{32}P or ^{35}S. *E. coli* cultures were exposed to the labeled viruses, which bound to the bacterial cells and injected their payloads. A blender was used to detach the viruses from the cells, and it was concluded that the shear force from the blender caused

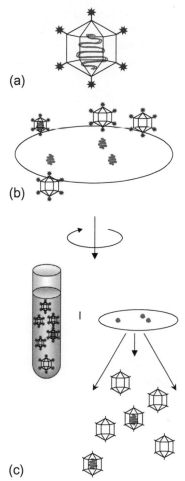

(a)

(b)

(c)

FIGURE 4.6 Summary of the Hershey-Chase experiments. (a) Hypothetical virus with ^{32}P (red)-labeled DNA and ^{35}S (blue)-labeled proteins. (b) When exposed to cells, labeled viruses will adhere via their coat proteins and inject their DNA. A blender was used to knock viral particles off of the cells to assess the location of the radioactive labels. Labeled proteins remained extracellularly, with viral ghosts. DNA, however, was found inside the cells. (c) Infected cells produced viral progeny containing some labeled DNA but no labeled proteins.

~75% of the ^{35}S but only ~15% of the ^{32}P to be released into the (extracellular) solution; most of the ^{32}P had been transferred into the cells. The viruses were injecting DNA, not proteins, into the cells.

After a period of incubation, new viruses were formed in the cells and the viral progeny were released. About 30% of the original parental ^{32}P was recovered in the progeny, but <1% of the ^{35}S was recovered. We now interpret this to mean the transfer of DNA was responsible for the viral coats in the progeny. While the newly formed viruses all had viral coats, the coat proteins were not constructed from the labeled proteins of the parent viruses. Instead, the DNA that had been delivered by the parent viruses had served as a message that led to the formation of protein products. DNA had been shown to be the genetic material.

4.1.4   Genomic DNA Is Double-stranded

Once it was established that DNA was the genetic material, the question arose as to how it could ever be replicated in a way that a mother could pass on a copy of her genes to her offspring. There must be some way in which the information is conserved from one generation to the next.

In the 1950s, a scientist named Erwin Chargaff reported results from his analyses of DNA from various species, finding that the composition of DNA varied from one species to another. However, he also noted that while the percentage of DNA nucleotides containing adenine might vary between species, the percentage of DNA bases containing thymine was roughly equal to the percentage of bases containing adenine. The same was true for guanine versus cytosine. These relations, now referred to as *Chargaff's rules*, were early indications of the double-stranded nature of DNA and the regular pairing of bases between the two strands. Combined with the X-ray crystallographic data of Rosalind Franklin, which indicated that DNA has a helical structure, Chargaff's data led to the notion of a double-stranded DNA helix with complementary strands due to A-T and G-C pairing.

Using the structures presented in Figure 4.7, it can be seen that hydrogen bonding can occur simultaneously at two sites within adenine/thymine pairs.

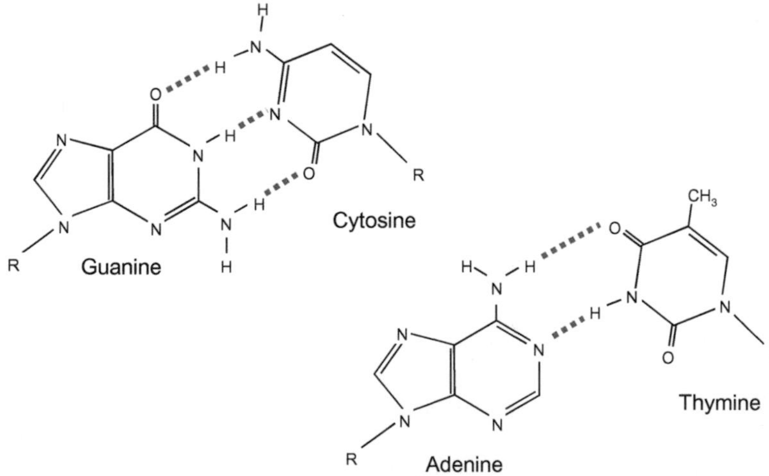

FIGURE 4.7   Hydrogen bonding can occur between the DNA base pairs as shown by the dotted lines.

Likewise, three hydrogen bonds can form in G-C pairs. Because the bases in double-stranded DNA generally adhere to these pairing rules, knowing the sequence of one strand in a double-stranded polynucleotide provides enough information to deduce the sequence of the complementary strand. While a sequence could be written as ATCGT, it is understood that this written sequence actually refers to

5′-ATCGT-3′

3′-TAGCA-5′

Consider two100 base pair stretches of DNA, one comprising solely G-C base pairs and the other consisting of only A-T base pairs. Because each G-C pair is held together by three hydrogen bonds, the bases in such a pair would adhere to one another more stably than those of an A-T base pair, which only has two hydrogen bonds. Extending this idea to our two hypothetical DNA fragments, one could guess that it would be more difficult to separate the double-stranded DNA (dsDNA) fragment made completely of G-C base pairs into two single-stranded DNA (ssDNA) fragments. The higher requirement for energy is applicable to melting temperature as well. Because of the extra 100 hydrogen bonds, it would take a greater amount of heat energy to melt the poly(G-C) strand into its single-stranded components. The temperature at which a dsDNA fragment is melted into two ssDNA fragments is known as the melting temperature (T_m) of the strand. T_m is directly dependent upon the base composition of the dsDNA in question. Figure 4.8 illustrates this correlation. If we took a solution of several identical dsDNA fragments and we then added heat to the system and monitored when they separated into single-stranded fragments, we would be able to determine the (G + C) content of the DNA fragments based upon this

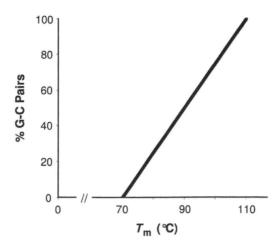

FIGURE 4.8 Because of increased hydrogen bonding, G-C base pairs require more energy to separate than do A-T base pairs. As a result, the greater the percentage of G-C pairs in a piece of dsDNA, the greater the melting temperature (T_m).

relation. Notice that the %(G+C) would be the same for the dsDNA and each separate ssDNA fragment.

4.1.5   DNA Replication Is Semiconservative

The fact the DNA replication takes place in a semiconservative fashion was shown in 1957 by equilibrium sedimentation experiments performed by Meselson and Stahl. Suppose we have a bacterial culture grown in a controlled environment. The bacteria, of course, have DNA in them. Before a given bacterium can divide into two daughter cells, DNA must be manufactured and an entire genome must be constructed. The materials for the new DNA will ultimately come from the cellular environment; in the laboratory, this will be growth medium. If we were to grow the bacterial culture for multiple generations in an environment that contained only the heavier ^{15}N form of nitrogen atoms, then (through subculturing techniques) we could obtain a culture in which eventually every nitrogen atom of the bacterial DNA was ^{15}N. If we were to extract the DNA from these cells and centrifuge it in a cesium chloride gradient, the genomic DNA would migrate a distance of x through the gradient (Figure 4.9). Because of the ^{15}N, the DNA from these bacteria would have a higher density than the DNA from a bacterial culture grown in normal (^{14}N-containing) medium, which would migrate a shorter distance (say, z) under the same conditions.

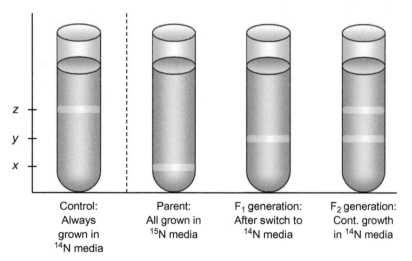

FIGURE 4.9   Cells grown in ^{15}N-containing media will have DNA of a higher density than untreated controls, as seen following centrifugation in a CsCl gradient. After moving the culture to ^{14}N-containing media, the first generation of daughter cells will have genomes that migrate a lesser distance (y) than did the genomes from the parent culture (x). The second generation of cells grown in the ^{14}N-containing media will have genomes that migrate upon centrifugation either the same distance as the F_1 generation or the same distance as untreated controls (z).

Next, some of the bacteria grown in the heavy medium were transferred to an environment containing normal (^{14}N-containing) medium and allowed to proliferate for some time (one generation). The results of DNA extraction and centrifugation in the CsCl gradient would produce a band at *y*, a distance somewhere between *x* and *z*. Perhaps even more interesting, if this newly transferred bacterial culture were allowed to proliferate for even more time (perhaps one additional generation), then two bands would be seen upon centrifugation. The bands would correspond to the distances *y* and *z* (Figure 4.9).

The explanation offered by Meselson and Stahl to explain the above results was that during DNA replication, the parental dsDNA strand was separated into two single-stranded templates, and a single strand of newly synthesized DNA was created to match up with each of the two templates. The DNA of each daughter cell would then contain dsDNA composed of one parental strand and one newly synthesized strand that contained ^{14}N. This would explain the single, hybrid band of DNA that migrated the intermediate distance of *y* in the above centrifugation experiments. For the second generation of replication, each hybrid strand would separate in the same fashion into two ssDNA templates. This time, however, one of the single-stranded templates would contain the heavy nitrogen, and the other template would contain ^{14}N. New DNA strands would be created to pair with the template strands, yielding two daughter dsDNAs composed of ^{15}N-^{14}N and ^{14}N-^{14}N. Half of the parent DNA would be conserved in each of the daughter cells, hence the term *semiconservative replication*.

DNA polymerase produces DNA polymers. This enzyme is responsible for producing the daughter strands made during replication. Similarly, when the cell uses genomic DNA to produce RNA, such as the mRNA produced during transcription, the cell uses the enzyme RNA polymerase.

4.2   FROM GENES TO PROTEINS

4.2.1   Introduction to the Genetic Code

The act of *transcription*, where RNA polymerase traverses genomic DNA to produce an RNA transcript that will be further processed to yield messenger RNA (mRNA), proceeds in the 5′ to 3′ direction. The mRNA is also produced in the 5′ to 3′ direction. This may at first seem contradictory, until one considers that the genomic DNA is double-stranded. We may write a code like ATGATGATGTGA but recall that such a code refers to

5′-ATGATGATGTGA-3′

3′-TACTACTACACT-5′

The strand that is written on top is referred to as the *coding strand*. Think of this as the strand that you would talk about if you were discussing the genetic code of a gene. The other strand, written in the 3′ to 5′ direction above, is called the *template strand*. RNA polymerase will bind both strands. They will be separated and the template (bottom) strand is the one that's going to be used by the enzyme to create a primary RNA transcript. The enzyme will move toward the right (with respect to the diagram), in the $5' \rightarrow 3'$ direction. We refer to the movement of the enzyme as the $5' \rightarrow 3'$ direction because that is the direction the enzyme is moving in terms of the coding strand, and it is the direction in which the enzyme is making an RNA sequence. The new RNA sequence will be complementary to the template strand and identical to the coding strand (with the exception that uracils will be used in place of thymines). The messenger RNA sequence corresponding to the dsDNA above would be 5'-AUGAUGAUGUGA-3'.

Transcription takes place in the nucleus of eukaryotic cells (and in the cytoplasm of prokaryotic cells) because that is where the genomic DNA is. *Translation*, or the conversion of an mRNA sequence into a polypeptide sequence, takes place in the cytoplasm. In eukaryotes, this necessitates the transport of the mRNA from the nucleus to the cytoplasm through nuclear pores. Translation is then carried out by organelles referred to as *ribosomes*.

When ribosomes process a polynucleotide, they handle the nucleotide bases in groups of three. These groups are referred to as *codons*. The mRNA sequence produced in our example above can now be thought of as AUG AUG AUG UGA. Refer to Figure 4.10 to determine the amino acids that are coded for by a codon sequence. The polypeptide is produced beginning with the N-terminus and ending with the C-terminus, or the $N \rightarrow C$ direction.

4.2.1.1 Degeneracy and Wobble

Again refer to Figure 4.10 (the genetic code) and note that an amino acid such as valine is pretty easy to decode once the first two bases have been located. GUU, GUA, GUC, and GUG all code for valine. The fact that the genetic code often has more than one codon that codes for a single amino acid is termed *degeneracy*. Also notice that when an amino acid is represented by more than one code, which is the case for every amino acid except methionine and tryptophan, the first two bases are identical in the cases where there are four or fewer redundant codes for the given amino acid. This has been hypothesized to be due to a condition known as "wobble," which is a straightforward concept, but to appreciate it, we must first discuss ribosomes and their interactions with *transfer RNA (tRNA)*.

4.2.1.1.1 Ribosomes and Translation

We were first introduced to ribosomes in this book during a discussion of antibiotics and the class that targeted the act of translation in bacteria through the targeting of specific ribosomal subunits. Ribosomes attach to messenger RNA and travel down it in the $5' \rightarrow 3'$ direction, interpreting codons and building a primary amino acid sequence based on a codon sequence with the help of tRNAs. A tRNA molecule is itself a polynucleotide, an RNA molecule that is typically

Second

		U		C		A		G		
		UUU	Phe	UCU	Ser	UAU	Tyr	UGU	Cys	U	
	U	UUC	Phe	UCC	Ser	UAC	Tyr	UGC	Cys	C	
		UUA	Leu	UCA	Ser	UAA	Stop	UGA	Stop	A	
		UUG	Leu	UCG	Ser	UAG	Stop	UGG	Trp	G	
		CUU	Leu	CCU	Pro	CAU	His	CGU	Arg	U	
	C	CUC	Leu	CCC	Pro	CAC	His	CGC	Arg	C	
		CUA	Leu	CCA	Pro	CAA	Gln	CGA	Arg	A	
First		CUG	Leu	CCG	Pro	CAG	Gln	CGG	Arg	G	Third
		AUU	Ile	ACU	Thr	AAU	Asn	AGU	Ser	U	
	A	AUC	Ile	ACC	Thr	AAC	Asn	AGC	Ser	C	
		AUA	Ile	ACA	Thr	AAA	Lys	AGA	Arg	A	
		AUG Met (Start)		ACG	Thr	AAG	Lys	AGG	Arg	G	
		GUU	Val	GCU	Ala	GAU	Asp	GGU	Gly	U	
	G	GUC	Val	GCC	Ala	GAC	Asp	GGC	Gly	C	
		GUA	Val	GCA	Ala	GAA	Glu	GGA	Gly	A	
		GUG	Val	GCG	Ala	GAG	Glu	GGG	Gly	G	

FIGURE 4.10 The genetic code. To read this chart, begin with the left-hand column to find the first base of your codon. This will determine the four rows of interest. Next, use the column headings to find the second base of the codon. This will determine the four columns of interest. The intersection of the four rows and four columns will yield four entries, a set which can be narrowed to the correct entry by using the base indicated on the right side of the chart that corresponds to the third position in your codon. Degeneracy is shown by colors: Amino acids with two codes are highlighted in orange or pink, the amino acid Ile has three codes highlighted in purple, amino acids with four codes are highlighted in yellow, and amino acids with six codes are highlighted in light green or blue. The three stop codons are highlighted in red. The single codons of Trp and Met (which is also the Start codon) are highlighted in white.

73-93 nucleotides in length. The 3′ end of the tRNA molecule can be covalently bonded to a specific amino acid. A tRNA molecule will pair with a codon via three bases in the tRNA structure known as the *anticodon* (Figure 4.11). If the mRNA codon is CUG, the tRNA that will be used in translation of this codon will have the anticodon CAG (verify this for yourself, remembering that the sequence of the anticodon is written in the 5′ → 3′ direction).

The large ribosomal subunit has three places that can bind tRNA: the A site, the P site, and the E site. The A site typically holds an aminoacyl-tRNA, meaning a tRNA carrying an amino acid. The P site typically holds the peptidyl-tRNA, meaning the tRNA that is holding the growing peptide chain. The E site can be thought of as a temporary holding site for empty tRNAs (Figure 4.12).

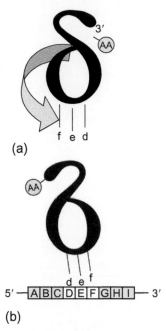

(a)

(b)

FIGURE 4.11   (a) The structure of a tRNA molecule is similar to the lower-case Greek letter delta. Here, an amino acid specific to the particular tRNA is attached to the 3' terminus. The anticodon, which will pair with an mRNA codon, is shown at the bottom of the panel. The arrow denotes the 5' to 3' direction. (b) The same tRNA interacting with an mRNA molecule (ribosome not shown). This panel illustrates the direction of the tRNA (3' to 5') relative to the mRNA molecule. The letters in the mRNA and anticodon represent a fictitious base-pairing system where A pairs with a, B pairs with b, *etc*.

FIGURE 4.12   Schematic of the ribosome, with the large subunit shown in light gray and the small subunit shown in darker gray. The three binding sites for tRNA are shown by the letters E (empty tRNA), P (peptidyl-tRNA), and A (aminoacyl-tRNA).

Translation of mRNA via ribosomes can be thought of as a three-step process. Figure 4.13a shows a ribosome in the process of translating an mRNA molecule, having the P and A sites already filled with tRNAs linked to the (poly)peptide and an amino acid, respectively. The first step of the process has an aminoacyl-tRNA associating with the ribosomal A site. The ribosome may

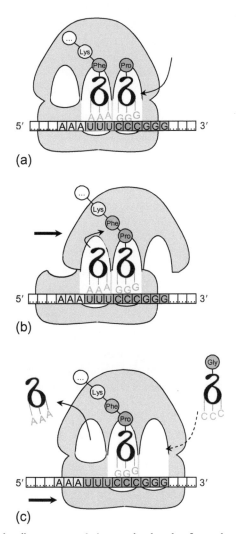

(a)

(b)

(c)

FIGURE 4.13 For the ribosome, translation can be thought of as a three-step process. (a) An aminoacyl-tRNA is loaded into the A site and checked for accuracy via anticodon hydrogen bonding with the codon. (b) The growing polypeptide chain is transferred to the aminoacyl-tRNA acquired in step 1. A conformation change keeps the peptidyl-tRNA in the ribosomal P site, and the now-empty tRNA is located in the ribosomal E site. (c) Another conformation change causes release of the old tRNA while the small ribosomal subunit realigns with the large subunit and is positioned three nucleotides (1 codon) downstream, ready to begin the process again.

process several different tRNAs until a proper match is found between the anti-codon of the tRNA and the mRNA codon lined up with the A site. Whether the proper tRNA is held in the A site is determined via base pairing (through hy-drogen bonding) between the anticodon and the corresponding mRNA codon. In step 2 of the process (Figure 4.13b), the growing peptide chain is cleaved at its C-terminus from the tRNA and then moved over to the α-amino group of the amino acid on the adjacent tRNA. A new peptide bond is formed to make the growing chain one residue longer. Note that, due to a conformation change in the ribosome, the growing peptide chain remains associated with the P site in the ribosome, even though the chain is now held by a different tRNA. Step 3 has the ribosome regain its original conformation but three nucleotides closer to the 3′ terminus of the mRNA (Figure 4.13c). The now-empty tRNA is released from the E site in this step, rendering the ribosome ready to start the process again. Note that while the ribosome is progressing in the 5′ → 3′ direction, the polypeptide chain is growing in the N → C direction.

4.2.1.1.2   Back to Wobble

The first two bases of the codon/anticodon pair must match up very well, but there is a little bit of play in the tRNA as it is held by the ribosome, so it can physically wobble into and out of contact with the third base of the codon. This implies that the third base in the codon is perhaps not as important in terms of coding because of this wobble. Refer once again to Figure 4.10 and note that GUU GUC GUA and GUG all code for the same amino acid—valine. This *wobble effect* helps to explain the degeneracy of the genetic code.

4.2.1.2   Mutations and Their Effect on Translation

If one base in the codon CUG were changed to CUA, this would indeed be a mutation. However, this would be an example of a *silent mutation* because the base change would not have any effect on the identity of the amino acid that is attached to the growing polypeptide by the ribosome, and therefore not affect the final polypeptide. Another, more serious type of mutation is the *frameshift mutation*. Such a mutation involves a shift in the groups of three that are read to make a codon. For instance, while shifting the reading frame to the right by one base has this mRNA sequence code for something completely different:

$$\text{ACG|AGC|AAC|GAA|UGA|AAA} \xrightarrow{\text{codes for}} \text{Thr-Ser-Asn-Glu}(\text{-Stop}),$$

$$\text{A|CGA|GCA|ACG|AAU|GAA} \rightarrow \text{Arg-Ala-Thr-Asn-Glu-}\dots$$

Note that not only are the amino acids different, but also the stop codon has been lost and the mRNA will continue to be translated into a larger polypeptide. A single frameshift mutation can therefore have a profound effect on the cell. These mutations can be brought about by *insertions* or *deletions*: adding an extra nucleotide or deleting one nucleotide from the exon, respectively.

4.2.2   Genes

That is the basis of how the process gets from DNA to a polypeptide. However, a stretch of DNA is more than just a coding region. A stretch of DNA also has control regions, such as regions that serve to activate transcription or even inhibit transcription. There may also be regions we don't know what they do (yet) but are important for the survival of the cell; maybe these regions serve as filler, not coding for anything, not necessarily activating or inhibiting transcription, but certainly fouling things up if they were to be removed. Let us therefore use the following definition for a *gene*: a stretch of DNA that functions as a unit to give rise to a polypeptide product via mRNA.

Consider a gene that codes for the green fluorescent protein. The part of the gene that specifically codes for the amino acids in the green fluorescent protein is known as the *exon*. An exon is a stretch of DNA that will be represented in a mature RNA transcript (such as mRNA, ribosomal RNA (rRNA), or tRNA). Before the gene is transcribed, RNA polymerase must first bind to the DNA. The spot at which it binds is referred to as a *promoter*. Other bases in the DNA sequence may serve to enhance the binding of the transcriptional machinery, hence the descriptive term *enhancer*. Enhancers can be upstream (meaning on the 5′ side of the coding region) or downstream (3′) to the exon. Enhancer sequences can also be backward and still serve to enhance transcription. Another DNA sequence that does not code for amino acids but still serves as a part of a gene is the eukaryotic silencer (known as an operator in prokaryotes). Figure 4.14 gives a general layout of some important genetic elements.

At one point in time, research showed that each step in a metabolic pathway was controlled by the product of a gene, and if one was able to destroy that gene somehow, then the function of the enzyme at that particular step in a metabolic pathway would be knocked out. This led to the "one gene-one enzyme" hypothesis, which we know now to not be entirely correct. It may seem, especially in light of the completion of the sequencing of the human genome, that we could now analyze the database and elucidate every gene in the human genome. After the sequencing had been completed, some hoped that for any malfunctioning protein an individual might produce, the corresponding genes that have mutations could be located and treatments be developed from there. The problem was not that simple, though. There is not a linear pathway from protein back to gene. This case in point is brought to light by the example of Gregor Mendel and his peas. Although Mendel published his famous work in 1866 (which went largely unrecognized until 1900), the gene responsible for wrinkled versus smooth peas was not discovered until 1990. (The gene codes for a starch branching enzyme. A defect in that gene would cause the peas to have wrinkled coats because they lacked appropriate branching to push out on the coat of the pea.) The moral is that it is easy to discover a phenotype, and it is relatively easy to discover the

FIGURE 4.14   Hypothetical setup of a gene, showing enhancers (Enh), a promoter (Prom), the transcriptional start point (arrow), the translational start codon (AUG), and a translational stop codon (UAG in this example). Note that enhancers can occur as a string of repeats, that a single gene can have different enhancers (denoted here by different fonts), and that some enhancer sequences can appear in either direction.

protein responsible for that phenotype, but it can be far more difficult to identify the gene or genes responsible for the phenotypic change.

It's not that a gene is producing a protein, it is coding for a polypeptide. And if we think back to when we were talking about the structure polypeptides (primary, secondary, tertiary, and quaternary structure, recalling that quaternary structure is the coming together of two or more tertiary structures) ... the example of hemoglobin was presented. Hemoglobin is a dimer of dimers, or it could be called a tetramer. Each of these subunits is a polypeptide, and four polypeptides come together to make the functioning protein hemoglobin. So, the "one gene one protein" hypothesis is more accurately modified to the "one gene one polypeptide" concept. In addition, having the primary sequence of a polypeptide does not mean we can backtrack directly to the genomic sequence. There is more than one codon that could code for a given amino acid for 18 of the 20 common amino acids. To make matters more complicated, having the primary sequence of a functioning polypeptide does not account for amino acids that were cut out during folding and processing by the cell to produce the functioning unit. Bases can even be cut out during RNA processing in producing a functioning mRNA. The point is that a gene codes for a polypeptide sequence, but so many modifications happen between gene and final polypeptide (which may only be one piece of a functioning protein) that the link between a phenotype and the associated gene can be difficult to establish, even with a sequenced genome.

4.2.2.1   How Many Genes Are in the Human Genome?

There are $\sim 3.08 \times 10^9$ bases in the human genome. For many years, it was believed that there were 100,000 genes in the human genome, based on the average number of amino acids in a protein. However, the completion of the human genome project revised that number down to about 33,000 genes. After further scrutiny of the human genome project results, the number was further revised down to about 25,000. At the end of 2010, four different counts had the total number of human genes between 18,877 and 38,621. In April, 2011, the RefSeq database, maintained by the National Institutes of Health in the United States, listed the number of human genes at 25,564 in April 2011. One might think that with the entire genome sequenced, it would be a straightforward matter to determine the exact number of genes in the human genome. The task is not so simple because, in addition to all of the noncoding regulatory regions we have discussed, introns (which also do not code for part of a polypeptide) can appear even in the middle of a gene. Coupled with the fact that if we say that the average gene utilizes 1000 nucleotides, 30,000 genes would be <1% of the total genome. Finding genes in the genome is the same as finding the proverbial needle in a haystack (Figure 4.15).

One of the reasons that the estimate of the total number of genes in the human genome has gone down is that it was found that certain polypeptides that are used to make large proteins are reused for other proteins. The fact that these

FIGURE 4.15 The number of genes in the human genome makes up a surprisingly small portion of the total DNA found in our chromosomes, as estimated by the above pie chart.

polypeptides can be used in more than one protein means the product of a gene can be used in more than one place. "One gene-one polypeptide" still holds, but it must be kept in mind that one gene can contribute to many different proteins.

4.2.2.2 Phenotypes

Why is it that one phenotype is dominant over another? It has to do with the proteins that are formed by an organism. Recall that each of our genes has two *alleles* (two copies), which occurs because we get one copy of DNA from our mother and one copy of DNA from our father. So, for every gene that we have in our bodies, we have two versions of it. If the mother and father of an organism each donated a dominant form of the gene to the offspring, then the offspring would display the dominant phenotype (also known as *wild type*). If, however, one allele coded for the wild type of the protein (perhaps the brown pigment seen in human eye color) and the other coded for a recessive form (perhaps blue, or *null*), the wild-type gene would be responsible for the cell producing a brown pigment while the other allele will fail to yield brown pigment. The offspring would still be walking around with the brown pigment in their irises—maybe not as much, but the brown pigment would still be present. So even having only one good version of the polypeptide coded for by a gene is enough to give one the wild phenotype (in most cases) and the only time we can get a mutant phenotype is when we have two mutant copies of the gene.

As an example of how the alteration of one gene can affect an entire organism and how that alteration can be passed to the future generations, consider blood types. Most people are aware that there are four main blood types: O, A, B, and AB. (We will not consider Rh factors in this example.) These four blood types indicate a specific antigen, known as the H antigen, that is attached to certain blood proteins (or a sphingolipid) on red blood cells. The gene responsible for the H antigen codes for a fucosyltransferase. This gene is one gene in a group of seven that constitutes the ABO locus found on chromosome 19. (A *locus* is a location, or where a gene can be found in the genome.) The three blood types arise from the presence or absence of a gene coding for an enzyme responsible for transferring galactosyl groups. The A type of the gene codes for an *N*-galactosyltransferase that will transfer an *N*-acetylgalactosamine to the H antigen, while the B type of the gene codes for a galactosyltransferase that will transfer a galactose to the H antigen (Figure 4.16). The O type of the gene does not code for such an enzyme, so the H antigen will not be further modified at the glucose end. The reason that the O form of the gene is inactive is that the gene is missing a single guanine in one codon, which causes that codon and all subsequent codons to be misread (a frameshift mutation). The result is that no active enzyme is made from that gene.

The H antigen is not interpreted as foreign, which means type O blood can be put into any individual. However, type A (or B) blood put into an individual with type O blood would be a problem because the red blood cells would contain glycoproteins that are not produced in the recipient's body and would therefore be recognized as foreign. Those with type O blood are known as universal donors, but they are certainly not universal recipients. Those with type AB blood can hold the title of universal recipient because, no matter what type of blood they receive, they should not mount an immune response to it because their body already makes matching glycoproteins.

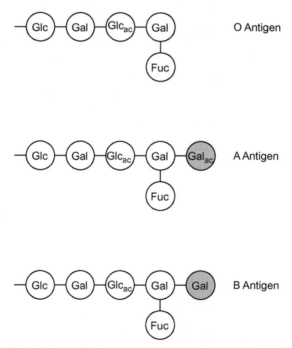

FIGURE 4.16   Structures of the H antigen, which determines ABO blood type. Glc, glucose; Gal, galactose; Glc$_{ac}$, *N*-acetylglucosamine; Gal$_{ac}$, *N*-acetylgalactosamine; Fuc, fucose. The sole difference between the A and B antigens from the O antigen is shown in pink.

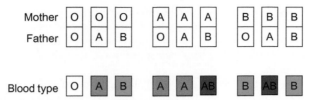

FIGURE 4.17 Blood type is determined by the H antigen, an allele of which is inherited from each parent. Each column above shows one of the nine possibilities for inheritance of the H antigen genes. Note that, for example, if a child inherits an A and an O allele, she is considered type A because of the immunological implications.

To make this idea mesh more with our idea of alleles, know that you have two alleles for every ABO locus: one from your mother and one from your father. Each of these alleles could contain the A, B, or O form of the gene just discussed. All of the possibilities are shown in Figure 4.17. While somebody with two type A alleles will have type A blood, somebody with both type A and type O alleles will also be said to have type A blood. If some of the type A oligosaccharide is made, whether a large or small amount, then the phenotype of the blood will reflect this because even a small amount of the A oligosaccharide would be recognized as foreign in an individual who did not have a type A allele. The same is true for type B alleles.

Karl Landsteiner discovered the ABO blood types in 1900, and 40 years later, he, along with Alexander Wiener, discovered the Rh blood types. The term "Rh" was short for rhesus macaque, the monkeys initially used in making the antiserum for blood typing. This blood typing system is very complex, as it involves 45 different red blood cell antigens. However, the entire system is essentially controlled by two genes, and inheritance has an even simpler binary form (either you have the antigens or you don't). A person with type O blood displaying the antigens is said to be O^+ ("O-positive"), as opposed to O^- ("O-negative").

Aside: Mother/Child Blood Issues

Consider for a moment a type O^- mother who is carrying a type O^+ fetus. If the blood of the fetus were to mix with that of the mother, the mother's immune system would mount a response against the Rh antigen of the fetus that would be devastating for both. This event is prevented *in utero* by a sort of blood-baby barrier. The blood of the mother and that of the fetus are two separate tissues (yes, blood is a tissue) that are kept separate, although close, in the placenta to ensure that they do not mix during gestation. However, during childbirth or the latter months of pregnancy, the placenta tears away from the uterus and some mixing of the mother and child's blood can occur. The amount of mixing is relatively low and does not present a significant problem for either. However, the mother's immune system will become educated against the Rh antigens of that child and be able to mount a quicker and more intense response against those Rh factors in the future, such as is possible if she carries another Rh-positive fetus. This could be disastrous during a subsequent pregnancy because fetal red blood cells would be destroyed, leading to anemia and a rise in unconjugated bilirubin levels, followed by several adverse conditions including brain damage or possibly death of the fetus or child shortly after birth.

To prevent this type of immunological attack, Rh-negative mothers-to-be may be given the drug RhoGam®. The drug contains antibodies against Rh antigens. The principal behind the mechanism of the drug is that Rh antigens will be bound by the antibodies in the drug before the antigens come into contact with the mother's active antibodies. The drug is given to the mother, and the inactive anti-Rh antibodies of the drug will circulate in her blood.

A type O mother giving birth to a second type A or B child can also pass some of her antibodies to the child during childbirth. The result will again be red blood cell destruction, which leads to anemia and heightened unconjugated bilirubin levels (jaundice). Such a condition will be treated with phototherapy ("bili lamps") to break down the bilirubin and blood replacement transfusions to clear out the maternal antibodies if the condition is severe enough.

The concept of typing can be further extended to include bone marrow or organs for transplantation. An in-depth matching must be performed because we don't want the recipient's body to reject the transplant. Rejection involves the recognition of certain molecules by the immune system. The fact that they are recognized by the immune system qualifies these molecules as *antigens*. Once the antigens are identified as foreign, cells of the immune system will be recruited to the area of transplantation and the "foreign" cells will be destroyed.

4.2.3   Transcription

4.2.3.1   The Start of Transcription: RNA Polymerase Binds to DNA

4.2.3.1.1   Prokaryotes

Our general discussion has taken us from DNA to RNA to protein. Let us now examine transcription at a greater depth. In human cells, with only about 25,000 genes in a genome of over 3 billion base pairs, one might wonder how RNA polymerase determines where to start transcription. Referring back to Figure 4.10 (the genetic code), we can see that AUG codes for START. AUG also codes for methionine, which indicates that at some point in the processing of polypeptides, they will all begin with a methionine. However, the initial polypeptide will often undergo modification, including the removal of some amino acid residues, as it matures into a functional polypeptide/protein.

The way that RNA polymerase determines where to start is different between eukaryotes and prokaryotes. The prokaryotic method is described here for simplicity, with the intention of giving the reader the general insight needed to appreciate transcriptional commencement.

Bacterial *RNA polymerase* contains a *core enzyme*, made from several polypeptide subunits, that performs the task of synthesizing an RNA strand from a DNA template. A separate subunit, known as *sigma (σ) factor*, can attach to the core enzyme and in so doing forms the RNA polymerase holoenzyme. The RNA polymerase holoenzyme will readily but weakly bind to DNA but is apt to fall off of a DNA strand as it quickly slides along the polymer. However, once it slides across a promoter region, the σ factor interacts with the bases of the promoter to add strength to the holoenzyme/DNA interaction and briefly halt the sliding. The strengthened binding causes a conformational change in the holoenzyme that serves to separate the dsDNA into two separate strands without hydrolyzing ATP. The polymerase is then able to synthesize an RNA polymer of ~10 nucleotides that is complementary to the DNA template strand.

At this point, the σ factor that served to halt the RNA polymerase holoenzyme at the promoter is now acting as a sort of anchor, preventing further transcription from taking place because the holoenzyme cannot move further downstream. The RNA polymerase core enzyme will then release σ factor and again continue its slide down the DNA, but this time, synthesizing complementary RNA as it goes at a rate of about 3000 nucleotides per minute.

At the end of the DNA coding region is a DNA sequence termed the *terminator* sequence, which contains a string of A-T base pairs. Perhaps, it is because A-U base pairs (formed as the RNA polymerase synthesizes a strand complimentary to the A-T-rich region of the terminator) are less stable, or perhaps, it is because self-pairing of As and Us within the final portion of the primary transcript causes a hairpin structure to form, but the result of the RNA polymerase traversing the terminator region is a dissociation of the core enzyme from the DNA template strand. This, in turn, causes a release of the newly formed RNA strand from the core enzyme, most likely due to a conformation change in the protein. The core enzyme is then free to associate with another σ factor and start the process again.

4.2.3.1.2 Eukaryotes

Eukaryotes, however, utilize three different RNA polymerases. RNA polymerase I is involved with transcribing genes encoding rRNA, RNA polymerase II transcribes genes that will become proteins (via mRNA) and some snRNAs (covered later, under Splicing), and RNA polymerase III mainly transcribes genes encoding tRNAs but also some snRNAs and one particular rRNA gene. The type of transcription most interesting to most biotechnologists will yield a product that becomes an mRNA that will be translated into a protein, so we will focus on RNA polymerase II in the following sections.

If the prokaryotic RNA polymerase core enzyme can be likened to eukaryotic RNA polymerase II, then the prokaryotic σ factor can be likened to general *transcription factors*, which is a large group of proteins that must bind to RNA polymerase II before it can start transcription. Transcription factors are abbreviated "TF," and those that bind to RNA polymerase II are abbreviated "TFII." We shall run into TFIIB and TFIID soon.

In eukaryotes, the product of RNA polymerase is indeed an RNA molecule, but it is not mRNA. Three modifications to the RNA are required to convert it into a mature mRNA ready for translation. Before we delve further into eukaryotic transcription, let us first discuss these modifications in greater detail.

4.2.3.1.2.1 The Eukaryotic 5' mRNA Cap

Even before RNA polymerase II has finished creating the entire primary transcript, the 5' terminus is modified through capping. The cap can be thought of as essentially an upside-down GTP molecule that is methylated at the nitrogen in position 7 in the purine base, sometimes with an additional methylation of 1-2 of the first two bases in the transcript (Figure 4.18). Creation of the cap is more complicated, though, as illustrated in Figure 4.19. The primary transcript will have a triphosphate at the 5' end. The first step in the capping process is the removal of one of these phosphates. Next, a GTP molecule is added in a 5'-5' linkage, with some of the energy for the reaction coming from cleavage of a pyrophosphate (think two phosphates) group. The third step entails methylation of the guanosine base at the nitrogen in position 7 in the purine rings. Occasionally, one or both of the ribose rings of the first two nucleotides of the primary transcript are then also methylated, at the 2' position.

Keep in mind that mRNA is not the only type of RNA in a cell; rRNA and tRNA are transcribed by RNA polymerases I and III, respectively. Other uncapped RNAs exist, too. RNA polymerase II is the only one of the three eukaryotic RNA polymerases that produces transcripts with a 5' cap, though, because it has a tail section that binds the three enzymes needed to perform the three capping steps listed above. The cap serves as a signal that the molecule is an mRNA molecule. By removing the 5' phosphate, it prevents degradation of the pre-mRNA by 5' exonucleases. The cap will also be bound by the cap-binding complex, which marks and mediates export of the mRNA from the nucleus into the cytoplasm. As we will soon see, the cap will also serve as an initial binding site for part of the ribosome to begin the translation process.

4.2.3.1.2.2 Splicing

The primary transcript will contain introns as well as exons. In becoming a mature mRNA, the cell will remove the introns through a process known as *splicing*. Figure 4.20 shows the process with two levels of detail. On the left side of the figure, it can be seen that a specific adenylate residue is brought to the 5' border of the intron. Hydrolysis at the 5' border is immediately followed by creation of a new phosphodiester bond between the exposed 5' intron phosphate and the 2' carbon on the ribose of the adenylate residue. This creates a loop structure in the intron. The newly created 3' hydroxyl at the end of the previous exon is then available to form a phosphodiester bond with the 5' phosphate of the next exon after another hydrolytic reaction frees the intron. The intron is released from the pre-mRNA as a loop structure with a tail, as structure known as a *lariat*.

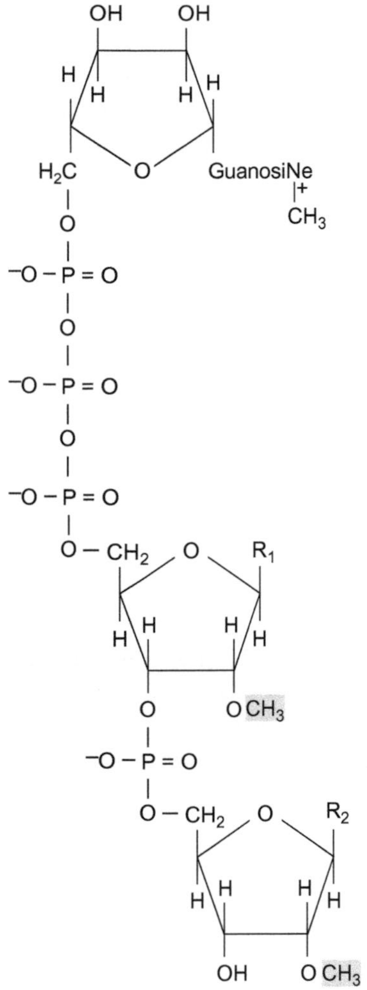

FIGURE 4.18   Structure of the 5′ cap added to primary transcripts created by eukaryotic RNA polymerase II. It consists of the positively charged 7-methylguanosine in a 5′-5′ linkage with the 5′ terminus of the RNA transcript. The first and/or second riboses of the primary transcript chain may also be methylated (shown by shaded CH_3 groups). The capital N in "guanosiNe" is meant to show the methyl group attaches to a positively charged nitrogen atom.

The act of splicing out introns as lariats is a very complex process, mediated by over 50 proteins and 5 additional RNA molecules. The five RNA molecules—U1, U2, U4, U5, and U6—are known as *small nuclear RNAs (snRNAs)*. Each snRNA is <200 nucleotides long and participates in recognition of intron/exon boundaries as well as remodeling of phosphodiester bonds. They do not work alone, though, and associate with several proteins to create complexes known as *small nuclear ribonucleoproteins (snRNPs)*. The snRNPs come together to

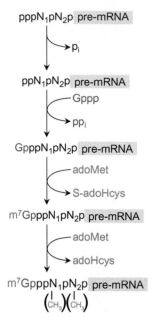

FIGURE 4.19   Order of steps in the creation of the 5′ RNA cap. (adoMet, S-adenosylmethionine, a methyl donor; adoHcys, S-adenosylhomocysteine). A different enzyme catalyzes each step.

form the structure known as a *spliceosome*. The right-hand side of Figure 4.20 shows some basic interactions between the snRNPs. First, the branchpoint adenine nucleotide is recognized by the branchpoint binding protein and a helper peptide (not shown), which recruit snRNP U2 to the site. At the same time, snRNP U1 identifies and associates with the upstream border between the intron and the adjacent exon. The other three snRNPs (U4/U6-U5) then become involved as a unit to help bring the 5′ splice site to the branchpoint and facilitate hydrolysis with the aid of additional proteins and ATP. During the process, there are rearrangements and dissociations of snRNPs in the spliceosome, resulting in only the U2, U5, and U6 snRNPs remaining with the lariat as it is released.

4.2.3.1.2.3   The Eukaryotic Poly(A) Tail   In prokaryotes, transcription of a gene halts when RNA polymerase passes a termination signal. The bacterial RNA polymerase will then release both the DNA and the RNA transcripts. Since prokaryotes lack a nucleus, no further processing is needed to allow the transcript to traverse a nuclear envelope.

Eukaryotic mRNAs must traverse the nuclear envelope, through nuclear pore complexes, to get to the cytosol for translation. The 3′ end of a mature mRNA consists of a string of about 200 adenylate residues, which play a role in the export of the mRNA from the nucleus as well as lending stability to the molecule once in the cytoplasm. The A residues are not coded for in the genome, though.

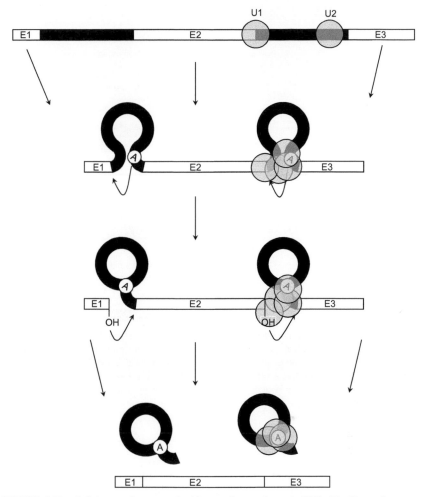

FIGURE 4.20   Splicing, or the removal of introns from eukaryotic RNA. The figure shows two introns being removed in different levels of detail. A specific adenylate residue is brought into the vicinity of the 5′ border of the intron, where hydrolysis and recreation of a phosphodiester bond are used to remove the intron as a lariat. The right-hand side of the figure shows the same process, but in the presence of snRNPs, which are essential in the formation of the spliceosome.

They are added after RNA polymerase II transcribes poly(A) and cleavage signals found in the DNA at the terminus of the gene. The DNA that encodes the signal to add the poly(A) tail has the consensus sequence AAUAAA-$[N]_{10\text{-}30}$—CA in mammals, and the cleavage signal immediately following is T- or G/T-rich and is <30 nucleotides in length (Figure 4.21). While the process of adding the poly(A) tail is far more complex than that of adding the 5′ cap, the progression of events boils down to the following:

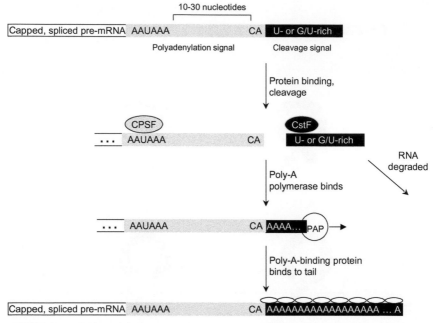

FIGURE 4.21   Eukaryotic mRNA has a string of A residues at its 3′ end, but these are not directly coded for in the genome. Rather, the gene has bases coding for RNA polyadenylation and cleavage signals that are transcribed and bound by three proteins, two of which are CPSF and CstF. These proteins help facilitate polyadenylation and cleavage, respectively. Once the cleavage signal is removed via hydrolysis, the poly(A) tail is synthesized by PAP and coated with multiple copies of poly(A)-binding protein.

1. Polyadenylation and cleavage signals in the DNA are transcribed by RNA polymerase II.
2. Cleavage and polyadenylation specificity factor (CPSF) binds to the AAUAAA signal, and cleavage stimulation factor F (CstF) binds to the G/U-rich signal.
3. RNA is cleaved by an endonuclease. The G/U-rich sequence will be degraded in the nucleus.
4. Poly(A) polymerase (PAP) binds to the pre-mRNA and produces the poly(A) tail. PAP operates like the conventional RNA polymerases, except that no DNA template is required.
5. Several copies of poly(A)-binding protein bind the newly created tail.

4.2.3.2   Regulation of Transcription

Proteins bind to DNA, and when they do, it is probably for a specific reason. RNA polymerase binding occurs because transcription is about to start. Still other proteins can bind to DNA and act as flypaper to help catch RNA polymerase (through mediators). Sometimes, it is important for a cell to reduce or completely shut off the expression of a specific gene, which is accomplished by another class of regulatory protein binding to a general (or specific) sequence of DNA. The portions of DNA that bind proteins for the purpose of transcriptional regulation are promoters, enhancers, silencers, and operators.

4.2.3.2.1   Promoters and Promoter Elements

4.2.3.2.1.1   TATA Box Gene regulatory elements play a significant role in determining how much of a gene is transcribed at any given moment. Perhaps the most important of these elements is the *promoter*, which serves as a binding site for transcription factors (in eukaryotes) or sigma (σ) factors (in prokaryotes), which serves to position RNA polymerase for transcription. Perhaps, the most famous promoter element is the *TATA box*. In eukaryotes, the TATA box is typically located 25-35 bases upstream of the transcriptional start site. The TATA box gets its name from its *consensus sequence* or the most common order of bases that make up this element. There is not a single sequence that is used everywhere as the TATA box, but some sequences are used far more than others. For instance, Table 4.2 shows the top TATA consensus sequences taken from a ranking of the 1024 possible sequences of [T A (A/T) (A/T) N N N N] (where N = any nucleotide) in yeast. From the list, it seems evident that an eight mer TATA sequence ought to adhere to four rules: it begins with "TATA," the sixth base is an A, the fifth and seventh bases are either As or Ts, and the eighth residue is a purine. So, what is meant by consensus sequence is not the same as the spelling of a word, which has only one correct order of letters, but rather is a reflection of the letters that are most commonly used to make up the ordered bases.

TABLE 4.2 The Consensus Sequence of the TATA Box in Yeast, Worked Out From All Possible Eight-Mers of [T A (A/T) (A/T) N N N N]

Position	1	2	3	4	5	6	7	8
	T	A	T	A	T	A	A	A
	T	A	T	A	T	A	A	G
	T	A	T	A	A	A	T	A
	T	A	T	A	A	A	A	G
	T	A	T	A	T	A	T	A
	T	A	T	A	A	A	A	A

Consensus: T A T A (A/T) A (A/T) (A/G).
The top six sequence candidates are listed in the table.

In prokaryotes, the promoter situation is a little bit different. In *E. coli*, there are two upstream sequences that often serve as promoters, being found in the -35 and -10 regions of the transcriptional start site. As already stated, σ factors bind to these sites. It just so happens that the transcriptional factor σ^{70} has the -10 consensus sequence of TATAAT, very similar to the TATA box of many eukaryotic genes. (For completeness, the -35 region consensus sequence for σ^{70} is TTGACA.) The -10 binding region in prokaryotes is sometimes referred to as the *Pribnow box*.

For eukaryotic genes that utilize the TATA box, alterations in the position or composition of this promoter sequence can have significant effects upon gene expression. If we were to change one base in the TATA box sequence, transcription of the corresponding exon(s) goes down, which should be intuitive.

If we were to change the base sequence between the TATA box and the normal transcriptional start site, not much will be changed in the transcription rate or product of the gene. However, if we were to remove bases between TATA and the usual transcriptional start site, RNA polymerase will begin transcription at a new site. How can changing bases in a region not have any effect while removing bases from the same region yield a new transcriptional start site? The answer lies in the fact that once assembled, the transcriptional machinery will have a fixed size. The transcriptional factor *TATA-binding protein (TBP)* will bind to TATA and aid in the binding of RNA polymerase to the gene. The active site of RNA polymerase, when bound to TBP, will be a set distance away from TBP; this distance will be about the same as the distance covered by a 25-35-base stretch of DNA. Whatever is next to the active site of RNA polymerase once the transcriptional machinery has been assembled will be where transcription starts. The identity of the bases between TATA and the active site of RNA polymerase is not of great importance, but the space that they take up helps to define the transcriptional start site.

4.2.3.2.1.2   CpG Islands The TATA box is not the only feature that can be used to identify the location of the gene. For instance, there is a class of genes that is constantly being expressed, known as *housekeeping genes*. Housekeeping genes code for proteins that are used for vital processes and are necessary at all times to keep the cell alive. Virtually, all of the cells in a given organism will express the same housekeeping genes. For example, consider that the cell always needs energy. One of the ways that the cell obtains this energy is through the process of glycolysis, which breaks a glucose molecule into two pyruvate molecules, two molecules of NADH, and a couple of ATP. The ATP will then be used by the cell for energy. Glycolysis is, for all practical purposes, always occurring in the cell, so it follows that glycolytic enzymes are always required, and therefore the genes for these enzymes must always be expressed. Housekeeping genes very often will be preceded by a stretch of DNA that is very rich in Cs and Gs. This stretch, often called a CG island, or more often a *CpG island* to distinguish it from a C-G base pair, is 1000-2000 base pairs long and remains unmethylated in all cell types. (Methylation, or the addition of a $-CH_3$ group to a nucleotide base, is a genomic modification often utilized by a cell to inactivate a gene.) CpG islands often surround the promoters for housekeeping genes. Unmethylated C-G base pairs are not so common in the genome, so having an extended stretch of unmethylated C-G appear in the genome is due to more than mere chance, serving to indicate the presence of an active gene in the vicinity.

4.2.3.2.1.3 GC Box Note that a CpG island is not the same as a *GC box*. A GC box is a transcriptional regulatory element containing the sequence GGGCGG. Assembly of the transcriptional machinery is enhanced by the binding of the protein Sp1, a transcription factor, to a GC box. Sp1 happens to be expressed in all eukaryotic cells, so having a GC box present in a transcriptional regulatory region increases the amount of transcription for that gene. GC boxes are often found within 100 bases upstream of the transcriptional start site, and although they may appear as a single instance, they are commonly repeated 20-50 times. Sometimes, GC boxes appear in genes that do not use a TATA box.

4.2.3.2.1.4 CAAT Box Another transcriptional regulatory element worthy of mention is the CAAT box. It has the consensus sequence of GG(T/C)<u>CAAT</u>CT and can be found ~75 base pairs upstream of the transcriptional start site. Keep in mind that the CAAT box, GC boxes, and even the TATA box are *promoter elements*; they are sequences that have been found within various promoters. Other common consensus sequences that are found in the vicinity of the transcriptional start site are shown in Table 4.3.

TABLE 4.3 The Consensus Sequences of Several Transcriptional Regulatory Elements

Element	Consensus Sequence	Binds Protein
TATA box	T A T A (A/T) A (A/T) (A/G)	TATA-binding protein (TBP)
GC box	G G G C G G	SP1 transactivator Sp1
CAAT box	G G (T/C) C A A T C T	CAAT-enhancer-binding protein (C/EBP)
BRE	(G/C) (G/C) (G/A) C G C C	Transcription factor IIB (TFIIB)
DPE	(A/G) G (A/T) C G T G	Transcription factor IID (TFIID)
INR	(C/T) (C/T) A N (T/A) (C/T) (C/T)	Transcription factor IID (TFIID)

The TATA box is a promoter element. There are some who will say that there's only one promoter, which contains the TATA box, and anything else that acts like a promoter should be termed a promoter-proximal element. Regardless of semantics, it is generally agreed that promoters are always upstream of the transcriptional start site, they occur within 200 base pairs of this start, and they must appear in the correct orientation. In addition, if they occur within 50 base pairs of the transcriptional start site, their location is fixed. (As with most rules, there are exceptions.)

4.2.3.2.2   Enhancers

Besides promoters, there exist other sequences in the DNA that bring about increased transcription levels of certain genes. These sequences are known as *enhancers*. Enhancers can be very far from the transcriptional start site—sometimes over 1000 nucleotides away. They can appear upstream or downstream and are active in either the forward or reverse orientation. Often times, they are short sequences that are repeated several times.

An enhancer can show activity despite being so far away from the transcriptional start site because of the overall conformation of a stretch of DNA. DNA is not a rigid entity; it is flexible and can be folded, perhaps to bring an enhancer closer to where the action is (see Figure 4.22).

Enhancers work by binding *activators*. Activators are proteins that help to build the transcriptional machinery. One could think of the transcriptional machinery as being a car built out of Lego® blocks. When you get your kit of Legos, there might be a big flat piece, there might be some wheels, or there might be some longer red blocks, some short, white blocks, and perhaps some intermediate-sized blue blocks. You could use these blocks to build a house, a car, a city-destroying monster, or something else such as a train with four wheels or even 12 wheels. By assembling these blocks, we could build all kinds of things or vehicles. We can think of the transcriptional machinery as a vehicle that's going to bind to DNA and roll down it, polymerizing an mRNA molecule as it travels. In some ways, a Lego train and the transcriptional machinery are similar. The enhancer sequence will bind some of these Lego pieces and the cell will start to build the locomotive. Since enhancers can be repeated, several of the same type of Lego block (activator protein) can be assembled. These pieces will help to bind other subunits of the transcriptional machinery—perhaps RNA polymerase II, the core of the locomotive. Other sequences in the DNA will bind still other proteins, such as TBP (TATA-binding protein). Still, other proteins can serve to collect these individual DNA-bound proteins into a functioning unit: the *mediator* protein complex serves to attach to activator proteins, general transcription factors, and RNA polymerase II. Still other proteins will bind during the assembly of a eukaryotic transcriptional machine, such as chromatin remodeling complexes and histone-modifying enzymes, which serve as a sort of cow catcher on the front of a train to clear the track (by unwinding the DNA and translocating histones) to allow the transcriptional train to pass and do its job.

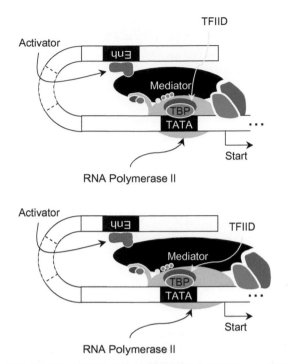

FIGURE 4.22 The eukaryotic transcription machinery. (Not all proteins are shown.) Note that the DNA enhancer is brought close to the transcriptional start site by an activator protein binding to a mediator that also interacts with RNA polymerase II. A gene may have many activator proteins and enhancers. The TATA box is bound by the TATA-binding protein (TBP), which is in turn bound by TFIID, one of many transcription factors involved with initiating eukaryotic transcription. Proteins involved with chromatin remodeling appear to the right, ahead of the polymerase.

Enhancers, promoters, and promoter-proximal elements all serve to increase the amount of transcription of a given gene, but they have differences between them such as promoters and promoter-proximal elements that are relatively close to the transcriptional start site (within 100 or 200 bases) and have a specific orientation, as opposed to enhancers that are more free-form, being found upstream *or* downstream, in the forward *or* backward orientation, and perhaps appearing as several repeated units.

Enhancers serve to enhance transcription levels (Table 4.4). One should not expect an enhancer to take the place of a promoter, though. Keep in mind that an enhancer will bind to activator proteins, which are to be bound by the mediator complex. A promoter, on the other hand, serves as the site for building the transcriptional machine that includes RNA polymerase II.

An example of an enhancer is the simian virus 40 (SV40) enhancer, which was the first enhancer to be adequately characterized. It has a sequence that is about 100 base pairs long, and in the virus itself, it appears about 100 base pairs upstream of the viral early transcription start site. This enhancer has turned out

TABLE 4.4 Hypothetical Plasmids and the Amount of Gene Expression Attainable with various Gene Elements

Plasmid contains	Gene Expression Level
Exon only	–
Promoter + exon	+++
Enhancer + promoter + exon	+++++
Enhancer + exon	–

to be a great tool for the genetic engineer. When constructing a gene, it has been found that the SV 40 enhancer stimulates transcription from all mammalian promoters. It can be inserted into a plasmid in either orientation, even thousands of base pairs from the transcriptional start site. Detailed research has shown that this enhancer is composed of several individual regions that act as protein binding sites, each of which contributes to the total activity of the enhancer.

As with promoter elements, there are also enhancer elements. The *E-box* is a transcriptional element that is found within many enhancers, hence the name *E*-box. It has the sequence CANNTG, where "N" stands for any nucleotide. E-boxes can be recognized by transcription factors that contain a basic helix-loop-helix structural motif (recall the discussion of supersecondary structure in proteins) in their folded conformation. The two "N" bases in the middle of the E-box sequence can determine specificity for certain transcription factors. Several E-boxes can appear in succession in a transcriptional regulatory region.

4.2.3.2.3   Silencers and Operators

The functional converse of an enhancer is the *silencer*. A silencer sequence, used by eukaryotes, will bind a *repressor* protein. A silencer shares most of the same properties with an enhancer (in that it can be very far from the transcriptional start site, can appear in either orientation, and can bind with a transcription factor protein), with the most notable difference being a lowering of the transcription level of a particular gene after a transcription factor is bound.

In prokaryotes, gene regulation is a little bit different. While prokaryotes do use promoters, they do not use enhancers. There are, however, *operator* sequences. Operators are bound by *repressor* proteins. (Note that the name of the protein here is the same as the name used for the eukaryotic protein that binds to a silencer.) Once a repressor has bound to its operator, the binding of RNA polymerase to the promoter is blocked. If an operator is bound, there's no transcription. If the operator is free, transcription is free to proceed. It's like an on/off switch.

4.2.4 Translation

4.2.4.1 Initiation of Translation in Eukaryotes

The transcriptional start site is not the same as the translational start site. Finding the location of a transcriptional start site is not as easy as looking for a specific DNA sequence. The location of the transcriptional start site is often determined by the physical size of the transcriptional machinery and the specific place where it is assembled on the DNA. We have discussed how the transcriptional start site might exist ~25-35 bp after a TATA box in a eukaryotic cell. There exist computer programs that can assist with predicting where the transcription of the gene will commence. Locating the translational start site, however, is much more straightforward. On the messenger RNA, it will be represented by the codon AUG. That makes the job of finding the translational start site very easy, unless there is more than one AUG. This problem can be eliminated by the cell by the use of a *Kozak sequence*.

An example of a strong Kozak sequence is GCCACC<u>A</u>U<u>GG</u>. Note the AUG. In the Kozak sequence, the AUG (underlined) is the start codon. The base found three bases upstream of the start codon must be a purine: an A or a G. Usually an A, it is considered crucial for efficient initiation of translation. In the absence of this purine, a G immediately following the start codon is essential. (Both of these important bases are shown in bold.)

The 11-base sequence	GCC[A/G]CC<u>AUG</u>G, for which the next base is not a U, is optimal
	[A/G]NN<u>AUG</u>G[not U] is strong ("A" at -3 is stronger than "G")
	Anything else is considered weak at best

If you are engineering a gene, knowing that eventually you want the gene to be translated, it is strongly advisable to have your start codon appear in the middle of a Kozak sequence.

The 5' cap of mRNA is used in eukaryotic cells as a site for the start of ribosomal assembly. Ribosomes consist of distinct subunits that must come together for translation of mRNA to occur. The 5' cap serves as an assembly point for the beginning of this process, which involves the small ribosomal subunit, the first tRNA, and several initiation factors. Table 4.5 lists the initiation factors, abbreviated eIF for "eukaryotic initiation factor," to give a glimpse into the complexity of the process. These initiation factors help to bring the mRNA to the ribosome for translation. Note that, through eIF-4G, the initiation complex also recognizes the 3' end of mRNAs. This is believed to be the reason that poly(A) tails have a stimulatory effect on translation.

Recall that the start codon, AUG, codes for methionine. This means that the translation-initiating tRNA will carry a Met residue. In eukaryotes, before

TABLE 4.5 A Shortened List of Initiation Factors and Their Roles in Ribosomal Assembly

Role	Eukaryotic Initiation Factors Involved
Bind to the small ribosomal subunit	eIF-1, eIF-1A, eIF-3
Binds to initiator tRNA (holding Met)	eIF-2 (plus GTP)
mRNA recognized and brought to the ribosome	eIF-4E (directly recognizes 5′ cap)
	eIF-4G (binds eIF-4E, poly(A) binding protein, and eIF-3)
	eIF-4A, eIF-4B

being paired with the mRNA molecule, the tRNA-Met molecule is loaded into a small ribosomal subunit together with the eIFs. It is interesting to note that the tRNA-Met molecule is the only one of the aminoacyl tRNAs that can bind tightly to the small ribosomal subunit without the large subunit being present. Still in eukaryotes, when the loaded small ribosomal subunit is being assembled on the messenger RNA at the 5′ cap, it recognizes the mRNA by virtue of initiation factors that have already bound to the cap. After some reorganization, the ribosome apoenzyme (the use of "apo" means the enzyme is not complete) starts rolling down the mRNA. When it gets to the Kozak sequence, progress will stall and the GTP bound to eIF-2 is hydrolyzed, enabling the release of all of the initiation factors and the binding of the large ribosomal subunit to yield a functioning ribosome for performing the task of polypeptide production. It's not that cells have functioning ribosomes floating around looking for mRNA—complete ribosomes must be assembled, just like the transcriptional machinery must be assembled. At the 5′ cap of an mRNA, the small ribosomal subunit and initiation factors assemble into—to continue a past analogy—a Lego locomotive that doesn't have a roof yet. The locomotive is able to roll down the assembly line until it encounters a Kozak sequence. It will stall, the initiation factors will fall off, and the roof will be put on. The roof happens to be the large ribosomal subunit. At this point, the cell will have a functioning ribosome that's going to translate the mRNA message.

The nucleotides immediately surrounding the start site in eukaryotic mRNAs influence the efficiency of AUG recognition during the above scanning process. If this recognition site differs substantially from the consensus recognition sequence (Kozak sequence), scanning ribosomal subunits will sometimes ignore the first AUG codon in the mRNA and skip to the second or third AUG codon instead. Cells frequently use this phenomenon, known as "*leaky scanning*," to produce two or more proteins, differing in their N-termini, from the same mRNA molecule. For some genes, this allows the production of a single protein that will be directed to two different compartments in the cell. The variable N-termini are removed after the protein is delivered to its destination.

Ribosomes will dissociate from the mRNA at the poly(A) tail. This is likely due to the poly(A) tail having only two sites available for hydrogen bonding per adenine (as opposed to three hydrogen bonds with guanine), which provides less force for holding the ribosome to the mRNA. The poly(A) tail also serves to increase the stability of mRNA by inhibiting the assembly of RNases that chew from 3' to 5'. Poly(A) tails are also involved in the export of mRNA from the nucleus to the cytoplasm.

4.2.4.2   Initiation of Translation in Prokaryotes

Another translational regulatory unit is the *Shine-Dalgarno* sequence. Simply put, it is the Kozak sequence for bacteria. The Shine-Dalgarno sequence is typically found around position -7 to -4 of the translational start codon, and it has the sequence AGGAGG. This sequence is complementary to part of the 3' end of 16S rRNA: ...GAUCACCUCCUUA-3' (the portion that is complementary to Shine-Dalgarno is underlined).

The way that bacteria determine the start codon differs from the method used in eukaryotes. Bacterial mRNAs do not have 5' caps, so ribosomal assembly by definition must be different. The small ribosomal subunit contains rRNA that is complementary to the Shine-Dalgarno sequence. Hence, while the Shine-Dalgarno sequence is similar to a Kozak sequence in that both help to determine the position of the start codon, the Shine-Dalgarno sequence is different because it allows the bacterial ribosome to be built at an interior position on the mRNA through direct binding to this sequence.

Since prokaryotic ribosomes can be assembled wherever there is a Shine-Dalgarno sequence and bacterial mRNAs can contain more than one start site, it is often the case that one mRNA can code for multiple polypeptides. A polynucleotide unit that codes for a polypeptide is called *cistron*. When an mRNA codes for multiple polypeptides, the mRNA is said to be *polycistronic*. Prokaryotes utilize polycistronic mRNAs via operons.

The bacterial genome, being very efficient, contains features known as *operons*. An operon is a contiguous stretch of DNA that codes for more than one polypeptide, the transcripts of which will be contained in a single mRNA. Just like an opera contains many different songs in a single continuous story, an operon contains many different messages that will be represented in a single mRNA. In *E. coli*, the *lac* operon codes for a single mRNA, but that mRNA encodes three proteins. Each of the protein-coding segments has its own translational start site and stop codon.

Aside

In eukaryotes, polycistronic mRNAs are very rare, but they do occur. This can happen because of the following:

(1) *Leaky scanning*. This is where ribosomes bypass an AUG codon that is located close to the 5′ cap or when the Kozak sequence is weak. This allows the ribosome to reach another AUG that is further downstream.

(2) *A lack of detachment*. After the translation of one part of the mRNA, a ribosome might continue scanning. In doing so, it might attach a new set of initiation factors and reinitiate at a downstream AUG or Kozak sequence.

(3) An *internal ribosome entry site*. In very rare instances, a ribosome might enter at an internal ribosome entry site, much like the bacterial model just presented.

QUESTIONS

1. Draw a graph to represent T_m versus %(A+T).]
2. Suppose there is a promoter for a gene encoding a nonessential protein that experiences a random mutation such that transcription of the gene can no longer occur. Hypothesize how such a nonlethal mutation might actually be beneficial to the species evolutionarily.
3. How does the wobble effect explain the degeneracy of the genetic code?
4. What helps position RNA polymerase?
5. One of the reasons it is supposed that DNA is evolutionarily favorable over RNA as genetic material is an inherent stability. However, it is thermodynamically more favorable to hydrolyze DNA than RNA. Resolve the paradox.
6.
    a. If my genome is 37% A, about what percent of it is G?
    b. If my genome is 26% A, about what percent of it is U?
    c. If Koco the Gorilla's genome 21% C, about what percent of Mighty Joe Young the Gorilla's genome is A?
    d. If a raccoon's genome is 21% G, about what percent of a giraffe's genome is C?
    e. If a wallaby's genome is 50% A, about what percent of its genome is G?
7. Imagine life evolving on an alien planet in such a way that the genetic code is binary; that is, only guanine and cytosine are used to code for a total of 12 amino acids. What is the minimum number of nucleotides required, per codon, to code for the 12 amino acids in this binary system?
8. How much energy does DNA polymerization require? Where does it code from?
9. How might we test to find a promoter? Give two examples.

AAUAAA
CA

10.
    a. Describe an experiment that showed the semiconservative nature of DNA

    b. What does semiconservative replication mean?

11. How can an enhancer be thousands of base pairs away from the promoter and still interact with the promoter?

12. Would it be a good or bad idea to make our genomes more efficient by deleting all of the introns? Explain your answer.

13. Explain why the sequence of the TATA box is not always TATA(AAT).

14. Below is an mRNA sequence. Indicate where the promoter is.

    CAAUGCGCGCGCGCGCGAGCAUUCAGGCGUAUAAAUAUCGGCAUUCACCGAUG

RELATED READING

Alberts, B., Johnson, A., Lewis, J., Raff, M., Roberts, K., Walter, P., 2007. Molecular Biology of the Cell, fifth ed. Garland Science, New York, NY.

Basehoar, A.D., Zanton, S.J., Pugh, B.F., 2004. Identification and distinct regulation of yeast TATA box-containing genes. Cell 116, 699–709.

Brockmann, C., Soucek, S., Kuhlmann, S.I., Mills-Lujan, K., Kelly, S.M., Yang, J.C., Iglesias, N., Stutz, F., Corbett, A.H., Neuhaus, D., Stewart, M., 2012. Structural basis for polyadenosine-RNA binding by Nab2 Zn fingers and its function in mRNA nuclear export. Structure 20, 1007–1018.

Cooper, G.M., 2000. The Cell: A Molecular Approach, second ed. Sinauer Associates, Sunderland, MA. Accessed via: http://www.ncbi.nlm.nih.gov/books/NBK9849/.

Ford, L.P., Bagga, P.S., Wilusz, J., 1997. The poly(A) tail inhibits the assembly of a 3'-to-5' exonuclease in an in vitro RNA stability system. Mol. Cell. Biol. 17, 398–406.

Hershey, A.D., Chase, M., 1952. Independent functions of viral protein and nucleic acid in growth of bacteriophage. J. Gen. Physiol. 36, 39–56.

Quigley, F., Martin, W.F., Cerff, R., 1988. Intron conservation across the prokaryote-eukaryote boundary: structure of the nuclear gene for chloroplast glyceraldehyde-3-phosphate dehydrogenase from maize. Proc. Natl. Acad. Sci. U. S. A. 85, 2672–2676.

Rubin, E., Farber, J.L., 1994. Development and genetic diseases. In: Pathology, second ed. JB Lippincott Company, Philadelphia, PA, p. 256.

Shatkin, A.J., Manley, J.L., 2000. The ends of the affair: capping and polyadenylation. Nat. Struct. Biol. 7, 838–842.

Shi, W., Zhou, W., 2006. Frequency distribution of TATA Box and extension sequences on human promoters. BMC Bioinformatics. 7 (Suppl. 4), S2.

Chapter 5

Cell Growth

5.1 THE EUKARYOTIC CELL CYCLE

Let's say we want to harness some eukaryotic cells to produce a product for us. The cells could be yeast that can produce ethanol under the right conditions, or they may be engineered islet cells that can produce recombinant human insulin. There exist many different cell types that can be harnessed to produce a wide array of products. As a biotechnologist, perhaps working in the industry for your own or somebody else's company, it will be important to have the cells produce the greatest amount of product for the lowest overall cost. If we have a cell type that produces a known amount of a given product per cell, we would like to have as many cells as possible simultaneously producing this product. There are, however, limits to the numbers of cells that can be grown in finite cultures. To understand the limits, we must first understand how cells grow, how they divide, and how we can get the maximum numbers of cells in our bioreactors to produce products for us. To gain such understanding, we must return to the cell cycle (Figure 5.1).

There are two basic events that occur during the cell cycle: cells double their DNA and they divide. The period of DNA synthesis is known as the *S* (synthesis) *phase*, while the period devoted to cell division is known as the *M phase* (with M standing for mitosis). Cells do not replicate their genomes and immediately undergo division, nor do they divide and immediately start making more DNA; there are gaps in between the S and M phases. For an actively growing cell, the gap preceding S-phase is referred to as G_1, while the gap between S-phase and mitosis is called G_2. In adults, the majority of cells in the body are not actively growing. Although cells in the adult human are growing and dividing for processes such as manufacturing new blood cells, growing new skin or mucosal cells that have sloughed off, or repairing lesions due to cuts or abrasions, most of the cells in the body are not engaged in growth and division. Cells can exist in this state of living without growing for an extended and indefinite amount of time, so this particular gap phase is drawn just outside the main cell cycle and is termed G_0 to distinguish it from G_1. Cells can remain in G_0 virtually forever (meaning for the life of the adult organism). A cell may exit M phase and sense that there is no demand in that particular part of the body for more cells of its kind, or maybe there is a lack of nutrients in the cell that would

An Introduction to Biotechnology. http://dx.doi.org/10.1016/B978-1-907568-28-2.00005-8
Copyright © W T Godbey, 2014.

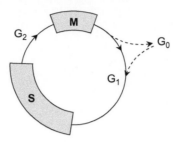

FIGURE 5.1   The cell cycle, including synthesis (S), mitosis (M), and gap (G) phases. G_0 serves as a holding point for the cell, allowing them to function normally without cycle progression.

allow for the production of a duplicate genome, so the cell will enter G_0. It can still perform its normal life functions, such as respiration or the production of a certain product (depending on the type of cell it is), but the cell will not double its genome or divide until certain conditions have been met.

5.1.1 Phases of Mitosis

Let us now consider mitosis in a little more detail. Looking at the cell cycle solely in terms of mitosis, there are five stages: interphase, prophase, metaphase, anaphase, and telophase. Prophase, metaphase, anaphase, and telophase are used to describe the process of cell division. Interphase refers to the rest of the cell cycle, including S-phase and the three gap phases (Figure 5.2).

Prophase This stage of mitosis involves the condensation of DNA into chromatids. Looking at an unstained cell in interphase, while it might be easy to discern the location of the nucleus it would be difficult to visualize the DNA within. The nucleus of an unstained cell in prophase might look like it contained a lot of dark spaghetti, because the DNA has condensed into chromatids (Figure 5.3b).

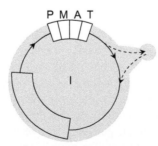

FIGURE 5.2   Another perspective of the cell cycle. Four of the five phases—prophase (P), metaphase (M), anaphase (A), and telophase (T)—are associated with M phase of the previous figure. The remaining phase—interphase (I)—is shaded, and includes everything that remains, including the S and gap phases.

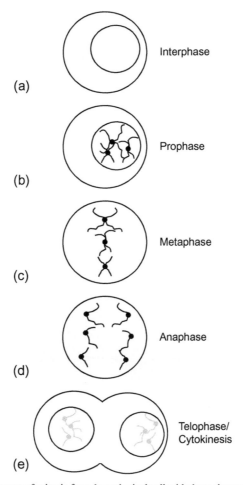

(a)

Interphase

(b)

Prophase

(c)

Metaphase

(d)

Anaphase

(e)

Telophase/
Cytokinesis

FIGURE 5.3   The stages of mitosis for a hypothetical cell with three chromosomes.

Metaphase In this stage, the chromatids will line up along an axis through the middle of the cell by way of a newly formed *mitotic spindle*. Also during this phase, the cell will dismantle its nuclear envelope. Recall that, since the cell has already performed S-phase, it contains two complete copies of its genome. That is why the chromosomes look like the letter X in Figure 5.3c. Each X represents both copies of the single chromosome, which are held together at the *centromere* (Figure 5.4a). The centromere is a region of chromosomal DNA. Not only is it where the two copies of the chromosome are held together, it is also where a structure known as the *kinetochore* forms. The kinetochore is made of layers of protein, serving as insertion points (maybe 10-40 of them in mammalian cells) for microtubules. These microtubules, also known as spindle fibers, connect the chromosome to *centrioles* (also known as spindle poles), which will

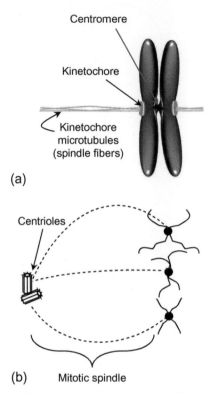

(a)

(b)   Mitotic spindle

FIGURE 5.4 (a) Location of the centromere, kinetochore, and kinteochore microtubules following metaphase. (b) Attachment of the mitotic chromosome to centrioles (spindle pole).

serve as anchors as the chromosomes are pulled apart in the next step of mitosis (Figure 5.4b). The transition from prophase to metaphase is sometimes termed pro-metaphase, and is when we see the mitotic spindle start to form and the chromosomes attaching to the spindle. The point in time when they line up defines metaphase.

Anaphase During anaphase, the sister chromatids are pulled apart, thus completely separating the two daughter chromosomes. The pulling is accomplished through the shortening of the kinetochore microtubules. At the same time, the spindle poles start to move further apart. Each chromosome is slowly pulled toward one of the two spindle poles (centrioles) (Figure 5.3d).

Telophase This step is marked by the reformation of the nuclear envelope around each set of daughter chromosomes, which decondense after reaching to the spindle pole (Figure 5.3e). At this point the cell will contain two separate, self-contained nuclei. This bi-nucleated cell will typically then divide into two daughter cells through a process known as *cytokinesis*.

Cytokinesis involves the division of the cytoplasm through the closing of a contractile ring, and is considered to be separate from mitosis. In most animal cells, cytokinesis begins during anaphase and ends slightly after telophase and the completion of mitosis. As the ring becomes smaller and smaller, there will be a need for additional plasma membrane to supplement the newly forming daughter cells. (The surface area of the plasma membranes of the two daughter cells will be greater than the surface area of the parent cell.) This new plasma membrane comes from the fusion of intracellular vesicles with the existing plasma membrane.

5.1.2 Control of the Cell Cycle

Although Figures 5.1 and 5.2 represent the cell cycle as a continuous process, the cycle does not proceed in a steady fashion like the hands on a clock. The lengths of the gap phases are variable, partly due to control points through which the cell monitors its own status and prevents execution of certain events when the time is not yet right. For example, replication of the entire genome is a very expensive event, requiring at least $2 \times (3.08 \times 10^9)$ ATP equivalents of energy just for polymerization. If the cell does not have sufficient energy stores to complete this task, it will not even begin S phase. This particular control point in the cell cycle is called *Start* (Figure 5.5).

Some questions that must be favorably satisfied for the cell to proceed beyond Start and into S phase are: (1) Is the environment favorable for replication? An energy-depleted environment is not conducive for the formation of large numbers of the dNTPs. (2) Does the cell have sufficient mass for dividing into two daughter cells? (3) Are there enough dNTPs present to undertake the task of replicating the entire genome? (4) Are there enough nutrients? The parent cell must ensure that there will be enough mitochondria and other

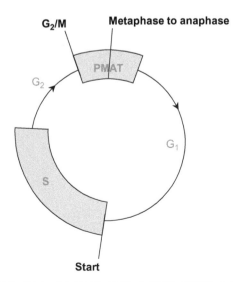

FIGURE 5.5   Cell cycle checkpoints (in black): Start, G2/M, and M to A transition.

organelles, proteins and other macromolecular products, and raw materials for forming enzyme proteins, structural proteins, and DNA by the two daughter cells. Synthesis phase is such a taxing process for the cell that if the cell is not prepared for the task then it will die. However, if the answer to all of the above questions is "yes" then the cell will progress past Start and commence DNA replication in the S phase.

There are a couple of other check points that are vital to successful cell division with the production of two viable daughter cells (Figure 5.5). One occurs at the transition from G_2 to M phase, where the cell again determines whether conditions are favorable for progressing into the next phase of the cell cycle—this time, mitosis. The questions are similar to those of Start: (1) Is the environment favorable? Perhaps the cell is facing a large metabolic challenge. Perhaps during S phase the environment was just fine, but now every other cell in the region is going through S phase or mitosis so the number of cells competing for nutrients has greatly increased. These reasons can render the environment unfavorable because nutrients and other resources have been depleted. (2) Has the replication of the genome been completed? It would not be advisable for the cell to split in half if it did not have two complete genomes for the daughter cells, so the cell must ensure that S phase has been completed.

The other remaining checkpoint is located within the M phase. It occurs at the transition from *metaphase to anaphase*. Recall that in anaphase chromatid pairs are separated. On a macro scale, one major thing that could go wrong is that the separation commences before all of the chromatids have been attached to microtubules via kinetochores. If this were to happen, one or more of the chromosomes would be lost during the separation, resulting in a daughter cell with an incomplete genome. Such a cell would be severely challenged or die. This checkpoint asks the question: is every chromatid attached to a spindle fiber?

Consider the amount of DNA per cell with respect to the cell cycle. Beginning with G_0/G_1, there will be a constant amount of genomic DNA per cell (Figure 5.6a). During S phase this amount will increase until it reaches a point that is double the starting value. From the end of S phase, through G_2, there will again be a constant amount of DNA in the cell. During M phase, although the cell is forming mitotic spindles and separating chromatid pairs, there will still be the constant, double amount of DNA in the cell, even though at the end of telophase the cell will contain two nuclei. It is only after cytokinesis, at the point when the contractile ring completely closes and the parent cell is cleaved into two daughter cells, that the amount of DNA per cell is reduced back to the original constant value.

Consider the number of mammalian cells in a given culture that are in each phase of the cell cycle at a given time (Figure 5.6b). The majority of the cells will be in G_0/G_1. (If all of the cells are well fed and actively growing, as should be the case in a culture plated yesterday, then there will be little need for G_0 and the majority of cells will be in G_1.) The figure also tells us that the majority of the rest of the cells are in G2. Since we can't distinguish G_2 from M in terms

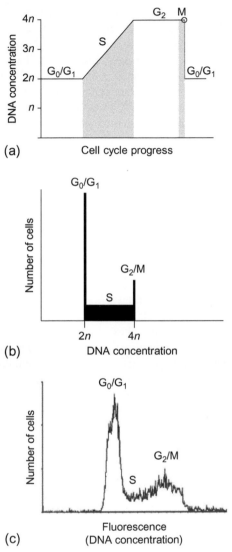

FIGURE 5.6 (a) The amount of polymerized DNA in a cell during the mitotic cell cycle. n = an arbitrary amount of DNA, dependent upon the type of organism. For humans, it would be 23 chromosomes' worth of DNA. (b) The theoretical relative number of cells in each phase of the cell cycle. (c) Measured number of cells in each phase of the cell cycle, as determined by PI fluorescence (indicating DNA concentration).

of DNA content per cell, the two phases are often combined and referred to as G_2/M. Everything between G_0/G_1 and G_2/M is S phase. Note that there are relatively few cells in S phase. S-phase costs so much energetically that the cell is committing a lot to go through it, which is why passing Start is such a key event in the cell cycle.

The relative amount of dsDNA per cell can be determined using *propidium iodide*, a dsDNA intercalating dye that brightly fluoresces red upon stimulation. The number of cells in a particular stage of the cell cycle can be determined experimentally using this dye, with typical results similar to what is shown in Figure 5.6c. Comparing panels (b) and (c) in the figure, the difference between theory and practice should become evident. While panel (b) shows discrete points for the beginning and ending DNA concentrations, panel (c) shows that these values are represented by approximately normal distribution curves.

Using propidium iodide and cell cultures that have had their cell cycles synchronized, the length of the cell cycle can be determined through a time course experiment. Human cells in culture typically have cell cycles of ~24 h. Although relatively complex, mitosis only takes about 1 h. S phase, on the other hand, usually lasts for about 8 h. Other types of cells can have quite different cell cycle lengths. For example, *E. coli* and budding yeasts can make it through the entire process in 20-30 min.

5.2   GROWTH CURVES AND THEIR PHASES

Now that we know how an individual eukaryotic cell grows and divides, we will move on to discuss how populations of cells grow. We have mentioned that the length of the cell cycle for microbes is very quick in comparison to human cells. Because of their short generation time, microbes are often used as cellular factories to produce molecular products because their numbers can be amplified greatly in a matter of hours. Although we will focus most of this section on prokaryotic cells, the principles for eukaryotic cell cultures are the same.

A graph of cell number versus time, commonly called the *growth curve*, is shown in Figure 5.7. There are several stages to this curve, both mathematical and physical, that warrant further discussion. These stages are lag, log (early and late), plateau, and death.

Lag phase Every cell culture begins with the inoculation of a medium with at least one cell, although typically many cells are used to start a culture. This is true for cultures in the laboratory as well as cultures growing on the food in your refrigerator. When the cells are first put into the growth medium, they do not immediately begin to replicate because they need time to acclimate to their new surroundings. This is called the lag phase because there's a period of time that must pass before the cells begin to replicate and populate the medium.

Log phase This is a period of exponential growth. There are two distinct regions within the log phase: early log and late log.

- Early log is a period of accelerated cell growth where cell numbers increase exponentially. Another name for this phase is the *exponential* phase.
- Late log is characterized by certain restrictions to cell growth. Although the slope of the curve is positive, it's second derivative is negative, meaning the shape is concave down. Another word for the late log phase is the *deceleration phase*.

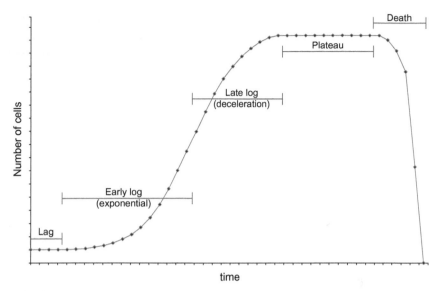

FIGURE 5.7   A hypothetical growth curve showing the progression of cell numbers in a culture over time.

Plateau phase While cells continue to divide, the division rate is greatly hampered. Any increases in cell numbers are offset by cell death, leaving the number of cells in the plateau phase roughly constant.

Death phase This phase is typical for cells grown in a vessel with a fixed amount of medium (a *batch culture*) for too great a time.

The reason for the lag phase is that once we inoculate our medium with cells, it takes a while for them to acclimate to the new environment from where they originated. Perhaps they were frozen at $-80\,°C$, maybe they were taken from another established culture, or maybe they were on the hands of an unsanitary food preparer. Whatever the previous conditions of the inoculate, the new culture environment is going to be different.

The log phase is associated with plentiful nutrients and ideal growing conditions. The cells have warmth, a large amount of space, and plenty of food so they're going to divide very quickly, which explains the exponential nature of this portion of the growth curve. However, in the deceleration phase, nutrients are beginning to become less plentiful. Referring back to the cell cycle, when the cycle comes around to the START checkpoint, individual cells may sense lower nutrient availability and delay progression beyond START to allow for the necessary buildup of dNTPs and energy stores. The deceleration in the rate of cell number increase is due to a gradual reduction in nutrient availability, an increase in waste concentration, as well as a reduction in the amount of free space in the culture container. Most cells (with cancer cells being a notable exception) adhere to the principle of contact inhibition, meaning they will not be as likely to divide if they bump into other cells or the sides of the culture container. As a cell culture contains more and more cells, more and more contact between cells will occur, leading to greater contact inhibition and a decrease in proliferation rates.

In the plateau phase some cells will be dividing, but there will also be some cells that are dying. The total number of cells in the culture will remain approximately constant. In the death phase, conditions are horrible. There will be no discernible cell growth, cell numbers will drop precipitously because of the environment that is virtually devoid of nutrients and may contain toxic waste products or secondary metabolites.

One can think of growing cells in culture as a chemical reaction. The reaction will begin with cells and substrates. One can inoculate some growth medium with cells on one day, and on the next day the culture vessel will contain something different. There will (hopefully) be a greater number of cells, cell products, plus unused substrates.

Original cells + Substrates → Original cells + New cells + Products + Unused substrates

While the concept is straightforward, different mathematical formulae are required to describe cell growth in each of the phases.

We can talk about how much *cell mass* is in a culture during exponential cell growth by describing the specific growth rate, designated by μ_{net}:

$$\mu_{net} = \frac{1}{x}\frac{dx}{dt}, \tag{5.1}$$

where x is the cell mass concentration, t is the time.

(Note that the above equation is not entirely rigorous. The term μ_{net} is often presented as a function of (the amount of growth) − (the amount of death). You will always have cells dying off for one reason or another, even during exponential growth. To keep the discussion straightforward we will omit the death parameter here.)

Sometimes it's convenient to talk about total cell mass in a culture, but at others one might be more interested in the *number of cells*. The above equation can be modified to reflect the rate of replication, designated by μ_{rep}.

$$\mu_{rep} = \frac{1}{n}\frac{dn}{dt}, \tag{5.2}$$

where n is the cell number concentration.

Calculations involving cell mass concentration are more convenient to work with because x represents a real number, whereas n is a positive integer. (It is infeasible to have half of a cell.) This implies that the function for μ_{rep} is not a continuous function. Again, μ_{net} represents the change in cell mass while μ_{rep} describes the change in cell number.

Equations (5.1) and (5.2) are not always interchangeable.

We cannot always quantify cell mass and convert it to a number of cells with accuracy. This is because μ_{rep} does not always equal μ_{net}. In the case where the cells have a lot of nutrients and a lot of space, they will be able to grow to a certain size and then divide freely, as is the case during early log phase. This will provide a condition where all cells are roughly the same size, so $\mu_{rep} = \mu_{net}$. However, when nutrients are scarce, some cells may become larger because of growth but they will not replicate; *i.e.*, they will not pass Start. This is the case in plateau phase. The mass per cell will be greater, so x/n (cell mass/cell number) will increase. In other words, the cells will tend to be bigger. We must therefore be cognizant of whether we are concerned with the amount of cell mass or the actual number of cells in a culture.

5.2.1   Growth Curve State—A Biotech Company Example

Suppose you have a biotech company that uses engineered cells to produce a specific biological product. Let's also suppose that the cells only make this product when they are not stressed, meaning the supply of nutrients and space are plentiful. When there aren't enough nutrients the cells will not have the luxury of making any extra products. They will only be making what they need to keep themselves alive. This implies that when the cell culture is in the plateau phase your company will not be harvesting much (if any) product. Also, when the culture is in the lag phase or in the early part of log phase, although there will be bountiful nutrients and space for growth, there will not be as many cells around to make your company's desired product. As a business person you will want to maximize production, so you should strive to keep the greatest number of cells as productive as possible. Profits can be maximized, in this case, by keeping cells in late log phase.

Once this particular culture enters the late log phase we know that we will have to feed the cells because they would not yield product after entering the plateau phase. The plateau phase can also be considered a warning phase because if the cells do not attain nutrition soon then they will die. Before this happens the cells could be passed to new cultures, meaning the cells in one flask can be divided among several new flasks and permitted to grow with new growth curves. If we divide one flask of cells among eight flasks, the culture is said to be undergoing *expansion*. However, because any bioreactor facility will have a finite amount of space, at some point most of the cells of a culture being *split* must be frozen to create a *stock*, or intentionally destroyed. Instead of undergoing expansion, the culture is now said to be in a *maintenance* situation.

Students are often bothered by the fact that when passing mammalian cells, over 75% (usually 7/8 or 15/16) of them will be intentionally destroyed. This is a reasonable concern in light of the fact that these cells may be producing a product that the company sells for profit. More cells would mean more profit. However, every laboratory has physical space limitations—there is only so much incubator space, and rooms are only so large and can house only so many incubators. Other costs must be considered, such as reagents, disposables, and personnel. Even if a given setup generates profit, doubling the size does not guarantee increased success. One must also consider market size and other economic factors. Having business-savvy professionals in your company's employ is an important part of a successful biotech company.

5.2.2   Be Aware of the Lag Phase

Consider a case where a batch cell culture has incubated for too long, and a small aliquot of cells is transferred from this plateau phase culture and placed in a new bioreactor with fresh culture medium. Referring to the graph in Figure 5.8, we see that the new growth curve will have the same elements we've already covered: it will have the same general shape and the same phases as before, but it will have a new y_0 and a different lag phase length. The length of the lag phase is related to how different the old culture environment was from the new one. For instance, taking cells from the end of lag phase, or the very beginning of the log phase, and starting a new culture with them should yield a relatively short lag phase for the new culture because the difference between the media should be very small. However, starting a new culture with cells from a plateau phase culture should yield a longer lag phase because the old and new culture media

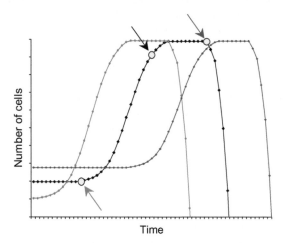

FIGURE 5.8   Growth curves to consider. Starting with a parent culture of *E. coli* cells (black curve), if we were to take a standard aliquot at the point indicated by the black arrow and start a new culture, the new culture would have a growth curve like the black line. Taking the standard aliquot at an earlier time (blue arrow) would yield a new culture with a growth curve like the blue curve. Note the lower starting concentration, shorter lag phase, and slightly longer plateau phase. Taking the standard aliquot at a later time (red arrow) would yield a new culture with a growth curve like the red curve. Note the higher starting concentration, longer lag phase, and slightly shorter plateau phase. The maximum number of cells in each culture should be the same because the results of contact inhibition should be unchanged for these cells. (Note that the curves do not represent actual values, but are drawn to illustrate the principles mentioned here.)

are very different. The cells may have to turn on or off many genes because they are being transferred from a nutrient-depleted environment to an environment rich in resources. The transferred cells first must detect the new surroundings then begin transcribing genes that are appropriate for the new surroundings. Since the cells will begin growing and dividing again, they will have to produce enzymes that will aid with the production of dNTPs, DNA polymerase, transcription factors, cytoskeletal proteins, *etc*. The further away the parent culture was from the lag phase on its own growth curve (every culture has its own growth curve), the longer the next lag phase will be. That's why, instead of taking our culture from late in the plateau phase, which might make sense because there will be a maximal number of cells, one might instead choose to pass the cells to a new culture during late log phase to shorten the time needed to again bring them into exponential growth. (Again, this discussion is for cells grown in batch.)

5.2.3 Cryptic Growth

There does come a point in the life of some cell cultures when, once a large number of cells begin to die, other cells will cannibalize them. This small increase in nutrient availability might allow for a brief period of cell growth at the end of the plateau phase, a period called *cryptic growth*. It is called "cryptic" because, on the surface, the medium should be virtually depleted of nutrients, so it would not be obvious why there would be an increase in cell numbers right before the death phase. We know now that nutrients are made available by the death and lysis of cells. An example of such a growth curve is given in Figure 5.9.

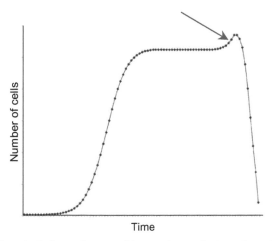

FIGURE 5.9   A hypothetical growth curve with a cryptic growth phase, denoted by the arrow.

5.2.4   Diauxic Growth

Consider the growth curve shown in Figure 5.10, which illustrates an example of *diauxic growth*. Note that there are two log phases. Diauxic growth is characterized by cells that utilize two carbon sources, where there is a preference for one over the other. For instance, cells that have the *lac* repressor can use the sugars glucose *or* lactose for energy. When in a medium that contains both of the sugars, they will preferentially use glucose. When glucose becomes scarce (as evidenced by the first plateau phase in the growth curve in the figure), the cells can alter which genes they transcribe to allow them to utilize lactose for energy. This switch to a new carbon source will allow for a second exponential phase of growth.

Returning to the example of a biotech company using cells to produce a commodity, we originally reasoned that passing cells to a fresh culture environment during the deceleration phase was preferable. This is not always the case. There are many occasions when the cells in culture do not produce a product of interest until they are in the plateau phase. This is the case for some antibiotic-producing bacteria. By the standards of human society, the microbial world can be very brutal. Some microbial cells will freely kill other cells when times get tough. When resources are limited and microbes are struggling for their own survival, certain microbes will start making antibiotics to kill other cells, especially of other cell types. If you happen to run a biotech company that produces antibiotics, then you might choose to let your cell cultures progress into the plateau phase to maximize product generation. Similarly, if you work with a type of cell that only expresses the gene that you're interested in when the cells are utilizing an alternative carbon source, such as during the second phase of diauxic growth, you might start your cultures with glucose to rapidly increase

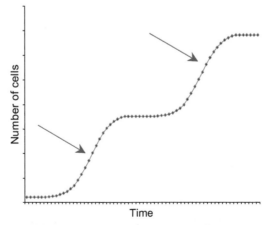

FIGURE 5.10 A hypothetical growth curve representing diauxic growth, where the culture switches to a secondary carbon source to achieve a second exponential growth phase. The two exponential phases are denoted by red arrows.

cell numbers. After an adequate population density has been generated, you might then let the cells metabolize the remaining glucose in the culture and then introduce the second carbon source. At this point the culture will be producing the product of interest. We shall see in a later chapter that this idea of rapidly growing cells and then changing culture conditions to obtain a desired bioproduct has been used for thousands of years.

5.3   MATHEMATICS OF THE GROWTH CURVE

5.3.1   Exponential Phase (Early Log)

We have already mentioned exponential cell growth, and that this phase of the growth curve is made possible because nutrient concentration is reasonably high, so there's relatively little to impede the cells from growing as fast as they can. Cells will progress unimpeded through the cell cycle and divide; one cell will become two, two will become four, four will become eight, etc. If we were to model that mathematically we would get an exponential curve. In the exponential phase, since nothing is being inhibited (the cell is growing and dividing as quickly as possible), $\mu_{net} = \mu_{replication}$, which means the equations that describe cell number and cell mass are interchangeable.

$$\mu_{net} = \frac{1}{x}\frac{dx}{dt},$$

Where x is the cell mass concentration, t is the time (hours).
The differential can be split up to allow integration of both sides:

$$\mu_{net}\,dt = \frac{1}{x}\,dx$$

$$\int_{0}^{t}\mu_{net}\,dt = \int_{x_0}^{x_t}\frac{1}{x}\,dx$$

$$\mu_{net}\,t = \left(\ln x_t - \ln x_0\right)$$

$$e^{\mu_{net}t} = \frac{x_t}{x_0}$$

$x_0 e^{\mu_{net}t} = x_t \rightarrow$ The cell mass at time t is equal to (the initial amount of cell mass) $\times e$ raised to (the rate of change in cell mass concentration) $\times$ (the amount of time that has elapsed).

5.3.1.1   Doubling time, indicated by $t_{1/2}$ *is it the amount of time that it takes for cell mass (or cell number) to increase to twice the starting concentration. Another way to say that is* $x_t = 2x_0$

Plugging this relation into the equation we just derived,

$$x_0 e^{\mu_{net} t} = 2x_0$$
$$\Rightarrow e^{\mu_{net} t} = 2$$
$$\Rightarrow \mu_{net} t = \ln(2)$$

$$\Rightarrow t = \frac{\ln(2)}{\mu_{net}} \rightarrow \text{The amount of time needed to double cell mass is equal to}$$

0.6931/(rate of cell mass increase).

Sometimes $t_{1/2}$ is written τ_d. Keep in mind that we are still discussing the exponential phase, and in the exponential phase we can go back and forth between cell mass and cell numbers, so the time needed to double cell mass is equal to the time needed to double cell number.

$$t_{1/2} = \tau_d = \frac{\ln(2)}{\mu_{net}} = \frac{\ln(2)}{\mu_{replication}} = \tau'_d$$

As a simple example, suppose a certain cell population has a doubling time of 20 min., and we start with 200 of these cells in culture. How many cells are in the culture after 1 h? First, determine the number of doublings: ($t/\tau_d = 60$ min/20 min $= 3$ doublings), then plug into:

$$n_0 \cdot 2^{(\# \text{ of doublings})} = 200 \cdot 2^3 = 200 \cdot 8 = 1600 \text{ cells.}$$

5.3.2   Deceleration Phase (Late Log)

Left unattended, the growing culture will undergo deceleration because nutrient supply is no longer limitless from a cell's point of view, and waste products are beginning to build up. This is a phase of unbalanced growth, and x and n are no longer interchangeable. As a result, the doubling time is different for cell mass versus cell number. Physically, the cells will be larger before they start to divide. Part of the reason for larger cells is that, when the cell is replicating its DNA it is using up a lot dNTPs, which represents a great amount of energy. Keep in mind that every base pair that's replicated requires two dNTPs: one for each new base that is being added to the growing double-stranded chains. This energy is in addition to the energy required to manufacture new deoxynucleotides and dNTPs. The ever-decreasing availability of nutrients in the medium is enough to hinder cells from passing the START checkpoint as easily as before, hence the larger average cell size. Another reason that $\mu_{net} \neq \mu_{replication}$ is that the cells are no longer focused on replication, but rather they are starting to focus on their own survival.

Refer again to Figure 5.7. As it is drawn, the time spent in the exponential phase is roughly equal to the amount of time spent in the deceleration phase. This is not necessarily the case for a given bacterial population. As the energy source (carbon source) present in the medium is metabolized by the cells, their growth will eventually be limited. The relation between μ_{net} and the concentration of the carbon source is given by:

$$\mu_{net} = \frac{\mu_{max} S}{k_S + S},$$

where S is the concentration of the energy source (the substrate), k_s is a constant describing the affinity of the microorganism for the energy source.

If we set the substrate concentration to a value equal to k_s, the equation becomes:

$$\mu_{net} = \frac{\mu_{max} k_S}{k_S + k_S} = \frac{\mu_{max}}{2}.$$

This implies that k_s can be defined as the substrate concentration at which that will produce a growth rate equal to half of the maximum possible growth rate. This constant serves as a measure of an organism's affinity for an energy source.

During the early log phase, when there is a very large relative concentration of S, the rate of growth (μ_{net}) will be as fast as it can be (*i.e.*, when $S >> k_s$, μ_{net} will mathematically approach the asymptotic value μ_{max}).

Another use for k_s has to do with where the transition from the exponential to deceleration phases appears on the growth curve. A low k_s implies that the microorganisms have a high affinity for the substrate, so they will latch onto it efficiently enough that deceleration will not occur until a greater percentage of the energy source has been used. The deceleration phase will be relatively short. If k_s is high then affinity for the substrate will be low, meaning the binding efficiency to the substrate is low, and the microorganisms will begin to sense that the energy source is becoming scarce when the absolute concentration of substrate is higher. The deceleration phase would last longer (Figure 5.11).

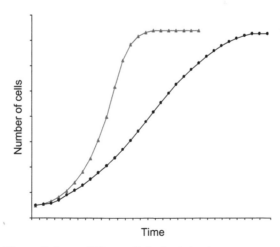

Time

FIGURE 5.11 When cells have a different affinity for their primary energy source, the shape of the growth curve will be altered. When there is high affinity (low k_s), the substrate will be bound efficiently, even at relatively low concentrations. leading to a short deceleration phase (red curve). When the binding affinity is lower (high k_s), the cells will sense a lower substrate concentration earlier. The deceleration phase will take longer and be more gradual.

5.3.3   Plateau Phase

Recall that the net growth rate in this phase is equal to zero. There are enough nutrients for cell maintenance, but continued large-scale replication is not supported. There may still be growth of some individual cells but, as pointed out earlier, net growth = (total growth) – (death). The bioreactor and medium cannot support any more population growth.

There are two scenarios that may occur regarding cell mass concentration in relation to the total number of cells. Either:

$$\mu_{net} = \mu_{rep} = 0 \, (\text{there are no cells dividing}), \text{ or}$$

$\mu_{net} = 0$ and $\mu_{rep} < 0$ (growth rate = death rate, with cell mass concentration staying constatnt while there are fewer viable cells). However, there will necessarily be a change in cell mass due to the maintenance of life. This change will be proportional to the initial number of cells:

$$\frac{dx}{dt} = -k_d x ,$$
$$\Rightarrow x = x_0 e^{-k_d t}$$

where x_0 is the cell mass concentration at the beginning of plateau phase, k_d is a constant that describes the rate at which the cell metabolizes its own reserves for the maintenance of life.

Often during the plateau phase the cells will begin to produce secondary metabolites. These metabolites may have value to the laboratory that is controlling the culture. Therefore, the plateau phase is where some companies get their money. In the exponential phase, cells are dividing and are focused on replication. In the plateau phase cell behavior is more concerned with survival. Antibiotics, for example, that are produced by the microbes are generated to help with their survival, with the possible goal of killing off some of the other microbes that are challenging them for the limited food supply. If your laboratory is engaged in harvesting antibiotics then the plateau phase is where the payoff is.

Secondary metabolites are different from primary metabolites. Primary metabolites are produced during the exponential phase, and are the natural result of the normal processes of the cell. For instance, your cells (even as you read this) are converting glucose into pyruvate, and the pyruvate is probably being converted to CO_2 and water. The pyruvate, CO_2 and water are all primary metabolites. Secondary metabolites are not directly involved in the chemical processes that keep an individual cell alive. In the case of the mold *Penicillium chrysogenum*, penicillin is a secondary metabolite that is produced as nutrients become scarce for the microbes. Under laboratory conditions, even though a culture of *P. chrysogenum* might be pure (meaning the only living things in the culture are *P. chrysogenum*), the cells will still produce the antibiotic as nutrients become scarce regardless of whether they are truly competing against other bacterial strains. This is because the genetics of the bacteria in that culture dictate that the antibiotic genes are expressed in times of limited resources.

It is now appropriate to reintroduce the question, "if we have a biotech company that deals with cell cultures, during what stage of the growth curve should cells be passed into a fresh culture?" Earlier, when we were only concerned with cell numbers or the production of primary metabolites, the answer was the deceleration phase. In the case just described, where the cells are producing a product for us in the plateau phase, we would want to maintain them in the plateau phase. The means by which this can be accomplished will be discussed shortly.

Consider a case where we have a microbe that produces something that's toxic to itself, such as yeast producing alcohol. Recall from earlier in this book, from the discussion on sterilization, sanitization, and aseptic conditions, that alcohols can be toxic to cells. Yeast are commonly used to produce the alcohol (ethanol) in fermented products such as beer, even though high concentrations of ethanol will kill the yeast cells themselves. Initially, there will be little to no alcohol in the cell culture, but it will increase slowly as the yeast metabolize glucose in the absence of oxygen. As the ethanol concentration increases to a certain level, the yeast will stop net growth. This marks the beginning of the plateau phase. One might ask how we can we get around this to obtain a greater concentration of alcohol from the yeast culture. One method could be to extract the alcohol from the culture, perhaps through distillation.

In general, there are several ways to extract secondary metabolites from a given culture. One way is to complex the product with a *non-metabolite*. A non-metabolite is a molecule that will not be used by the cell during metabolism. Such a molecule could be used to bind the secondary metabolite, allowing us to pull the complex out of solution. This form of harvest will allow the cells to continue to produce the secondary metabolite without toxic effects to themselves.

5.3.4   Death Phase

Death phase is very similar to the plateau phase. Plateau phase is zero order on a logarithmic scale (N is a constant), but it could also be considered a special case of first order where $dN/dt = c = 0$, meaning the graph is linear but the slope is 0. Death phase is also a linear phase on a logarithmic plot, but it has a negative slope, equal to $-k_d$.

$$\frac{1}{x}\frac{dx}{dt} = -k_d$$

$$\int \frac{1}{x}dx = \int -k_d dt \;\leftarrow \text{Equation for a line, on a logarithmic scale.}$$

$$\ln x = -k_d t + c$$

$$x = x_0 e^{-k_d t}$$

5.4 COUNTING CELL NUMBERS

In light of Equations (5.1) and (5.2), one must be able to quantify cell mass and cell numbers. Let us first discuss methods for determining cell numbers.

5.4.1 Hemacytometer

One common and relatively inexpensive way of counting cells is with a hemacytometer (Figure 5.12). In the figure, note that the main square is divided into nine regions, of which five have been labeled. The etched lines on the hemacytometer are very precise. In the figure, the lines inside the area denoted E are 0.2 mm apart, demarcating an area of $1 \, mm^2$. The other noted areas, A-D, also

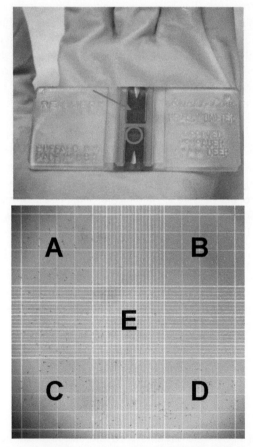

FIGURE 5.12 (top) A standard hemacytometer. The etched lines of the two counting grids are barely visible with the naked eye, being denoted here by a blue arrow and circle. (bottom) The area in the circle above is viewed with a microscope. With a 4× objective, nine distinct sub-areas are clearly visible.

denote standard areas of $1\,mm^2$. The subdivisions of these areas permit easier counting. The cover slip specific for the hemacytometer is such that when $10\,\mu l$ of cell suspension are loaded into one of the chambers, the height of the liquid will be $0.1\,mm$. This means that the volume of liquid over any of the areas A-E is $1\times1\times0.1\,mm$

$$= 0.1 mm^3$$
$$= 1\times10^{-4}\,cm^3.$$

A good reason to convert to cm^3 is that $1\,cm^3 = 1\,ml$, so we can now say that the volume is $1\times10^{-4}\,ml$. This means that if one were to count the number of cells in area E, the total could be multiplied by 1×10^4 to obtain an approximate number of cells per milliliter. Counting the cells in areas A through D and dividing by 4 will give an even better representation of the concentration of cells in the original suspension.

Counting cells using a hemacytometer is a very straightforward process, with one possible exception. Referring to Figure 5.13, notice the cells denoted by the arrows. Each one could be considered to lie inside or out of the region

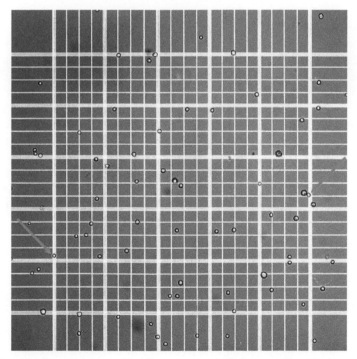

FIGURE 5.13 The area in area "E" (Figure 5.12) magnified using a 10× objective lens. Note that this level of magnification is needed to be able to count cells—the same cells are present in Figure 5.11, bottom. Also note that sometimes cells will fall on left or right boundaries (blue arrows), as well as top or bottom boundaries. A simple rule of counting cells on left and top borders, but not counting cells on right and bottom borders, easily removes this quandry.

being counted. When considering the number of cells that are being counted, it would be inefficient to also keep track of fractions of cells, so instead either the cell will be counted or will not. Assuming that the probability of a cell touching the left-hand border of the counting area is the same as that of a cell touching the right-hand border, a common practice is to count the cells that touch the left border but not to count cells that touch the right border. The same reasoning applies to counting cells that touch the top border but not counting cells that touch the bottom border of a region being considered.

Using the above rules and scale-up factor, you should be able to count 43 cells in the region shown in Figure 5.13 (verify this for yourself). This would lead us to approximate the concentration of cells in our original suspension as 430,000 cells/ml. If the total volume of the original suspension were 8 ml, this would imply that we have 3,440,000 cells on hand.

In terms of microbial cell growth that's fine. You write down the number and you decide whether the concentration is good enough to accomplish the next thing you need to do (such as isolate a desired mass of DNA). If we are working with a mammalian cell culture, we might need to use a specific number of cells to, for instance, seed a porous scaffold for tissue engineering. Determining the volume of cell suspension that will contain a specific number of cells is a common practice in the laboratory.

Consider a gene delivery experiment that will involve the transfection of 100,000 canine skeletal muscle cells. We would typically grow such a culture to late log phase and trypsinize it, meaning we use an enzyme (trypsin) that will cleave proteins, including those used for cell adhesion, so that the cells will no longer be stuck to the culture flask and will float in solution. The cell solution would then be centrifuged into a solid pellet to allow us to pour off the trypsin-containing solution, and the cells would then be resuspended in a known volume of medium. Suppose that after performing these steps, 10 μl of the cell suspension were loaded into a hemacytometer and subsequent counting yielded an average of 77 cells per counting area. The question is now, "how much of this cell suspension should be plated for a 100,000-cell transfection experiment?"

$$77 \, \text{cells} / \text{counting area} \Rightarrow 770,000 \, \text{cells} / \text{ml}$$

$$\frac{770,000 \, \text{cells}}{1 \, \text{ml}} = \frac{100,000 \, \text{cells}}{x \, \text{ml}}$$

$$x = 0.1298 \text{ml} = 129.8 \mu\text{l}.$$

We would plate 129.8 μl into one cell well, let it grow overnight, and the next morning we would be able to transfect roughly 100,000 cells. One might ask why the cells do not multiply during this incubation period. First, after taking the cells from the flask and transferring them into a new culture environment, there will be a lag phase. Second, since the cell cycle for mammalian cells is about 24 h and the incubation is overnight, there will not be enough time for the population to double. Third, some cells will die and some will possibly make it to cell division, so the actual number of cells that are transfected the next day will probably not be exactly 100,000, but having a standard number of cells plated per well will help ensure that roughly the same number is being transfected no matter how many wells, or days, we use for our experiments.

5.4.2   Agar Plates

For bacterial cultures, one can determine cell number with the aid of agar plates. If we were to evenly plate a given volume from a cell culture on an agar plate and allow it to grow overnight, the next morning we should be able to count the number of colonies on the plate with the naked eye (Figure 5.14). Each of these circular bacterial colonies originated from a single cell which was plated the night before. So, if we plated 100 μl of a bacterial culture and the next morning we saw that we had 100 colonies growing on the plate, we could infer that the culture contained one *colony forming unit* (CFU) per μl when the cells were originally plated. It may be easy to think of a CFU as being a bacterium, but this would not necessarily be correct. Dead and most dying bacteria will not form colonies, so CFU does not necessarily equal the number of cells in an aliquot, it indicates the number of cells that were able to grow on the agar to form colonies.

We can take a sample of cells from our culture, dilute the sample if we wish, take a known volume (such as 100 μl) and evenly distribute the volume on the surface of an agar plate. After an overnight incubation at an optimal temperature (37° for *E. coli*, 30° for yeast), the number of colonies can be counted and divided by the volume of culture that was initially plated to yield the CFU concentration at the time the cells were plated.

5.4.3   Cell Counters and Flow Cytometers

Consider two electrodes connected to a power source with an ohmmeter in the circuit. If the electrodes are placed into an electrolyte solution, then some current

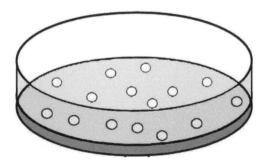

FIGURE 5.14   Agar plates can be streaked with a known volume of bacterial cell suspension to get an idea of cell concentration. Presuming that they do not overlap, the resulting circular colonies come from individual cells able to proliferate, known as colony forming units.

will be able to pass between the two electrodes and the current (lack of resistance) can be noted. If the electrodes are very small and very close, and we were to place a cell between them, an increase in resistance would be detected because the plasma membrane, made of phospholipids, is a poor conductor of electricity. Figure 5.15 is an illustration of how these principles have been applied to produce a cell counter, which is a special case of a particle counter. Cell medium, regardless of the type of cell for which it was designed, will contain salts (or else the cells would face a hypotonic environment and swell/burst), and therefore will conduct electricity. Negative pressure will be used to pull cells into and up a small cylinder that contains an electrode. The cylinder will contain an aperture that should ideally allow no more than one cell to pass at a time. A second electrode will reside on the other side of the cylinder (shown outside the cylinder in the figure). As a cell enters the aperture, a momentary increase in resistance will be detected on the ohmmeter and the signal will be interpreted as one cell. At the same time, the amount of fluid that has been pulled through the cylinder is controlled and monitored or so that the number of cells per unit volume can be determined. This concentration is usually output in the form of cells/ml.

Flow cytometers work in a fashion similar to cell counters, with the exception that cells are detected optically instead of via changes in resistance. A laser is directed across the inlet tube, and cells are detected when the laser light is scattered away from the detector on the other side.

5.5 COUNTING CELL MASS

5.5.1 Packed Cell Volume

The preceding three methods are means by which one can determine cell numbers, but it can be faster or less expensive to determine cell mass. One straight-forward means of obtaining such values is via *packed cell volume*. Starting with cells in suspension, the culture is centrifuged in a special graduated conical tube. The resulting pellet is measured against the graduations to determine the packed cell volume. If the density of the specific cell type is known, then cell mass can be calculated from the volume.

5.5.2 Wet and Dry Weight

Perhaps a more precise application of the above method is the determination of the *wet weight* of the cells. Once again, cells will undergo centrifugation to form a pellet. The supernatant is poured off and the pellet weighed. The weight of the tube plus pellet will be determined, and the tare weight of the empty tube subtracted to yield the wet weight of the cells. There is no need to utilize an average cell density as mass will be determined directly via a balance. This method is very straightforward and simple to perform. It is routinely used to produce an estimate of cell mass. While determination of the wet weight is simple, there is error associated with the variable amount of liquid remaining over the pellet and

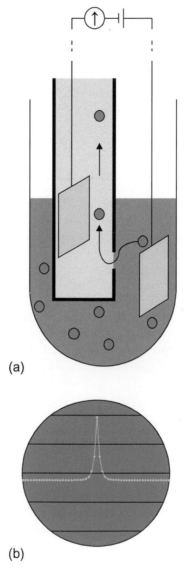

(a)

(b)

FIGURE 5.15   (a) Schematic of a cell counter. Cells (green) and medium are drawn up the collection tube via a vacuum. When a cell passes between two electrodes (gray), (b) a momentary increase in resistance will be detected.

trapped between individual cells after the supernatant has been removed. A more accurate estimate of cell mass can be made using the *dry weight* of the cells. The procedure for determining the dry weight is essentially the same as above, except that after the first centrifugation the cells will be washed and undergo a second centrifugation, followed by drying of the pellet. The wash step is very important.

Without it, salts, proteins, carbohydrates, and other constituents of the cell medium will contribute to the weight of the pellet, even after it has been dried.

5.5.3   Optical Density

The spectrophotometer can be used to determine cell mass for Gram negative bacteria such as *E. Coli*. Optical density measurements are taken at 600 nm (red light) in plastic cuvettes, and the reading compared to a calibration curve. The calibration curve must be generated for the specific spectrophotometer that is being used because optical density relies, in part, on light scattering, and readings will vary depending on the distance of the photodiode detector from the sample (which may vary with machine). Generation of the calibration curve is time-consuming and often utilizes other methods such as agar plates to determine the cell counts of the standards. However, once a reliable curve has been generated, it can be repeatedly used to determine cell mass concentration for many experiments to come.

It might be interesting to note that plastic cuvettes are used for determining the optical density of bacterial cultures instead of the quartz cuvettes used to determine the absorbance of DNA/RNA-containing solutions because (1) quartz cuvettes are over 100 times more expensive than plastic, and (2) at 600 nm, the light that is being used is of relatively low energy and will not induce autofluorescence, unlike the 260 nm (ultraviolet) light that is used to determine DNA concentration.

5.6   SCALE-UP

Although a process might work on a very small scale, making it work on a larger scale involves more than just using bigger tubes. This issue of scale-up is very important for the biotechnologist, whether it be used for producing viruses for transduction on an industrial scale, for producing large amounts of bacterial cell culture for production of recombinant protein, or for producing large amounts of herbicide to selectively protect genetically modified plants. To more fully appreciate the significance of scale-up, let us address the problem in terms of increasing the volume of bacterial cell cultures. Let us begin with some basic geometry.

Consider two cylinders, the first with radius=height=1 (arbitrary units), and the second with radius=height=2. Using $v=\pi r^2 h$, the volumes of the two cylinders are found to be equal to:

$$\text{Cylinder 1:} v_1 = \pi(1^2) \cdot 1 = \pi$$
$$\text{Cylinder 2:} v_2 = \pi(2^2) \cdot 2 = 8\pi$$

While the ratio of height to radius has been held constant, doubling the radius and height produces an eightfold increase in volume.

We will now look at the general case. Consider two more cylinders, with r/h being held constant at a value of k_1. Because of this relation, we will be able to simplify the volume equation by removing one of the variables:

$$r/h = k_1$$
$$\Rightarrow r = k_1 h$$
$$\Rightarrow v = \pi \left(k_1 h \right)^2 h$$
$$= \pi k_1^2 h^3$$

Now, if we wish to scale-up the dimensions of our first cylinder by a factor of c, still holding $r/h = k_1$, we can determine a general relation for the change in volume:

$$h_2 = ch_1$$
$$\Rightarrow v_2 = \pi \left(k_1 h_2 \right)^2 h_2$$
$$= \pi k_1^2 h_2^3$$
$$= \pi k_1^2 c^3 h_1^3$$

So, although the heights are scaled by $h_2/h_1 = c$, the volumes are scaled by

$$v_2/v_1 = \frac{\pi k_1^2 c^3 h_1^3}{\pi k_1^2 h_1^3} = c^3$$

Check this against the original problem, where the scaling factor was equal to 2. We found that the cylinder volumes $v_1 = \pi$ and $v_2 = 8\pi$, so $v_1/v_2 = 8 = $ (the scaling factor)^3.

Consider the same problem of scale-up for a pair of cylindrical culture vessels, except this time we will deal with adherent cells (cells that are attached to the walls of the culture vessel). For adherent cells, surface area is of key interest. As before, we will impose the restriction that the ratio of radius to height is constant:

$$r/h = k_1$$
$$\Rightarrow r = k_1 h$$

$$\text{Surface area} \equiv SA = 2\pi rh$$
$$= 2\pi \left(k_1 h \right) h$$
$$= 2\pi k_1 h^2$$

Again, scaling up by a factor of c (still holding $r/h = k_1$) we can determine a general relation for the change in surface area:

$$h_2 = ch_1$$
$$\Rightarrow SA_2 = 2\pi \left(k_1 h_2 \right) h_2$$
$$= 2\pi k_1 h_2^2$$
$$= 2\pi k_1 \left(ch_1 \right)^2$$
$$= 2\pi k_1 c^2 h_1^2$$

$$\Rightarrow SA_2 / SA_1 = \frac{2\pi k_1^2 c^2 h_1^2}{2\pi k_1^2 h_1^2} = c^2$$

We now see that, holding the ratio of radius to height constant, volume will vary as a third order relation with the scaling factor while surface area will vary as a second order relation. This may become more intuitive if we recall that volume is a 3-dimensional descriptor while surface area can be reduced to 2 dimensions. These facts take on greater importance in matters of cell culture, where the cells of interest can grow both adherently and non-adherently. Certain bacterial cultures are one example. Scale-up of the non-adherent cells is a matter of volume, while scale-up for adherent cells is largely a matter of surface area.

Consider a culture where you have adherent cells growing on the bottom and the sides of the vessel, but also non-adherent cells growing in suspension. Scale-up becomes a more interesting problem in this case because the amount of surface area scales with r^2 while the volume scales with r^3. If the adherent cells produce more product per cell than the non-adherent cells, scaling up a cylindrical vessel without a change in geometry will be a losing proposition: the cost of medium and laboratory spare requirements will increase faster than the amount of product. A change in geometry should be regarded.

Now consider the bioreactor shown in Figure 5.16, which utilizes a stir bar to constantly mix the medium to bring oxygen to the cells. If we were to scale-up the volume of this bioreactor while leaving the delivery of oxygen to the cells unaffected, the rate of mixing will have to be increased, perhaps by increasing the RPM of the stir bar. However, as the RPM of the stir bar is increased, the amount of shear stress upon each cell will also be increased. Increasing the volume while keeping the vessel shape, rate of mixing, and shear stress constant cannot be achieved. The biotechnologist should be aware of which parameters are most important and try to control them, realizing that other parameters will not scale at the same rate.

There is a variety of relations that must be considered during the scale-up of cell cultures, keeping in mind that not all can be controlled simultaneously.

surface area	S.A.	$\propto r^2$
volume	V	$\propto r^3$
pump rate	Q	$\propto nr^3 (N = \text{revolutions/minute})$
energy input	P	$\propto n^3 r^5$
energy input/volume	P/V	$\propto n^3 r^2$
pump rate/volume	Q/V	$\propto n$

Mathematically we can see there's a difference in all of these parameters. While we can control any single one of them, we cannot simultaneously control all of them.

FIGURE 5.16   A spinner flask, commonly used for tissue engineering applications. *Photo courtesy of Prof. Taby Ahsan, Tulane University.*

We just saw RPM comes down to n, and P/V comes down to n, so RPM and P/V can be controlled concomitantly. Since P/V controls oxygen availability, if we increase the pump rate while keeping the volume (and radius) constant then we will have more oxygen available. If we have a culture mixed slowly versus a culture mixed quickly, which one will be mixed with the air above it more quickly? The one with the faster mixing. P/V gives you that. Alternatively, you can control the Reynolds number. Consider a bioreactor where oxygen is being bubbled up through the culture, and the bioreactor has sieve plates that allow the oxygen to get through, but as the oxygen bubbles strike the plate there will be some turbulence which will aid mixing. If we were to scale this bioreactor, we could do so in a way that would preserve the geometry of the bioreactor, but O_2 availability would be altered. We could try to preserve oxygen concentration by increasing the flow rate of oxygen with the increase in volume, but then we would be increasing the amount of turbulent force at the sieve plate, which could damage the cells. Once again, you can control some but not all of the scale-up parameters.

Example 5.1

In case you are still not a believer, let us take a sample problem and work it two ways:

Consider a 2 l batch fermentation system whereby 75% of the (intracellular) target product was associated with attached cells and 25% was associated with cells in suspension. The total output of this reactor was 2 mg of product/liter/day.

Now consider a scale-up to a 20,000 l reactor that has the same height-to-diameter ratio as the original reactor: 2-1. Assuming that both tanks are cylindrical, that cells will bind to the walls AND bottom of the tanks, that binding to other internal components of the bioreactor is negligible, and that no cells bind to the tops of the tanks, what will be the yield of the 20,000 l system?

Answer:

Before we get to the scale-up portion of the problem, let's take a look at just what is going on in the system. In the first bioreactor, 2 (mg/l/day) · (2 l) → the total output of the first bioreactor is 4 mg/day.

The problem also states that 25% of the product is due to cells in suspension. We can think of this amount as being directly related to the volume of the bioreactor (V). So, the amount of product in terms of reactor volume = 25% (4 mg/day) = *1 mg/day*.

We also know that 75% of the product is due to cells attached to a surface, such as the walls of the bioreactor. It is common for cells to behave differently when in suspension versus when they are attached. We can think of the amount of product resulting from attached cells as being directly related to the surface area of the bioreactor (SA). So, the amount of product in terms of reactor surface area = 75% (4 mg/day) = *3 mg/day*.

Note that the 25/75 split of product coming from volume and surface area is unique to the small bioreactor. When we move to the larger bioreactor, V and SA will scale differently, so the 25/75 split will be irrelevant.

Next, let us utilize the relation between height and diameter. We will use this relation in different ways for solutions (a) and (b).

h: $d = 2 = h/d$

$\Rightarrow h = 2d = 2(2r)$ (where r = radius)

$\Rightarrow h = 4r$

Let us now move on to determining the yield of the 20,000 l system:

(a) Solution using a geometric method:

2 l Reactor	*20,000 l Reactor*
We can easily find the basis for scaling the amount of product due to volume:	
$V = 1$ mg/day for a 2 l reactor	
$\Rightarrow V = 1$ mg/day/2 l	
$\Rightarrow V = 0.5$ mg/l/day $\longrightarrow$	Output from $V_2 = (0.5$ mg/l/day$) \cdot (20,000 l)$
	= 10,000 mg/day
We will use the volume to solve for r:	
$V = \pi r^2 h$	
$= \pi r^2 (4r)$	
$V = 4\pi r^3 \longrightarrow$	$20,000 = 4\pi r_2^3$
	$\Rightarrow r_2 = 11.6754$ dm
$2 l = 2$ dm³ $= 4\pi r_1^3$	
$\Rightarrow r_1 = 0.5419$ dm	
$SA = 2\pi rh + \pi r^2$ (only one area of a circle because no cells attach to the top)	

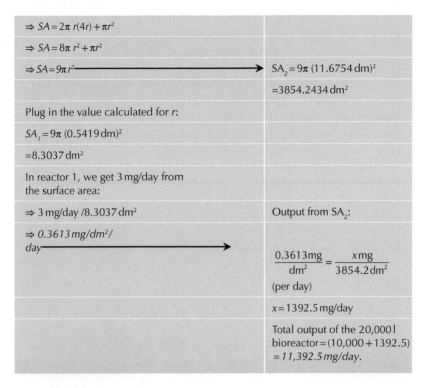

$\Rightarrow SA = 2\pi\, r(4r) + \pi r^2$	
$\Rightarrow SA = 8\pi\, r^2 + \pi r^2$	
$\Rightarrow SA = 9\pi\, r^2$ ⟶	$SA_2 = 9\pi\,(11.6754\,\mathrm{dm})^2$
	$= 3854.2434\,\mathrm{dm}^2$
Plug in the value calculated for r:	
$SA_1 = 9\pi\,(0.5419\,\mathrm{dm})^2$	
$= 8.3037\,\mathrm{dm}^2$	
In reactor 1, we get 3 mg/day from the surface area:	
$\Rightarrow 3\,\mathrm{mg/day}\,/\,8.3037\,\mathrm{dm}^2$	Output from SA_2:
$\Rightarrow 0.3613\,\mathrm{mg/dm^2/}$ day ⟶	$\dfrac{0.3613\,\mathrm{mg}}{\mathrm{dm}^2} = \dfrac{x\,\mathrm{mg}}{3854.2\,\mathrm{dm}^2}$ (per day)
	$x = 1392.5\,\mathrm{mg/day}$
	Total output of the 20,000 l bioreactor $= (10,000 + 1392.5)$ $= 11,392.5\,\mathrm{mg/day.}$

(b) *Solution using the scale-up factor* c:
The volume scaling factor $= 20,000\,l/2\,l = 10,000 \Rightarrow c^3 = 10,000$
The surface area scaling factor is c^2, which $= (c^3)^{2/3} = (10,000)^{2/3} = 464.159$
Now, simply plug in the original outputs due to V and SA from the original reactor and multiply by the appropriate power of c:
1 mg/day $(10,000) + 3$ mg/day $(464.159) = 11,392.5\,mg/day$.
This is the same answer as in part (*a*), but was a whole lot easier to obtain!

QUESTIONS

1. Name and describe 3 checkpoints in the cell cycle.
2. Can one tell the difference between G0 and G1 by using a microscope?
3. What is the difference between absorbance and optical density?
4. Explain what diauxic growth is and why it happens.
5. If a cell suspension is so concentrated that individual cells in a hemacytometer cannot be resolved for counting, how can one still use the hemacytometer to find the concentration of cells?
6. Suppose a particle counter collects medium at 40 μl/min for 30 s and counts 23 cells. What is the cell concentration of the cell suspension in cells/ml?

7.
    a. Describe why it might better to pass batch cell cultures in the late log phase rather than in the early log or plateau phases.

    b. Is it always best to pass cells in the late log phase? Explain.

8. Dr. Richtofen puts a 100 μl of cell sample through a particle counter and gets a cell count of 735 cells/ml. He then puts a 100 μl aliquot of the same sample (Dr. Richtofen is obsessive compulsive) onto an agar plate and counts only 55 colonies the next day. Why did the agar plate produce fewer colonies than there were cells counted by the cell counter? (Assume Dr. Richtofen didn't kill any of the cells between the tests.)

9. When do cryptic growth and diauxic growth occur? What is the difference between the two?

10. Can a cell population grow exponentially for an infinite amount of time? Explain your answer, and include a labeled plot of ln population growth kinetics.

11. Is the net change in the number of cells zero, negative, or positive from the beginning of the lag phase to the end of the death phase? Explain your answer.

12.
    a. If we have a population of cells in early log phase, in what phase of the cell cycle are most of them in?

    b. If we have a population of cells in plateau phase, in what phase of the cell cycle are most of the cells?

13. You are asked to quantify the concentration of *E. coli* that are growing in a culture. You remove 200 μl and streak a plate. After 6 h the plate looks like this:

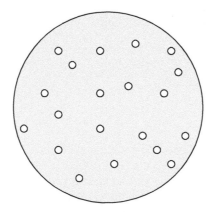

    a. What was the concentration of bacteria, per ml, at the time you streaked the plate?

    b. Assuming that the culture has been in (early) log phase since you streaked the plate and that doubling time $\tau_d = 20$ min, what is the concentration of the culture now? (Please leave your answer in the form of an expression: *do not simplify.*)

c. Assume instead that the culture has been maintained in the plateau phase since yesterday, what is the concentration of the culture now?

d. Forget you ever saw parts *b* or *c* of this problem. Suppose that your culture was behaving strangely, so you decided to streak a second plate 2 h after you streaked the plate above. After 5 h of incubation, your second plate had 160 colonies on it. What is the doubling time for the culture (in minutes)?

e. Assuming that the culture has been in (early) log phase since yesterday and that $\tau_d = 1\,h$, what is the concentration of the culture now? What is the value of μ_{rep}?

14. For a transfection experiment, suppose that you want to plate 100,000 cells per well. Also suppose that you loaded a hemacytometer with $10\,\mu l$ of a sample of the cells that you want to use and obtained the pictured result:

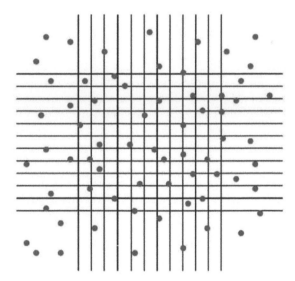

a. How much of the solution will you plate into each well for your transfection experiment? (Show your work.)

b. Given that the area of the (magnified) counting region above is 1×1 mm, determine the height of the liquid based on the scale-up factor presented in this chapter.

15. Examine the three growth curves on the right. Each curve, A, B, and C, represents a separate strain of microorganism grown in identical culture conditions. (You may consider the lengths of all three lag phases to be equal.)

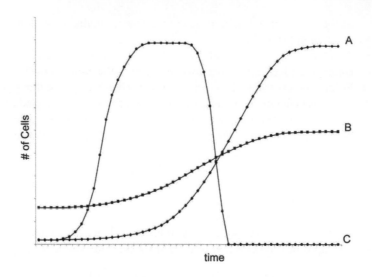

a. If the cells in question break down oil, such as what might be seen following the explosion of a drilling platform in the Gulf of Mexico, which of the three strains would be best-suited for the job? Give two reasons for your answer.

b. Consider the three curves in a different setting: Suppose the cells produced a desired product during plateau phase. (Assume that the same amount of product per cell is produced regardless of strain.) Which strain would you use for your business that produces that product? Give one reason for your answer. State all of your assumptions.

16. If we start two new cultures by taking a set volume out of an existing batch culture at times t_1 and t_2, where t_1 is in the early exponential phase and t_2 is in the plateau phase, what will the new growth curves look like? Express your answers on a single graph. (Think about how many cells are in the new cultures at $t=0$, and also think about the lengths of the lag phases. Also keep in mind how many cells there are at the two new plateau phases, and when the death phase might occur.)

17.

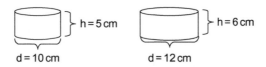

In going from stirred bioreactor 1 to stirred bioreactor 2, we could keep RPM constant and hold O2 availability steady by bubbling oxygen up through the second tank in addition to stirring. However, another parameter would change because the geometry of the flow patterns changed. Name the parameter.

18.

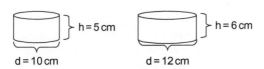

What is the scale-up factor for volume in going from the first cylinder to the second?

    a. $(6\text{-}5)^3$
    b. $(5/6)^3$
    c. $(6/5)^3$
    d. $[(6/6)\text{-}(5/5)]^3$
    e. 5^3
    f. 6^3

19.

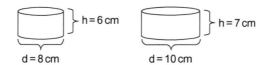

What is the scale-up factor for volume in going from the first cylinder to the second?

    a. $(5\text{-}4)^3$
    b. $(4/5)^3$
    c. $(5/4)^3$
    d. $[(7/5)/(6/4)]^3$
    e. $(5/4)^2 \cdot (7/6)$
    f. $[(7/6)/(5/4)]^3$

20. In Example 5.1, what percentage of the product comes from volume and what percentage comes from surface area in the 20,000 l bioreactor? Why is your answer not 75%/25%, the percentages of the system before scale-up?

21. Suppose that you have two reactors in the shape of cylinders that have the same relative geometry, with the height being 1.5 times the diameter. The second reactor has a volume of 100,000 l, and the volume of the first reactor is 12.5 l.

    a. What is the scaling factor, c, between the two reactors?
    b. What is the ratio of the radii of the two reactors (r_2/r_1)?
    c. If the surface area of the first reactor is 30.33885 dm^2, what is the surface area of the second reactor?
    d. What is the height of each reactor, in inches?

22. Consider a 1 l batch fermentation system whereby 90% of the (intracellular) target product is associated with attached cells and 10% is associated with cells in suspension. The total output of this reactor is 6 mg of product/liter/day.

Now consider a scale-up to a 64,000l reactor that has the same height-to-diameter ratio as the original reactor: 2-1. Assuming that both tanks are cylindrical, that cells will only bind to the walls of the tanks because of the stirring that is taking place, and that binding to other internal components of the bioreactor is negligible, what will be the yield of the 64,000l system?

RELATED READING

Alberts, B., Johnson, A., Lewis, J., Raff, M., Roberts, K., Walter, P., 2007. Molecular Biology of the Cell, fifth ed. Garland Science, New York.

Cooper, G.M., 2000. The Cell: A Molecular Approach, second ed. Sinauer Associates, Sunderland, MA.

Shuler, M.L., Fikret, Kargi F., 2001. Bioprocess Engineering: Basic Concepts. Prentice Hall, Upper Saddle River, NJ.

Zwietering, M.H., Jongenburger, I., Rombouts, F.M., van 't Riet, K., 1990. Modeling of the bacterial growth curve. Appl. Environ. Microbiol. 56, 1875–1881.

Chapter 6

Microbial Killing

6.1 THE GRAM STAIN

While the outermost portion of the mammalian cell is the plasma membrane, prokaryotic cells have 1-2 additional layers on the exterior of the lipid bilayer that surrounds the cytoplasm. A cell wall surrounds the cytoplasmic (or inner) membrane of the prokaryotic cell, and in some species of prokaryotes, an outer membrane surrounds the cell wall (Figure 6.1). When identifying bacteria, the first delineation that is often used is whether or not the cell wall is exposed or if it is hidden by the additional outer membrane. The determination can be made by the use of the Gram stain.

The Gram stain was developed by Christian Gram, a researcher in Berlin in the 1800s. He developed a procedure, using a dye, a mordant, and a counterstain to determine whether or not the bacterial cell wall was exposed. If it's exposed, the cell would be called Gram-positive, and if it is not, then the cell is termed Gram-negative. Interestingly, the Gram stain was first described by Carl Friedlander, not by Christian Gram. Friedlander worked with Gram and alluded to the staining procedure used by his colleague in an 1883 paper concerned with *pneumococci*. The famous paper by Gram was published in 1884. Although Friedlander used the technology and was the first to publish with it, history has since corrected the situation by crediting Gram with the original procedure, now known as the Gram stain.

The Gram stain is simple and only takes about 10 min to perform. The first steps are to place a bacterial sample onto a slide and fix the cells in place with heat. The sample might require dilution to help ensure a monolayer of bacteria, which is desired to allow for complete penetration of the dyes that will be used. The extracellular fluid (medium and diluent) must be removed without removing the cells, and the cells must be attached to the slide to prevent them from being washed off during the stain and rinse steps of the procedure. Heat is generally used for such fixation. The slide is gently waved over a Bunsen burner to dry the slide, and the heat will cause evaporation of the fluid as well as *denaturation* (unfolding) of proteins. The unwound proteins will bind to the surface of the slide, resulting in the adherence of the cells to the slide. An alternative method of fixation is via alcohol, which acts to denature proteins, again leading to cellular attachment.

An Introduction to Biotechnology. http://dx.doi.org/10.1016/B978-1-907568-28-2.00006-X

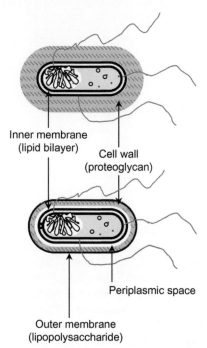

Inner membrane
(lipid bilayer)

Cell wall
(proteoglycan)

Periplasmic space

Outer membrane
(lipopolysaccharide)

FIGURE 6.1   Proteotypical bacteria. (Top) Gram-positive and (bottom) Gram-negative. Both cells have cytoplasms surrounded by an inner membrane made from a lipid bilayer, and both inner membranes are surrounded by a cell wall made of proteoglycan. Gram-negative cells also have an outer membrane made of lipopolysaccharide. The cell wall in Gram-negative bacteria is generally thinner than that of Gram-positive bacteria, and there is a periplasmic space between the cell wall and the inner membrane.

Currently, as opposed to Gram's time, the dye most commonly used to stain the cells is crystal violet. Crystal violet has the structure of that shown in Figure 6.2. Note that the dye is a carbocation. The positive charge will readily interact with anions, such as I^-. Gram's iodine, a solution made up of iodine and potassium iodide, will supply the I^-. Once iodide ions have been attached to crystal violet molecules, a precipitate will form. Because it is responsible for forming a precipitate, Gram's iodine is known as a *mordant* (Figure 6.3).

Because the cells have been fixed, crystal violet will be able to penetrate the cells. The incubation time for staining the slide with crystal violet is only about one minute, after which the slide is rinsed with tap water. Next, a solution of Gram's iodine is placed upon the slide for one minute, followed by a rinse with *decolorizer* (methanol, ethanol, or acetone). Grams iodine will cause the dye to form a bulky precipitate, and the decolorizer will act as a solvent for the dye. However, certain cells will retain the purple dye because they have a thicker cell wall. The proteoglycan network of these cell walls will serve to trap the dye molecules within the cells. However, in other cells, there are an outer membrane and a thinner cell wall. The decolorizer will serve to severely damage the outer

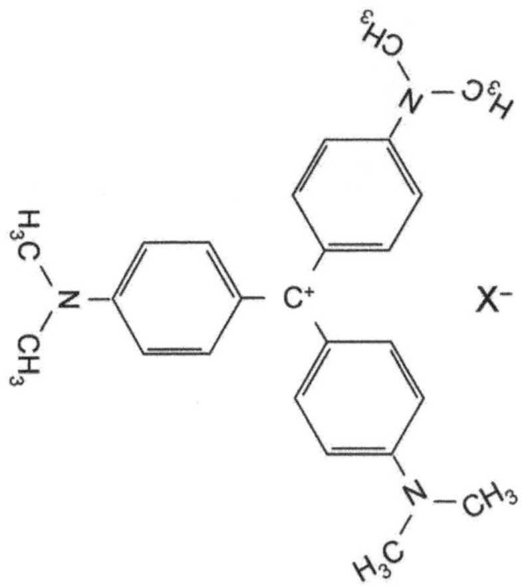

FIGURE 6.2   The structure of crystal violet.

FIGURE 6.3   Crystal violet without (left) and with (right) the mordant Gram's iodine. Notice that the mordant causes the dye molecules to precipitate.

membrane and the thinner cell wall will be unable to effectively trap the precipitated dye. The presence or absence of dye molecules after this step is the basis for labeling cells as Gram-positive or Gram-negative.

The next step is to *counterstain*. The counterstain is performed because, at this point, cells will be either purple or unlabeled, and it is desirable to have every cell labeled. The counterstain is a dye such as safranin or basic fuchsin, each of which carries a positive charge. The counterstain is able to get through the cell membrane(s) because the cells have been fixed and are especially permeable after exposure to the decolorizer. The dye will enter the cytoplasm and adhere to negatively charged components, thus providing a pink color to the inside of all of the cells. An example of staining results is given in Figure 6.4.

At this point, one might ask, "Why do we care about the Gram stain? What good is it, and can you use it on all bacteria?" The answer will be revealed on the following pages. We will discuss phospholipids, proteins and glycoproteins, and the cell wall. The reason we might want to perform a Gram stain is because we want to determine whether an infection in a patient or a cell culture is Gram-positive or Gram-negative, because that's going to help us determine

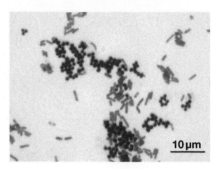

FIGURE 6.4   Gram's staining results of a mixture of *E. coli* (pink; Gram-negative) and *S. aureus* (purple; Gram-positive).

which antibiotic to use. But before we can talk about antibiotics, we need some more background on microbial cell killing.

6.2   MICROBIAL RESISTANCE TO KILLING

If we want to kill microbes, there are various levels of cleanliness. You've probably heard the term "germ" before. The word "germ" is like the word "stuff." It's a catchall term. There are different "germs" (microbes) to consider, and they have different resistances to different cleaning methods (Table 6.1).

TABLE 6.1 General Resistance of Contaminants to Physical and Chemical Methods of Control

Highest resistance

 Bacterial endospores

Moderate resistance

 Protozoan cysts

 Some fungal sexual spores

 Naked (non-enveloped) viruses

 Some vegetative bacteria

Least resistance

 Most bacterial vegetative cells

 Fungi (spores and hyphae)

 Enveloped viruses

 Yeast

The hardest things to destroy are bacterial endospores. You're not going to be able to destroy bacterial endospores by washing with soap and water, scalding water, or 5.25% bleach (the concentration of regular Clorox®). You need to go through some high-end processes to get rid of bacterial endospores. Luckily, bacterial endospores are not huge plagues in our lives. When you work in a science laboratory, especially one that performs cell culture, you need to worry about spores, but for your everyday life, they are not so much of a concern.

Microbes with moderate resistance to killing methods include protozoan cysts, some fungal sexual spores, naked (non-enveloped) viruses, and some vegetative bacteria. Have you ever had a staph infection? Such a condition involves contamination with *Staphylococcus aureus* ("staph") in the vegetative state. Staph is on your pencil. It's on your desk. It's on your skin and in your respiratory tract. Staph is virtually everywhere, and if you get cut, staph will commonly be present in the cut itself. (Cleaning the wound with soap and water, with the help of a functioning immune system, will typically be enough to protect the body from further invasion by these bacteria.) Other microbes that you may have heard of in the "moderate resistance" classification include *Mycobacterium tuberculosis* and *Pseudomonas*. *Pseudomonas* is the one that can live in soap dishes. (Yes, one should periodically clean the soap dish or the inside of the liquid soap dispenser!)

Microbes with the lowest resistance to destruction include most bacterial vegetative cells, fungi (spores and hyphae), enveloped viruses, and yeast. Vegetative bacteria are not in the protected state and are relatively defenseless. This partially explains why, if you get a cut and it gets infected, the infection is commonly a staph infection. If you wash your hands after you get a cut, it is easier to remove or destroy the types of microbes listed as having the lowest resistance. Of those that are left, *Staphylococcus aureus* is one of the most common. The staph are a little more difficult to get rid of, especially in comparison to fungi and yeasts.

Aside: Vegetative Versus Spore Forms

When we refer to "vegetative bacteria," we are pointing out a bacterial state—in this case, the growing and reproducing form. Many bacteria can also enter into a spore form, where they remain dormant until conditions improve. The spore form is a defensive state taken in response to adverse temperatures, pressures, or nutrient availabilities. Thick cell walls and dormant metabolisms make spores resistant to hostile environments—sometimes for millions of years. Spores have been isolated and cultured from the stomachs of extinct bees embedded in amber for 25-40 million years. Spores have also been ascribed some of the mysterious sicknesses and deaths of those who first entered King Tut's tomb, rumored by many to have succumbed to the "The Curse of the Pharaohs" (Figure 6.5).

FIGURE 6.5   (Top) Fly embedded in amber (from http://www.icr.org/article/a-45-million-year-old-brewers-yeast-still/). (Bottom) Tomb paintings from the Valley of the Kings (from http://www.smithsonianjourneys.org/blog/2010/08/17/the-curse-of-king-tuts-tomb/).

6.3   STERILIZATION, DISINFECTION, AND SANITIZATION

Now that we know what we're up against, let us now turn to the different levels of "clean" (Figure 6.6).

6.3.1   Sterilization

There is a difference between sterile, disinfected, and sanitized. *Sterile* means all the viable microorganisms and viruses have been destroyed or removed. There are different methods that can be used to achieve this level of cleanliness, such as the autoclave or ethylene oxide gas. An *autoclave* works like a pressure cooker. For sterilization, it's not enough just to boil things. While boiling water will kill most bacteria, it will not kill bacterial endospores. As a liquid, water cannot get any hotter than boiling, and the boiling point at standard pressure (1 atm.) is 100 °C. We could use steam to sterilize instruments or surfaces, but in many situations, that is both inconvenient and dangerous. If we are to achieve sterilization while keeping water in the liquid phase, we will have to incorporate increased pressure, which will result in higher boiling temperatures. This can be achieved with both pressure cookers and autoclaves. Using Henry's law,

$$PV = nRT \Rightarrow P = T\left(nR / V\right),$$

where P is the pressure, V is the volume, T is the temperature, n is the number of moles of a substance, and R is the gas constant.

We see that for a given amount of H_2O held in a fixed volume, raising the temperature will increase the pressure, and vice versa. At a pressure of 2 atm. (1 atm above standard pressure), water will boil at ~121 °C. These conditions

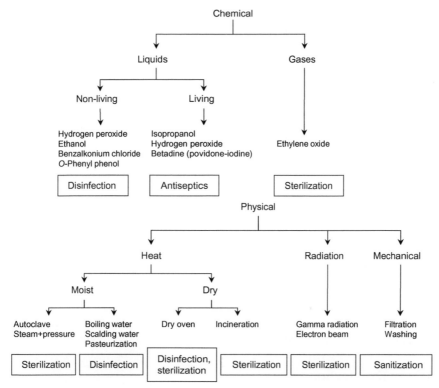

FIGURE 6.6   There are many different means available to remove or destroy microbes. Methods have been grouped into chemical (top) and physical (bottom), with the resulting levels of cleanliness shown in boxes. Refer to the text for discussions on what each level of cleanliness means.

are enough to kill all life forms within 15 min and are the conditions used in most autoclaves.

Ethylene oxide (EtO) is a flammable, colorless gas at temperatures above 51.3 °F (10.7 °C) that smells like ether at toxic levels. Its structure is shown in Figure 6.7. EtO is used in the production of solvents, antifreeze, textiles, detergents, adhesives, polyurethane foam, and pharmaceuticals. Smaller amounts are used in fumigants, sterilants for spices and cosmetics, and during hospital sterilization of surgical equipment. The term *cold gas sterilization* refers to sterilization via ethylene oxide. There are specific reasons why one might want to use ethylene oxide instead of an autoclave. Suppose we have some tissue culture dishes that we wish to sterilize. We cannot put them in the autoclave because they're made of plastic that will melt. Keep in mind, however, that ethylene oxide gas is expensive, can be dangerous, and has many side reactions. For example, ethylene oxide plus water yields antifreeze (ethylene glycol).

Irradiation is an alternative to cold gas sterilization. It is the method of choice for companies producing large scale like plastic laboratory tubes and

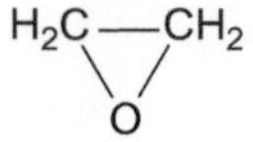

$$H_2C \overset{\displaystyle}{\underset{\displaystyle O}{\diagdown}} CH_2$$

FIGURE 6.7 The structure of ethylene oxide.

culture ware. There are two techniques that can be used: gamma irradiation and electron beam (E-beam) radiation. Gamma irradiation uses ^{60}Co to produce gamma rays that will disrupt or destroy microorganisms. Such radiation can easily penetrate packaging and an enclosed cluster of plastic items several layers thick. Required doses for irradiation are rather high, with a minimum of 2.5 megarad needed for items packaged in air. E-beam radiation can deliver high doses of radiation much more quickly than gamma irradiation, but the treatment volume will be much smaller and the penetrating power is much lower. For comparison, gamma irradiation will require minutes to hours to deliver the dose needed for sterilization but can penetrate ~50 cm into a sample. E-beam radiation can deliver the same dose in seconds but can only penetrate ~5 cm.

"Sterile" does not mean "safe to consume." One could run a cup of bleach through the autoclave, but it will still be a cup of bleach and quite toxic to drink. It is a common misconception that "sterile" means "safe." So, just because we sterilize something does not mean that it's safe to consume. It just means that we've removed all the viable microorganisms or viruses (see Figure 6.8).

6.3.2 Disinfection

Disinfected means all of the vegetative pathogens have been destroyed. This term is typically used to describe surfaces. Lysol®, 5% bleach, alcohol, and boiling can all be used to disinfect items or surfaces, but bacterial endospores will not necessarily be killed by the disinfection process.

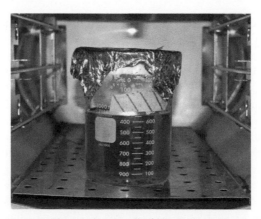

FIGURE 6.8 Mmmmmmm, this attractive green liquid has been autoclaved, so there are no viable microorganisms in it. Is it therefore safe to drink?

6.3.3   Sanitization

One level lower on the cleanliness scale is *sanitization*. Sanitization refers to any process applied to inanimate objects including air, liquids, and surfaces such that microorganisms are removed mechanically. Examples of this include certain air purification systems, water filtration, dishwashing, and washing your clothes. The process is not the same as disinfection, which utilizes a chemical agent, and it is certainly not the same as sterilization in that viable microorganisms are left behind after the process. When you wash your dishes at home, there will still be bacteria on them after you finish. This may seem disgusting, but generally, the residual microbes should not present a problem if you have a functioning immune system. In many restaurants, the cleaning of dishes is taken a step further, with the incorporation of high heat and steam to disinfect dishware, which is analogous to boiling. Keep in mind that sometimes in a restaurant, you may get a fork or a knife that has still got some food on it. Although you can see the debris, it has still been disinfected if it was exposed to the steam. These terms—sanitized, disinfected, and sterilized—refer to the number of viable microorganisms on a surface, but the terms do not necessarily infer anything about other molecules on the surface, such as toxins. So, you can have a sterile cup of bleach and not be able to drink it. You can have a disinfected fork with a piece of dried shrimp on it that is safe to eat (icky, but safe to eat, although it is probably safer just to send the fork back). You can also have a sanitized plate that looks clean but has who-knows-what growing on it, left behind by the previous user. In this latter case, we have to trust in our immune systems and that the restaurant used sufficient cleaning methods to bring the number of microorganisms down to a safe level.

6.3.4   Antiseptics

To understand antiseptics, one should first understand the root of the word. Consider the word "*sepsis*," which is a medical term meaning there are microorganisms alive in the blood and/or the tissues of the body. If a person is septic, then he is in big trouble because the infection is often global, meaning it is throughout the body, having been carried by the circulatory system. Sepsis must be treated very quickly and aggressively. "*Aseptic*" simply means "not septic." It is sometimes used as an adjective to imply the prevention of sepsis. Consider the physician who is about to perform a procedure. The doctor will wash his hands very well. He may place a drape over parts of the patient that are near the area to be worked on. If the procedure is a surgery, the doctor will cover his hair and mouth and wear a gown that has been sterilized. These measures are known as aseptic technique. Similarly, when biotechnologists work with cells, we do so in a biological safety cabinet because it prevents room air from delivering dust, lint, spores, or what-have-you into the cell medium with which we are working. In addition, a person performing cell culture in a biosafety cabinet will not move his hands or arms over what is being worked upon, including any open containers, because particles can fall off of the lab coat, the skin, or possibly the gloves. The prevention of sepsis—the prevention of bacteria growing in our cell cultures, in our mice, or in our patients—is known as aseptic technique. *Antiseptics* are chemicals applied to body surfaces to destroy or inhibit the growth of vegetative pathogens. Their use is *similar* to disinfecting the skin. An example of sanitizing your skin would be washing your hands under a faucet using regular hand soap. You would be removing a great deal of bacteria. The use of an antiseptic would be different—perhaps using a hand soap with triclosan or swabbing the skin with an alcohol. Disinfectants and antiseptics are different. Disinfectants are used on inanimate surfaces, and antiseptics are for body surfaces like your skin. Disinfectants can potentially be harsher than antiseptics because one does not have to worry about the preservation of living tissue.

To extend our discussion, what would be the result of sterilizing your finger? Killing every cell in your finger! Your cells are microorganisms too, so to sterilize any part of your body would essentially mean to kill it.

Hydrogen peroxide is a very effective antimicrobial. In fact, what you buy from the store—3%—is very effective, killing a broad spectrum of microbes within 10-15 s. It is used as both a disinfectant and an antiseptic. However, it is not the best agent to put onto a healing wound because it can damage your own cells. Hydrogen peroxide works by making hydroxyl oxygen radicals, which can oxidize DNA, RNA, proteins, and membrane lipids. H_2O_2 will serve to help clean a fresh wound by killing microbes. When you first get an open wound, you should clean out any debris, which includes dead cells, tissue, dirt, and what-have-you. One could wash the wound with hydrogen peroxide. However, after the healing process begins, hydrogen peroxide will take away the body's work in wound healing. At the end of the day, you may have grown fresh granulation tissue to cover the wound. You wouldn't want to strip that away by killing those cells. The healing process is complex, and what might work well on day 0 might not be the best agent on day 2.

6.4 MICROBIAL CELL DEATH

We have been discussing how to kill microbes, but just what is microbial cell death? Since we as humans haven't adequately defined life, how do we define death? One popular definition is, "the permanent termination of an organism's vital processes." "Vital" means life. So why not just say, "the cessation of life?" Death is therefore "not life." It's pretty easy to tell when a person is dead—breathing stops, the body is not moving around, the heart is not beating, and there are no electrical impulses in the brain—but it can be difficult to tell when a microbe dies. When a microbe dies, it just sits there, but that's quite often what it did when it was alive, so in this case, it's not doing anything different as far as motion is concerned. The definition we're going to use for microbial cell death is *the permanent loss of reproductive capability, even under optimal growth conditions*. Such a definition is needed for miocrobes because of their different states. Consider spores. Some microorganisms, when they are under assault, armor up in a ball. Spores have a thick, protective proteinaceous coat that is hard to penetrate, even with most chemicals. There is no locomotion, and the spores do not respire. Many definitions of "life" state that metabolic function (respiration) is required. If we were to use the definition that death equals "not life," then "death" would mean the cessation of metabolic function. Spores are not performing metabolism, so does that mean that spores are dead? That's where the second half of the above italicized definition comes in. If you place a bacterial endospore into optimal growth conditions (perhaps a warm, moist environment with ample nutrients), then it would desporulate into the vegetative form of that bacterium, which is then actively respiring and replicating. So, a spore cannot reproduce; but when you put it under its optimal conditions, it will become something other than a spore and demonstrate features that we typically equate with life.

6.4.1   Death by Alcohol

Alcohols are germicidal because of their abilities to interact with membrane proteins and to disrupt lipid bilayers. The number of carbons on the alcohol will help determine how effective it is as an antimicrobial (Figure 6.9). For example, the one-carbon alcohol is methanol. Methanol is not considered to be very microbicidal. While it can kill some microbes in a population, it is not an effective microbicide because, with one carbon, it is difficult for the methanol to disrupt lipid bilayers such as those found in the plasma membrane.

Ethanol, the two-carbon alcohol, is far more effective as an antimicrobial agent. Ethanol is used in the laboratory as a disinfectant. For example, it will be sprayed in biological safety cabinets at a concentration between 50% and 90% (typically 70%). Anything <50% may be too dilute to kill certain microbes. Pure ethanol isn't desirable, either, because certain microbes (such as the Gram-positive strains *Staphylococcus aureus* and *Staphylococcus pyogenes*) are more resistant to 100% ethanol than to lower concentrations. Also keep in mind that when you spray down a surface, such as your desk, it's not that the surface is instantly "clean." You are compromising the integrity of the cell membrane so that when you apply some shear, you're more likely to lyse cells that have had their membranes destabilized by the alcohol. Simply spraying the surface with ethanol will not yield the same level of disinfection.

Using ethanol as a disinfectant is good in the laboratory. However, when you go into the doctor's office, they don't rub your skin before an injection with ethanol, they rub it with rubbing alcohol (isopropanol). The reason is ethanol evaporates too quickly. With two carbons, ethanol is more volatile than alcohols with three carbons or more. For an injection, the needle can carry microbes from the surface of your skin into your body. The skin is rubbed with the less volatile isopropanol because it will have a greater chance to interact with microbes that are hiding within the stratum corneum (the upper layers of the skin). In addition,

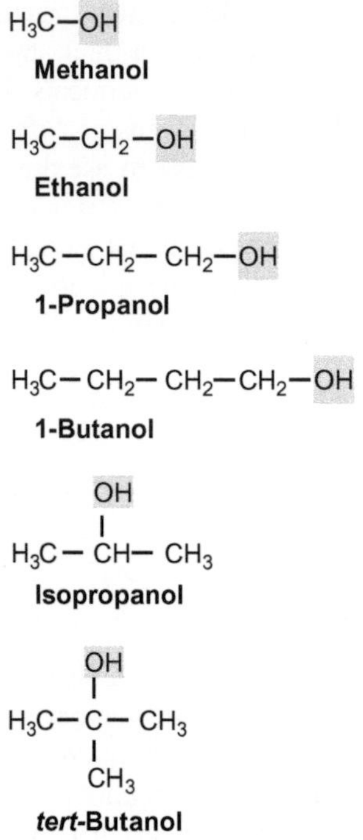

H₃C—OH

Methanol

H₃C—CH₂—OH

Ethanol

H₃C—CH₂—CH₂—OH

1-Propanol

H₃C—CH₂—CH₂—CH₂—OH

1-Butanol

$$OH$$
$$|$$
H₃C—CH—CH₃

Isopropanol

$$OH$$
$$|$$
H₃C—C—CH₃
$$|$$
$$CH_3$$

***tert*-Butanol**

FIGURE 6.9   The structures of several common alcohols. Considerations including hydrophobicity, volatility, and cost should be taken into account when deciding on a suitable alcohol disinfectant or antiseptic.

having three carbons renders the molecule somewhat more hydrophobic and allows it to interact with the hydrophobic portions of the plasma membrane (lipid tails, hydrophobic cores of folded proteins) more easily than an alcohol with only two carbons; therefore, the microbicidal activity of isopropanol is greater than that of ethanol.

Isopropanol also tends to be less expensive than ethanol. One of the reasons is that you can't drink isopropanol. "Pure" ethanol, because it can be consumed, can be subjected to taxation. The way that laboratories get around having to pay for the expensive ethanol tax is by purchasing ethanol that has impurities such as methanol in it. Drinking the typical laboratory ethanol will make one violently ill, but the good news is that it is fairly inexpensive.

One might think at this point that using alcohols with even greater numbers of carbons would serve as even better choices as microbicidal agents. However, such alcohols will have greater hydrophobicity, serving to defat the skin more easily leading to hardening and cracking of the skin. These alcohols also tend to be more expensive, and as the number of carbons is increased, the solubility of the molecules in water goes down. If you are working in the laboratory and need to spray down your biosafety cabinet but have run out of ethanol and isopropyl alcohol, you *could* use butanol, but it is not recommended to swab a patient's skin with butanol before giving an injection.

6.4.2   Antimicrobial Drugs

Let us now turn to antimicrobial drugs. This discussion will explain why we bothered to learn the Gram stain. When we take antimicrobial drugs ("antibiotics"), what we want to do is to capitalize on differences between the targeted microorganisms and humans. This presents a variety of potential targets, as outlined in Figure 6.10. If, for instance, we had a drug that completely destroyed all cell walls, then we could annihilate the pathogenic microbes that are inside us without harming ourselves because our cells do not have cell walls. In reality, sometimes, it's relatively easy to get to the cell wall and sometimes it is difficult. Recall that in Gram-positive bacteria, there is no outer membrane, so the cell wall is exposed. Such bacteria are more susceptible to drugs that target the cell wall. Penicillin is one example of such a drug.

6.4.2.1   Targeting the Cell Wall

Consider the prokaryotic cell wall. The cell wall is primarily made of the peptidoglycans N-acetyl muramic acid (NAM) and N-acetyl glucosamine (NAG) (Figure 6.11). These molecules are arranged in a very ordered fashion, with alternating layers of NAMs and NAGs connected as shown in Figure 6.12. They are also cross-linked between NAMs via a relatively short protein sequence. This cross-linking is structurally important. Think of a deck of cards. If you were to pick up a deck of cards and try to rip it in half, you might not be able to, but if you slide the cards off of the deck one by one, it will not take much force to displace the entire deck. However, if there were glue between each of the cards, you wouldn't be able to slide the cards off because instead of 52 individual sheets, you would have a single

> **Cell wall**
> > We don't have one
>
> **Translation**
> > Prokaryotes use different
> > ribosomes
>
> **Nucleic acid synthesis**
> > We don't manufacture all of
> > our own co-factors
>
> **Cell membrane**
> > Prokaryotes use different
> > components (*e.g.*, sterols)

FIGURE 6.10   Potential targets for antimicrobial drugs and a reason why they can be targeted.

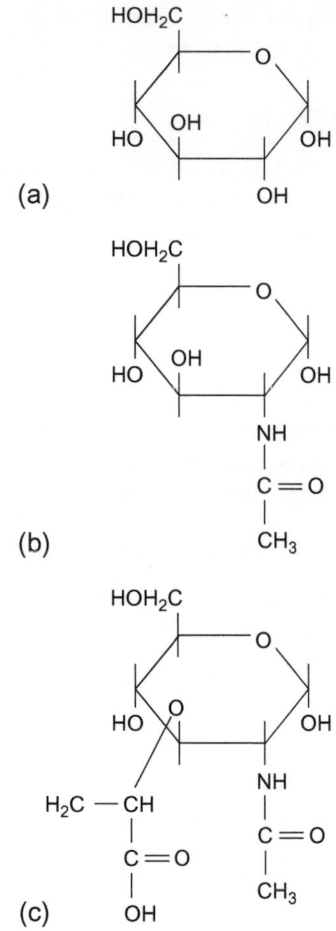

FIGURE 6.11   The structures of (a) glucose (included only as a reference for (b), (b) N-acetyl glucosamine, and (c) N-acetyl muramic acid.

block. That's similar to what the cross-linking between two NAMs is for. The building of the cross-links between NAMs involves pentapeptides. The sequence of the five constituents depends on the species of prokaryote, but there are some commonalities. The third member typically has two amino groups. The amino acid lysine is often used in the third position, although diaminobutyric acid and diaminopimelic acid are also common. The fourth member is often an alanine of some sort (either the D- or L-enantiomer). The fifth member is usually a D-alanine. During the cross-linking of two such pentapeptides, at least one of the fifth position, D-alanines will be removed as the fourth member is linked to the diamino member of the other pentapeptide (Figure 6.12b). Some microorganisms use a spacer (such as an

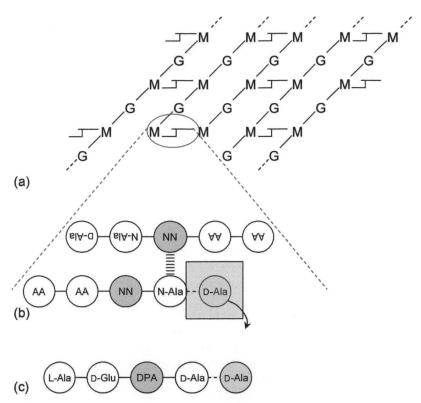

FIGURE 6.12   (a) The bacterial cell wall is composed of alternating layers of NAMs and NAGs. Interlayer cross-links are also formed between NAMs. (b) NAMs will contain a species-specific side chain, generally 4-5 amino acids in length, that will participate in cross-linking with a successive NAM. The third member of the NAM side chain typically has two amino groups (NN), such as the amino acid lysine. The fifth member is often a D-alanine, which will be lost during the cross-linking of the fourth member to the diamino-containing member of an adjacent NAM side chain. c) A common pentapeptide used in cross-linking NAMS in the cell walls of *E. coli* is [L-alanine – D-glutamic acid – diaminopimelic acid (DPA) – D-alanine – D-alanine (which may be lost during the cross-linking process)].

oligoglycine) to bridge the gap between the two pentapeptides, which will affect cell wall porosity and density. Note the use of D-amino acids. It is thought that the bacteria use D-amino acids because they help them to avoid enzymatic degradation by proteinases that only recognize the L-enantiomers.

In order for bacteria to grow in size, the cell wall must be broken (by autolysins), made larger by the insertion of new NAMs and NAGs (via transglycosylases), and then resealed via the formation of new cross-links (using transpeptidases). Penicillins and cephalosporins bind to transpeptidases to prevent the resealing of the cell wall. With the drug bound, the transpeptidases are no longer active so there will be no further cross-linking. Without the cross-linking, the cell wall will be weak, and a weak cell wall will make the cell prone to lysis (rupture), especially because of osmotic forces. The drugs penicillin G, ampicillin, amoxicillin, methicillin, and oxacillin all work in this fashion.

6.4.2.2   Targeting Translation

We, as eukaryotes, also differ from prokaryotes in how we perform translation. Both prokaryotic and eukaryotic cells use ribosomes for translation, but our ribosomes are fundamentally different from bacterial ribosomes. Bacterial ribosomes are made of one 50S and one 30S subunit. The 50S and 30S subunits come together to form 70S ribosomes. Eukaryotic ribosomes have a 60S and a 40S subunit, which come together to form an 80S ribosome. If we target 70S ribosomes with an antibiotic, we can knock out translation in bacteria, which means they can no longer make proteins. Without proteins, the bacteria will die because they will be unable to perform metabolic functions. (Metabolic enzymes are made from proteins.) Since we do not have 70S ribosomes, our cells would theoretically be unaffected by the drug. We can therefore (in theory) target bacterial ribosomes and destroy them without hurting any of the functions of our own cells. Things just aren't that simple, though. Mitochondria are organelles within eukaryotic cells that behave like symbiotic bacteria. They have their own genomes and carry out translation for some mitochondria-specific proteins. Drugs that target prokaryotic translation can also affect the mitochondria, which can be significantly deleterious since the mitochondria are responsible for most of the energy production in eukaryotic cells.

Examples of drugs that target translation include streptomycin and gentamicin, which bind to the 30S ribosomal subunit of prokaryotic ribosomes to cause misreading during the translation of mRNA. Erythromycin inhibits translocation of the 50S subunit during translation. Tetracycline blocks proper tRNA attachment to the ribosome.

Aside: When Does $60 + 40 = 80$?

S stands for the svedberg unit, which is related to the sedimentation coefficient. The large ribosomal subunit in eukaryotes has a sedimentation coefficient equal to 60×10^{-13}, or 60 svedbergs, which is where the name "60S subunit" comes from. Keep in mind that sedimentation in this case refers to how far something migrates through a fluid under centrifugal force. It is a function of relative density of the particle versus the fluid and frictional drag, which depends on the geometry of the particle. It should be intuitive that if we drop two rocks having the same shape and size but different densities into a pond, the one with the greater density will settle faster. Also consider, however, two pieces of sandstone having a mass of 500 g each, but one is roughly spherical and the other is rather flat. If you drop them into a pond at the same time, the round one will sink to the bottom first because it has less frictional drag. Svedberg units are a way to describe migration through a fluid, taking density and geometry into account.

When two subunits come together to make a protein, the final protein may have a geometry that affords it less frictional drag than either of the individual subunits. Consider the yin and the yang (in three dimensions). Individually, each would migrate through a liquid rather inefficiently, but together, they would make a sphere, which has a far lower coefficient of drag. This is similar to the case for ribosomes. In eukaryotes, although the two ribosomal subunits are of 60 and 40 svedbergs, bringing them together yields a ribosome of only 80 svedbergs. 60S + 40S = 80S. Aha—another way to win a bet! Knowing biotechnology pays.

6.4.2.3   Targeting Nucleic Acid Synthesis

As we have already discussed, DNA is made up of four bases, abbreviated A, G, C, and T. (Refer back to Chapter 4 if this does not make sense.) Half of the bases (A and G) are purines and half of the bases (C and T) are pyrimidines. The synthesis of purines requires N^{10}-formyl tetrahydrofolate, which is generated from tetrahydrofolate. Tetrahydrofolate synthesis, in turn, requires dihydrofolate, which requires *para*-amino benzoic acid (PABA) for synthesis in microbes (Figure 6.13). In humans, on the other hand, dihydrofolate is derived from folate, which is ingested as a vitamin (vitamin B9). *Vitamins* are molecules that are required for good health but must be consumed because the organism cannot

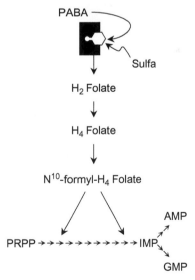

FIGURE 6.13   In microbes, *para*-amino benzoic acid (PABA) is required for the synthesis of purines (AMP and GMP in this figure). Sulfa drugs compete for the active site in the enzyme that uses PABA. When the synthesizing enzyme is occupied with a sulfa drug, it is unavailable to incorporate PABA into folate. (Two N^{10}-formyl-H4 folate molecules are involved in the *de novo* building of a purine.)

make them. If we were to take a drug that impairs the enzyme responsible for making dihydrofolate, we are not going to hurt ourselves (in theory) because we are not making our own dihydrofolate in the first place.

In microbes, folate synthesis requires PABA. The sulfonamides (sulfa drugs) are a class of drugs that have structures similar to PABA (Figure 6.14). The enzyme that converts PABA to folate will be led down a blind path as it binds to the sulfonamide. It is as if we throw in something else for the enzyme to do, taking some of the enzyme's action away from converting PABA to folate as it binds the sulfa drug. The enzyme therefore doesn't produce as much folate, so purine synthesis is ultimately impeded. This is an example of *competitive inhibition*. The enzyme's attention has been taken away from PABA, so less PABA will be converted to dihydrofolate, which, in turn, means the bacterial cells won't have as much tetrahydrofolate, which means they will be deficient in purine synthesis. The result will be that the bacterium cannot create a copy of its own genome, so cell division will be impossible because of a lack of necessary starting materials.

6.4.2.4   Targeting Cell Membranes

Drugs that target the cell wall are effective on most Gram-positive bacteria. However, what can be done to combat an infection of Gram-negatives? Apart

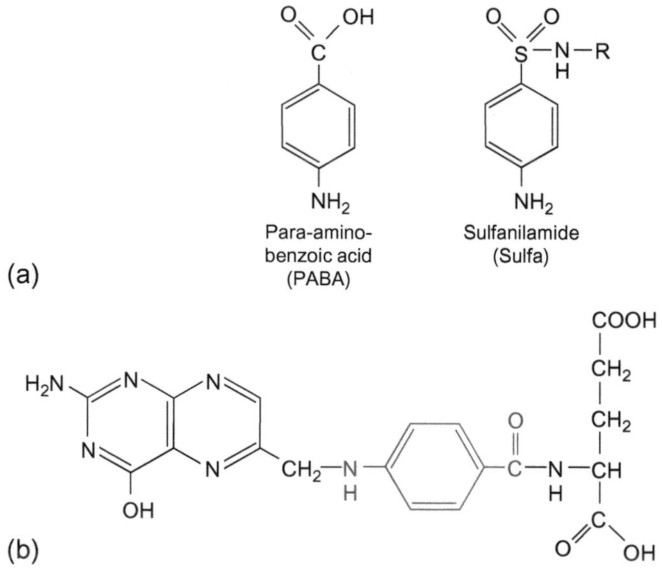

FIGURE 6.14   (a) The structures of PABA and sulfa drugs. Note the similarity between the structures. (b) The structure of folic acid. The portion derived from PABA is shown in red.

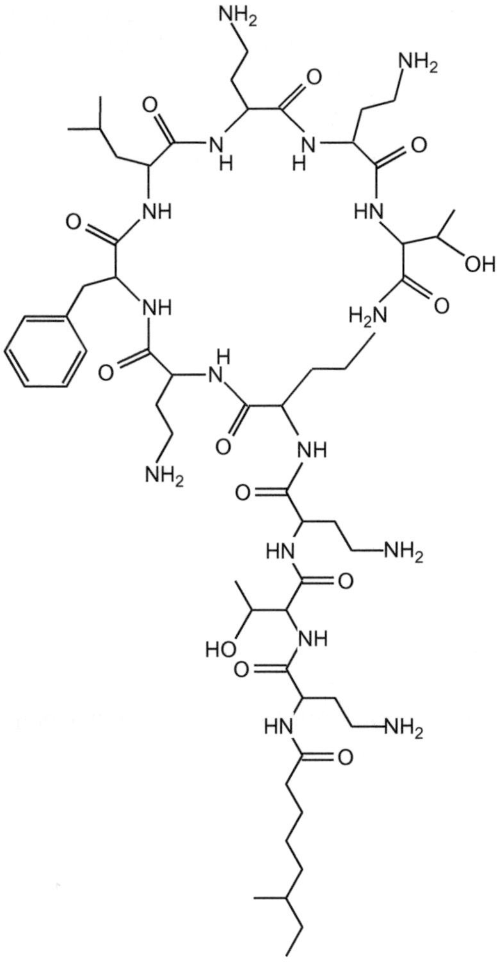

FIGURE 6.15 The structure of polymyxin B. Note the similarity in structure to membrane phospholipids, except that the head group of polymyxin B is much larger.

from broad-spectrum drugs (which will not be covered in this text), one could consider polymyxins. These drugs have a structure vaguely similar to phospholipids (Figure 6.15), but the polar head groups are so large that they disrupt membrane integrity upon insertion. They bind to the lipopolysaccharides (outer membranes) of the Gram-negative bacterial cell outer membrane to cause structural instability. Polymyxins are generally used as topical treatments.

Polyenes such as amphotericin B and nystatin are used as antifungal agents. Fungi have different sterols in their cell membranes than we do. These drugs bind to ergosterol in the fungal membrane, causing leakage of small ions.

Unfortunately, the drugs are not well targeted and can interact with cholesterol molecules (which are also sterols) in our own cell membranes, destabilizing our cells and causing significant side effects.

Aside: Penicillin Allergy

In class after class, students have asked for a reason to explain why some people are allergic to penicillin. Since there is space left on this page, the question will be addressed here. (Woohoo!)

Penicillin is a drug that, by itself, is too small to spark an immune response. However, the carbonyl carbon of the β lactam ring can covalently bond with an amino group in a protein to produce a penicilloyl group (breaking the β lactam ring). Not only will the penicillin lose its antimicrobial activity, but also the penicilloyl group on the larger molecule will have antigenic activity in some people, meaning it will be recognized by the immune system.

QUESTIONS

1. Why would one purposely place a vial of bacterial endospores in an ethylene oxide gas sterilizer along with instruments that were going to be used for human surgery?
2. Why would monochloro-*o*-phenylphenol or benzalkonium chloride be good disinfectants but poor antiseptics?
3. Because of fears of bacteria in the drinking water, a certain town has undergone a "boil water" order. However, boiling will not kill all bacteria.
   a. Why was the town told to boil its water, then? (Give an example of your answer.)
   b. What would survive the boiling?
   c. Why don't most people have to worry that the water they drink is not sterile?
   d. How could one go about sterilizing a plastic water bottle?
4. Which would be a better antiseptic for preparing a patient's arm for an injection, ethanol or isopropanol?
5. Jim is in the waiting room at the Muddy Hole Hospital, awaiting surgery. The doctor steps out and says that the operating room has just been sanitized. Jim is disgusted by the news and insists on having his operation performed at the hospital in the neighboring city. What could have possibly upset Jim?
6. If a bacterium were to stay in the spore state forever, why would it not be considered dead?
7. Looking at the following sets of three, indicate whether the members of the following sets are good antiseptics, disinfectants, or neither:
   - Bleach, quats, and boiling
   - H_2O_2, soaps, and isopropanol

 ‑ Methanol, Triton X-100, and DMSO
 ‑ Dry heat (500°) for 30 min, incineration, and application of electron beam

8. Consider the sulfa antibiotic class. What is the term used to describe when the enzyme's "attention" is drawn away from PABA and towards a sulfa drug?

9. Name two problems with using methanol as an antiseptic.

10. Consider a soap or detergent: Is a longer or shorter hydrocarbon chain the better solvent for the lipid tails found in membrane lipids? Is there a limit to how long or short the chain can be in the soap/detergent? Explain your answer.

11. Nystatin targets ergosterol in the fungal cell. We discussed side effects in humans due to nystatin also binding to cholesterol. Could cholesterol, therefore, be considered a competitive inhibitor of nystatin?

12. If I had a drug that would double the number of cross-links in the Gram-positive cell wall and make them unbreakable, would it serve as a good or bad choice of antibiotic to treat Gram-positive infections?

13. Name four targets of antibiotics, and give one specific drug for each target.

14. Why is it not advisable to use soap in your dishwasher?

15. In Chapter 2, it was stated that L-amino acids are used in our bodies. However, there are purposes in nature for which D-amino acids are used. Identify a very important one.

16. Explain why the polyene drug amphotericin B has harsh side effects when administered to humans.

RELATED READING

Beveridge, T.J., Davies, J.A., 1983. Cellular responses of *Bacillus subtilis* and *Escherichia coli* to the Gram stain. J. Bacteriol. 156, 846–858.

Cano, R.J., Borucki, M.K., 1995. Revival and identification of bacterial spores in 25- to 40-million-year-old Dominican amber. Science 268, 1060–1064.

Davies, J.A., Anderson, G.K., Beveridge, T.J., Clark, H.C., 1983. Chemical mechanism of the Gram stain and synthesis of a new electron-opaque marker for electron microscopy which replaces the iodine mordant of the stain. J Bacteriol. 156, 837–845.

Edwards, R.G., Youlten, L.J., Dewdney, J.M., 1986. Penicillin hypersensitivity: mechanism, diagnosis and management. Indian J Pediatr. 53, 37–44.

Friedlander, C., 1883. Die Mikrokokken der Pneumonie. Fortschr. Med. 1, 715–733.

Gram, C., 1884. Uber die isolirte Farbung der Schizomyceten in Schnitt-und Trockenpraparaten. Fortschr. Med. 2, 185–189.

Ingólfsson, H.I., Andersen, O.S., 2011. Alcohol's effects on lipid bilayer properties. Biophys J. 101, 847–855.

Koc, E.C., Koc, H., 2012. Regulation of mammalian mitochondrial translation by post-translational modifications. Biochim Biophys Acta. 1819, 1055–1066.

Linley, E., Denyer, S.P., McDonnell, G., Simons, C., Maillard, J.Y., 2012. Use of hydrogen peroxide as a biocide: new consideration of its mechanisms of biocidal action. J Antimicrob Chemother. 67, 1589–1596.

Morton, H.E., 1950. The relationship of concentration and germicidal efficiency of ethyl alcohol. Ann N.Y. Acad. Sci. 53, 191–196.

Murray, P.R., Rosenthal, K.S., Kobayashi, G.S., Pfaller, M.A., 2002. Medical Microbiology, fourth ed. Mosby, St. Louis.

Popescu, A., Doyle, R.J., 1996. The Gram stain after more than a century. Biotech Histochem. 71, 145–151.

Rutala, W.A., Weber, D.J., and the Healthcare Infection Control Practices Advisory Committee (HICPAC). Guideline for Disinfection and Sterilization in Healthcare Facilities, 2008. Centers for Disease Control and Prevention. http://www.cdc.gov/hicpac/pdf/guidelines/Disinfection_Nov_2008.pdf.

"Sterilization by Gamma Irradiation." Kátia Aparecida da Silva, Aquino Federal University of Pernambuco-Department of Nuclear Energy, Brazil. From: Gamma Radiation, Edited by Feriz Adrovic, ISBN 978-953-51-0316-5, Publisher: InTech, http://dx.doi.org/doi: 10.5772/2054. Chapter 9.

Sterilization of Plastics. Zeus Technical Whitepaper, pp. 1–7. http://www.zeusinc.com/UserFiles/zeusinc/Documents/Zeus_Sterilization.pdf.

Talaro, K., Talaro, A., 1996. In: Sievers, E.M. (ed.) Foundations in Microbiology. Wm. C. Brown Publishers. Dubuque, IA.

Chapter 7

Cell Culture and the Eukaryotic Cells Used in Biotechnology

7.1   ADHERENT CELLS VERSUS NONADHERENT CELLS

Animal cells are typically grown as an adherent culture. Consider a culture dish with medium in it. If you were to pour the medium out of the dish, the cells would remain where they were because they are stuck to the bottom of the dish—they are *adherent*. They are able to remain attached to the dish via adhesion proteins such as *integrins*. The integrins make up an entire family of proteins that are used by cells to attach to an extracellular surface. In the body, this surface might be extracellular matrix. In the laboratory, the surface might be a culture dish or a synthetic matrix covered with collagen. (To be complete, *cadherin* should be mentioned. Cadherin, which appropriately gets its name from "calcium-dependent adherence," is a similar family of adhesion proteins that are used by cells to attach to other cells.)

Proteins such as fibronectin, vitronectin, osteopontin, the collagens, thrombospondin, fibrinogen, and von Willebrand factor can be used to promote the attachment of cells to surfaces. These proteins all share a common characteristic: the tripeptide sequence arginine-glycine-aspartate, abbreviated *RGD* (from their one-letter amino acid codes; Figure 7.1). RGD sequences can be used to coat surfaces such as tissue engineering scaffolds to promote cell attachment. Attachment is achieved through the recognition of RGD by integrins, which are transmembrane proteins that act as receptors.

7.2   PRIMARY CELLS, CANCER CELLS, AND CELL LINES

We already know that not all cells are the same; differences between eukaryotic and prokaryotic cells have already been discussed. Looking at eukaryotic cells, there are further classifications that must be understood if one is to work with live cells, especially in a culture situation. These include primary cells, cancer cells, and cell lines. One type of primary cell, the stem cell, is a subject of Chapter 17 and will not be presented here.

An Introduction to Biotechnology. http://dx.doi.org/10.1016/B978-1-907568-28-2.00007-1
Copyright © W T Godbey, 2014.

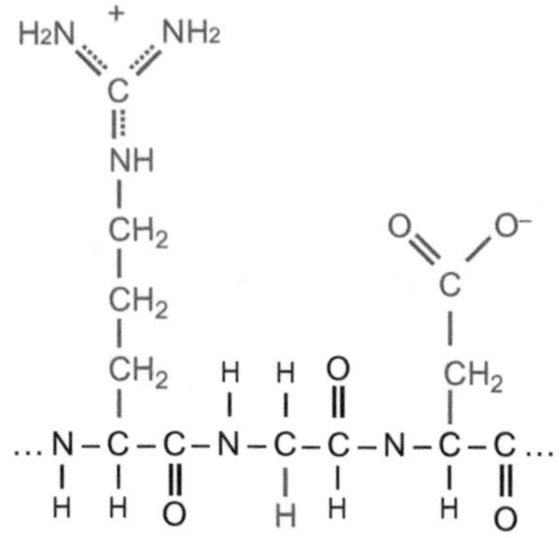

FIGURE 7.1   An RGD sequence. Amino acid identities are defined by side groups, which are shown in red. (See Chapter 2 for more information.)

7.2.1   Primary Cells

Cells that have been taken directly from a body or a tissue are known as *primary cells*. They can be obtained by biopsy, surgery, or autopsy. Primary cells can also be cultured for a finite amount of time as *primary cell cultures* (cells from primary cell cultures are slightly different from primary cells).

Suppose you have given permission for a sample of your own cells to be collected and grown-up in the laboratory for research purposes. Your doctor takes a punch biopsy and starts a culture via *explant*. An explant is where a tissue sample is taken and extracellular matrix is broken down. This can be accomplished via mechanical means such as mincing with a scalpel blade, by chemical means via digestion with an enzyme such as papain, or both. Following the mincing/digestion, the processed tissue is placed upon a prepared growth surface, covered with growth medium, and allowed to incubate undisturbed for several days. After the incubation, the tissue mounds are examined via light microscopy. If cells are observed on the plate in the area immediately surrounding the tissue mounds, the mounds are gently removed to prevent adverse effects from further tissue degradation or products of cell death. The remaining cells are allowed to proliferate as much as the current conditions will allow in a process known as expanding the culture. When cell numbers are great enough, selective media may be employed to prevent the growth of undesired cell types. Also, as the culture is expanded, it may be *passed* into one or more fresh culture vessels. Passing a culture means the cells have been removed (by chemical or mechanical means) from one culture vessel and placed into a new one. When a primary culture has been passed once, the new culture is called the *secondary culture*.

Back to our example, suppose that several years after you graduate, you return to the laboratory of your old doctor to visit your cells. You were told that there's not much to see because there are no active cultures at the moment. Your cells that are still in the laboratory have been frozen, but the active cultures had to be discarded because "they got old." "How can that be?" you exclaim, "I'm in the prime of my life. How could my cells possibly be old?" The answer is related (in part) to something known as the *Hayflick limit*, which helps to explain the mortality of primary cells and cultures.

In 1961, Hayflick and Moorhead published proof of the idea of limited cell division and the eventual *senescence* (old age) of cells in culture. The idea behind the Hayflick limit is that not all of an animal cell's genome is replicated during the cell cycle. As a result, each chromosome will be shortened with every cell division (actually, with each S phase—see Chapter 5). After a number of cell divisions, chromosomes will have shortened to the point that important parts of the genome are missing, so cells will not be able to perform as needed to continue to grow and divide.

Consider a human chromosome that is undergoing replication. When ds-DNA is replicated, the parental DNA strands are separated at the replication fork. Each of the daughter DNA strands is polymerized in the 5′ to 3′ direction. For one of these strands, the process is straightforward. For the other strand, however, the 5′ to 3′ direction of polymerization leads away from the replication fork, so this strand must be replicated in segments known as *Okazaki fragments* (Figure 7.2). As the replication fork reaches the end of the chromosome, the final Okazaki fragment will not be formed because its origin would be beyond the end of the chromosome. As a result, this daughter DNA strand will be shorter. Shortening will occur with every replication until eventually the DNA that is not replicated will be important to the survival of the cell. At this point, the cell either will cease replicating or will die. This is the concept used to justify the Hayflick limit.

The ends of eukaryotic chromosomes are made up of special DNA sequences that are repeated many times. These specialized ends are known as *telomeres* (Figure 7.3). In humans, the telomere sequence is GGGTTA, and it is repeated approximately 1000 times at the end of each chromosome. It is the shortening of telomeres that is responsible for senescence in some primary cells.

Immortal cells such as cancer cells or cell lines often have an agent called *telomerase*. Telomerase is an enzyme that has the function of adding telomere sequences to the ends of chromosomes each time the cell divides. To make matters more complex, not all immortal cells express telomerase. These cells get around the telomere-shortening problem by a telomerase-independent pathway referred to as *alternative lengthening of telomeres (ALT)*. It has even been shown that it is possible to switch between telomerase-positive and telomerase-negative telomere-lengthening processes. A general rule of thumb is that somatic (mortal) cells do not express telomerase, and telomerase expression is seen in many immortal cells.

The subject of telomere shortening was also relevant to the case of Dolly the sheep, the first successful clone of a large animal. Dolly was the result of a successful cloning exercise, but she only lived to be six years old. Although apparently a newborn lamb at birth, Dolly got old fairly quickly. The reason for

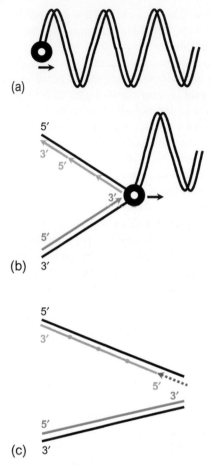

FIGURE 7.2 (a) Replication of dsDNA involves several enzymes, including a helicase, which separates the two strands and DNA polymerase. The replication machinery is shown as a ring. (b) As the replication fork progresses, the "leading strand" (in blue) is easily polymerized in the 5′ to 3′ direction. The "lagging strand" is also created in the 5′ to 3′ direction, but this must be done in segments known as Okazaki fragments (in orange). (c) Okazaki fragments require a primer to be laid down before polymerization can occur. At the end of the strand, there is no place for the primer, so the final Okazaki fragment will not be created (red and dashed) and the end of the lagging strand will not be replicated.

this was that the DNA from the somatic cell that was used for cloning Dolly had already undergone telomere shortening. When this DNA was transferred into the enucleated oocyte, although the oocyte could be considered a "new" cell, the transplanted genetic material had undergone some aging. When Dolly started growing and developing and her cells continued to divide, they continued to age from the point where the original somatic cell had been harvested.

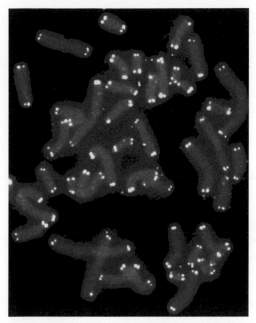

FIGURE 7.3 Chromosomes (in blue) have been stained with a probe to detect telomeres, shown in yellow. Telomeres appear on the ends of the chromatids.

7.2.2 Cancer Cells

There are two main factors that determine whether a cell can be considered a "normal" somatic cell or a cancer cell: mortality and *contact inhibition*. We have just discussed mortality as being a limit on the number of times a cell can divide in culture. Most somatic cells are mortal, although some cells such as stem cells can be propagated for extended periods and therefore seem to be immortal. This does not qualify stem cells as cancer cells, though. The second factor, contact inhibition, must first be addressed. For a typical somatic cell in proper culture conditions, it will grow and divide (and migrate) until it makes contact with something such as another cell or the edge of the culture plate. In terms of cell migration, making contact with an object causes the cell to change direction. As more and more cells populate the culture plate, there is an increasing likelihood that cells will make contact with each other. Eventually, the cell will be surrounded on all sides by other cells or the edge of the culture plate, and the culture has formed a *monolayer*. The cell will no longer migrate at this point and will also stop replicating. This monolayer of cells is going to remain a monolayer—there will not be continued cell division. In terms of the growth curve, the culture will have reached a plateau phase, not because of a lack of nutrients but because of a lack of space. Similar circumstances occur in 3-dimensional cultures and in the body and help to explain why (hopefully) we do not continually have large masses of tissue growing out of us. In short, a cancer cell is a cancer cell because (1) it is immortal and (2) it is not contact-inhibited.

In a 2-dimensional culture, after forming a monolayer (and sometimes before), cancer cells will begin to grow on top of each other. They might form a second layer, or they might begin to grow vertically and then branch out to form a structure that looks something like a mushroom or a ball and chain. It does not take much force to break these structures to release small balls of living cells, which can then relocate to another area to set up a new colony of cells. This is not an uncommon feature in cultures of metastatic cancer cells. Other cultures of metastatic tumor cells may have cells that do not adhere tightly to the culture support, rendering many of the cells easily displaceable and allowing them to detach to set up new colonies in areas with more room to grow. Metastasis occurs in similar ways in the body. Tumor cells will divide to form a small mass; then metastasis occurs when one or more of the tumor cells lets go and migrates through the bloodstream or the lymphatic system to a distant site, often a lymph node.

7.2.3   Cell Lines

A cell line can be considered to be a cross between a "normal" cell and a cancer cell, both figuratively and sometimes literally. Cell lines are created in the laboratory to display the key characteristics of a specific cell type while at the same time being immortal. Cell lines are of great value to the research community because they allow for the study of specific cell types without the necessity of returning to the same donor repeatedly as cells reach senescence. They also provide for a very large source of cells that can be used in multiple laboratories around the world with little variation from culture to culture.

Cell lines can be created in a number of ways. A primary culture can be passed into a secondary culture, which gets passed into a tertiary culture until the characteristics of the culture change. Eventually, one or more of the cells in the culture will undergo *transformation*, which is a switch from a mortal cell to an immortal cell. The transformation of a cell in the body may spell cancer for the individual, but the transformation of a cell in culture can be the birth of an immortal cell line. It is common for transformed cells to become *multinucleate* or contain more than one nucleus (Figure 7.4). Remembering that most primary cells have a limit to the number of times they can replicate; it is fairly easy to select for transformed cells as passage numbers become large—they will select for themselves. Untransformed cells will either die or be outcompeted as they enter senescence.

A second way to establish a cell line is by explanting a biopsy of cancer cells. The cells have already been transformed, so the need to let the cells divide beyond their expected senescence is not necessary. Such cell lines are appropriately referred to as cancer cell lines.

A third way to create a cell line is through the fusion of a primary cell type with a cancer cell type. Fusion can be accomplished, for instance, by placing the cells in contact with one another in poly(ethylene glycol) and administering an electrical current to cause membrane perturbation. When the current is stopped,

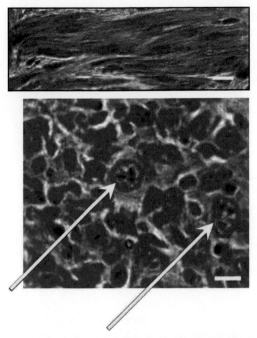

FIGURE 7.4   Stained sections of normal and cancerous cells. Each sample was stained with hema-
toxylin and eosin. Scale bars denote 10 μm. (a) Smooth muscle cells. Each (pink) cell has a single
(purple) nucleus. (b) Cells from a tumor of transitional cell carcinoma. These cancers stain quite
differently from normal tissue. Notice that cancer cells can have multiple nuclei. Two such cells are
denoted by the arrows.

the cells may have their plasma membranes permanently joined. Many cells will
die as a result, but in theory, it only takes one surviving cell to start a new cell line.
The goal of this method is to retain characteristics of the primary cell for future
study while imparting immortality, thus creating an inexhaustible supply of cells
that behave with a standard set of traits. However, let us consider a reality of this
method in greater detail. Human somatic cells have 46 chromosomes (23 pairs)
in their genomes. Let's say we are going to make a fusion cell line from some
endothelial cells that we obtained from a biopsy. The resulting cell line will not
have 46 chromosomes, since we are fusing two cells that have their own genomes
and their own nuclei. In fact, not only will the resulting cell not be restricted to 46
chromosomes, but also it will not be restricted to a single nucleus. These cells are
also multinucleate. This is why most attempts at fusing two cells do not produce a
viable hybrid cell. When a viable cell is produced, it will not carry with it all of the
properties of the parent somatic cell. That is a major drawback to using cell lines.
The data and results that are obtained from cell lines are not always applicable to
primary cells in a living organism. While cell lines are a great tool for research,
data obtained with them should be taken with a grain of salt.

   Although simple in principle, producing a new cell line can be quite a daunting
task. One might be able to earn at least part of a PhD by creating and characterizing
a novel cell line to study a specific problem.

QUESTIONS

1.

(a) Suppose that every time you passed cells, you always did so in a 1:8 split. (A 1:8 split means that 1/8 of a plate or flask of cells is passed to a new culture vessel. This is accomplished by removing the cells from the original container (typically via an enzyme such as trypsin), pelleting the cells, resuspending the cells in a known quantity of medium, and transferring 1/8 of the resulting cell suspension into a new culture vessel.) If your original cells undergo 11 doublings before the first split, you always pass your cells when they reach a certain density, and you pass them into the same size of culture vessel every time; how many passages can you expect to perform before the cells reach senescence? (You may use a Hayflick limit of 60 for this problem.)

(b) Perform part (a) again, but for a 1:16 split and a 1:4 split.

(c) Write a general equation to predict the number of passages one can expect to get from a primary culture.

2. How many passages can one expect to get out of a culture of cancer cells? Why?

3. Suppose you passed a primary culture of cancer cells, storing a frozen aliquot of them and propagating the rest. After the nonfrozen cells had undergone 50 additional passages,

(a) What is the passage number of the frozen cells?

(b) What is the passage number of the nonfrozen cells?

(c) How many doublings have the nonfrozen cells undergone?

(d) If you thawed the frozen cells and placed them into a new culture, would you expect this culture to behave the same as the culture that had undergone 50 additional passages? Why or why not?

4. What might be a problem with introducing active telomerase into all of your own cells?

5. Name one similarity and one difference between a cancer cell and a stem cell.

6. Give two opposing reasons why a biotechnologist might be concerned with cell adherence.

RELATED READING

Hayflick, L., Moorhead, P.S., 1961. The serial cultivation of human diploid cell strains. Exp. Cell Res. 25, 585–621.

Kumakura, S., Tsutsui, T.W., Yagisawa, J., Barrett, J.C., Tsutsui, T., 2005. Reversible conversion of immortal human cells from telomerase-positive to telomerase-negative cells. Cancer Res. 65, 2778–2786.

Ruoslahti, E., Pierschbacher, M.D., 1987. New perspectives in cell adhesion: RGD and integrins. Science 238, 491–497.

Chapter 8

Fluorescence

Apart from being just flat-out cool, the principles of fluorescence are important for the biotechnologist to learn because of the prevalence of fluorescence in both biotechnological research and application. Fluorescent molecules (fluorophores) are very often used as reporters or tracers. Fluorescent tags can be attached to other molecules, such as antibodies, to allow for detection (for quantitation) or tracking (for location). Fluorescent reporters such as the green fluorescent protein (GFP) can be used to verify that a cell is expressing a plasmid that has been delivered. The principles we will cover in this chapter are used in applications ranging from the determination of DNA concentration and purity, to qPCR, to the demonstration of cell type or the location of specific molecules inside the cell during certain processes and under certain conditions. As always, before we discuss the applications of fluorescence, let us first turn to a discussion of how fluorescence works.

8.1 STOKES' EXPERIMENTS

In the year 1852, a 102-page paper was published by G. G. Stokes that described some interesting experiments involving sunlight and a solution of quinine. You can perform these experiments yourself with a prism, some colored glass, and a tube of quinine (which is a component of tonic water).

In the first experiments, Stokes covered a test tube with black paper. A hole was cut into the paper to allow light to enter through the side of the tube, and the top was left uncovered (Figure 8.1). Sunlight was allowed to enter the tube, which contained a solution of quinine, and Stokes observed the solution from the top of the tube. A pale blue arc of light was observed in the quinine. When a smoke-colored piece of glass was placed between the sunlight and the quinine, the arc disappeared. When the glass was moved between the quinine and Stokes' eye, the arc reappeared. The glass was said to have the property of being able to absorb the "invisible rays beyond the extreme violet," or ultraviolet (UV) light as we would now say.

In a later set of experiments, Stokes placed boards in front of his laboratory window so that only a vertical slit of sunlight could enter the room. He then allowed the light to enter a series of prisms so that it was dispersed into a spectrum (Figure 8.2). Placing a test tube with quinine in the rainbow of light did very

An Introduction to Biotechnology. http://dx.doi.org/10.1016/B978-1-907568-28-2.00008-3

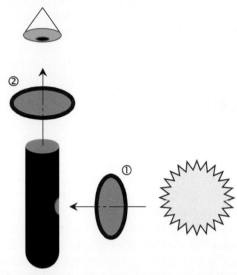

FIGURE 8.1 In this experiment, sunlight enters the side of a tube containing a solution of quinine, and the effect on the quinine is observed from over the top of the tube. Stokes placed smoke-colored glass in either position 1 or 2, as indicated. With the glass in position 1, no fluorescence was detected in the tube. In position 2, a pale blue arc of light was observed in the tube.

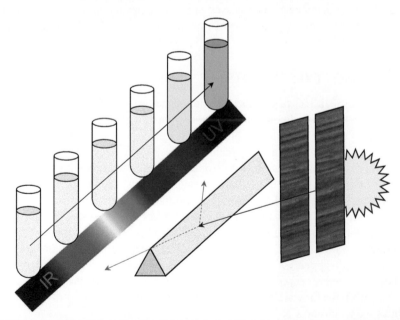

FIGURE 8.2 Stokes' laboratory was darkened, except for a slit of light that was allowed to enter between two boards he put in his window. The light was dispersed by prisms, and a tube with quinine was run through the resulting spectrum. The contents of the tube remained clear until it entered the extreme end of the violet light. When the tube was moved further still, beyond the visible violet color, the quinine became opaque and glowed even more intensely.

little, with the solution remaining clear like water as it was moved slowly from the red to the blue light. When it reached the extreme end of the violet light, a "ghost-like gleam of pale blue light shot right across the tube." However, when the tube was moved even further, past the violet region of the visible light spectrum, the quinine solution became opaque with pale blue light of greater intensity. In Stokes' own words, "it was literally *darkness visible*."

So now we know that sunlight can make quinine glow (fluoresce) and that it is a specific component of sunlight—UV rays—that is responsible for inducing the fluorescence. In another set of experiments, Stokes was able to get a better idea of the nature of the light that came out of the fluorescing quinine. First, he had sunlight go through a solution of water before it hit the prisms. As expected, the complete visible light spectrum was observed. However, when the water was replaced by quinine, the light hitting the prism was split to reveal only colors in the violet or extreme violet range (Figure 8.3).

It would seem that light in the UV spectrum served to make the solution of quinine glow, and the light coming out of the glowing quinine was of a different color

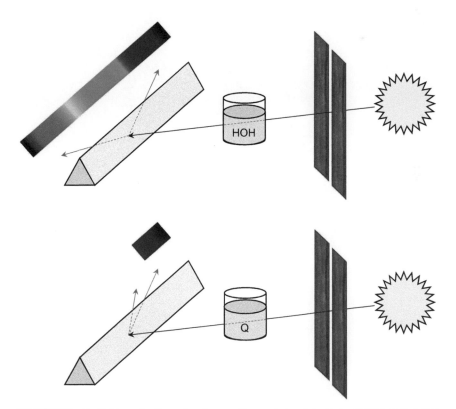

FIGURE 8.3 Sunlight passing through water was dispersed via prisms to reveal the typical visible light spectrum. However, when the light passed through a solution of quinine, only colors in the deep violet portion of the spectrum were observed.

than what was going in. To further verify these observations, Stokes repeated the experiments described in Figure 8.1, but this time, he used cobalt-colored glass (Figure 8.4). Cobalt glass was selected because it "is highly transparent to the chemical rays." (Once again Stokes is referring to what we would now call UV rays.) The results obtained were in sharp contrast to what was obtained when smoke-colored glass was used: When the cobalt glass was placed between the sunlight and the quinine, the amount of fluorescence remained strong, but when the cobalt glass was placed between the quinine and the eye, virtually all light from the quinine was blocked. Stokes referred to the phenomena as

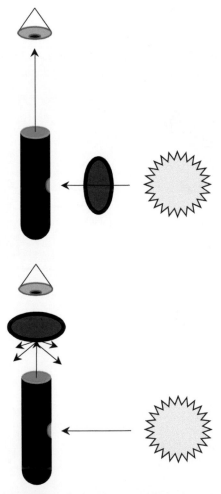

FIGURE 8.4   The experiment from Figure 8.1 was repeated using cobalt glass, which allowed UV rays to pass. There was a distinct difference in the light that was needed to make the quinine fluoresce (UV), and the light that was emitted by the quinine.

a change in the refrangibility of light. Today, we describe his observations in terms of excitation and emission wavelengths of a fluorophore.

8.2   FLUOROPHORE PROPERTIES

8.2.1   Excitation and Emission

It is not uncommon for objects to absorb light—we see it every day. An apple appears red because it absorbs green and blue light while reflecting red. Fluorophores are special in that, not only do they absorb light, but they emit some of the absorbed light at a different wavelength. Recalling Newton's law on the conservation of energy, it should be intuitive that if a fluorophore absorbs energy from a photon and the energy is then released as a photon, the latter photon will have energy that is less than or equal to the amount of energy that was held by the photon prior to absorption. Looking at the visible light spectrum, the photons in the violet region carry more energy (and have shorter wavelengths) than do the photons in the red region of the spectrum. This means that if a fluorophore absorbs at any given color, the photons emitted will be the same or to the right of that color (as pictured).

We can consider what is going on with fluorescence on the subatomic scale. A fluorophore is a molecule, be it GFP or quinine. Within the molecule are individual atoms, each with nuclei and electron clouds. Let us consider a key electron within the fluorophore. We can illustrate the energy state of that electron with the aid of a Jablonski diagram, shown in Figure 8.5. In the diagram, the lowest energy state of the electron—the ground state—is denoted by S_0. Lower limits of higher electronic energy states are shown by bold lines and the letter "S" with a higher subscript (e.g., S_1 and S_2). There also exist vibrational energy states within the electronic energy states, which are indicated by thinner lines; the vibrational energy states for S_0 are indicated in the diagram by the letter V. When energy is transferred from a photon to an electron, the electron will move to a higher energy state in the diagram.

An electron cannot absorb energy from photons of every single wavelength. The diagram in Figure 8.5 shows discrete locations for the possible energy states of an electron, which implies that an electron will only absorb photons that have wavelengths (energies) that will move the electron exactly to one of the higher energy states. In other words, an electron will only absorb photons that have wavelengths (energies) equal to the difference in energy between the current energy state and a higher one.

Once energy has been absorbed, the electron will eventually return to the ground state. This can be accomplished in several ways, but the most common are *vibrational relaxation*, *internal conversion*, and *fluorescence*. Vibrational

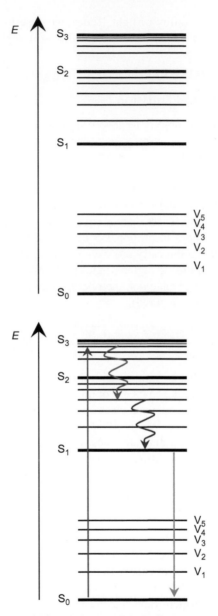

FIGURE 8.5   *(Top)* A typical Jablonski diagram of singlet electron energy states. *(Bottom)* When an electron is elevated to a higher energy state (purple line), it will return to the ground state by releasing the energy in various ways. Red squiggle: internal conversion, which crosses an electronic state (heavy line). Blue squiggle: vibrational relaxation, where the energy descends through vibrational states within the same electronic state. Green line: the large energy drop from a higher electronic state to the ground state is achieved via the emission of a photon. This emission is observed as fluorescence.

relaxation is where an excited electron gives some of its vibrational energy to another electron in the same or a different molecule, in the form of kinetic energy. Because this strictly involves vibrational energy, the energy drop is small: the energy drop will be between vibrational levels but will not go into a new electronic (S) level.

On a strictly mechanical level, internal conversion is the same as vibrational relaxation. The difference is that internal conversion converts the electron to a lower electronic energy state, meaning the energy level can drop below the current S level. Internal conversion happens at higher energy states because the distances between vibrational energy states, as well as in electronic energy states, get smaller and smaller (the lines in the Jablonski diagram get closer and closer).

Vibrational relaxation and internal conversion will serve to move the electron down to one of the electronic energy states (S_n, but typically not S_0 if the electron is in S_1 or higher). When the electron moves from an elevated electronic energy state back to S_0, a photon is emitted and the light energy is observed as fluorescence.

Not every molecule can serve as a fluorophore. Certain geometries are more amenable to absorbing photon energy than others. In addition, just because a molecule absorbs a photon does not mean that it will emit the energy in the visible light spectrum. In general, ring structures are more able to support the absorbance (and emission) of photons. The structure of quinine is no different in this respect (Figure 8.6). Proteins that serve as fluorophores will also contain at least one aromatic amino acid in the active fluorescent region.

Fluorophores are described, in part, by two key wavelengths: the excitation wavelength (E_x) and the emission wavelength (E_m). The distance between these two values is, rightly enough, referred to as *Stokes' shift*. As we shall see soon when we discuss filters, fluorophores with larger Stokes' shifts provide greater utility to the researcher.

Nothing in nature is perfect, including E_x and E_m. These values are used to indicate wavelengths of maximal absorption and emission. However, there is a distribution of wavelengths at which a fluorophore can be excited or will emit.

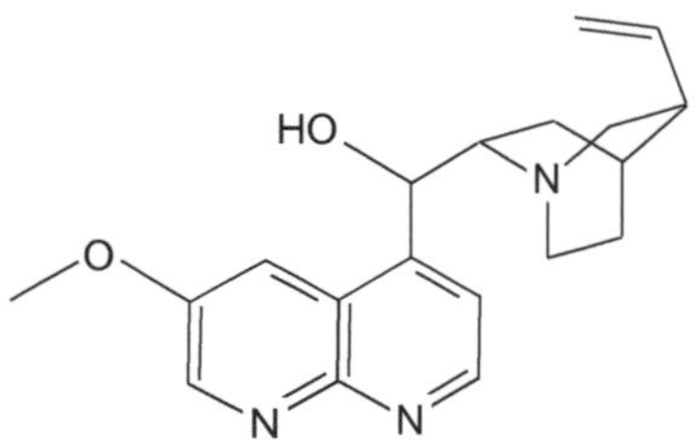

FIGURE 8.6   The structure of quinine. Note the rings in the structures, which contribute to the fluorescent potential of the molecule.

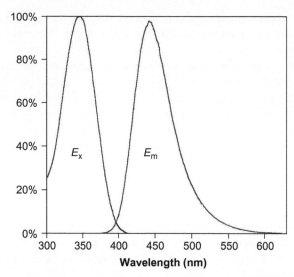

FIGURE 8.7 A hypothetical fluorescence curve. The optimal excitation wavelength for this fluorophore is 346 nm, and the greatest amount of emission is at 442 nm. Stokes' shift is 96 nm. Note that, even though $E_x = 345$ nm, you could use 350, 360, or even 370 nm light to excite the fluorophore.

Figure 8.7 is the fluorescence curve for GFP. Note that E_x, E_m, and Stokes' shift are easily determined from the graph. Also note that these parameters do not tell the whole story. The heights and widths of the peaks are also important pieces of information to consider in determining how ideal a fluorophore will be for your purposes.

8.2.2   More Descriptors of a Fluorophore

Simply knowing E_x and E_m of a fluorophore does not provide the entire picture of the fluorescence characteristics of a fluorophore. For instance, if you shine light on a certain amount of fluorophore, how much light will be absorbed? This value is known as the (molar) *extinction coefficient*, ε. It can be determined empirically by noting the absorbance of a solution of the fluorophore and applying Beer's law: $A = \varepsilon l c$, which states that the absorbance of a molecule is equal to the extinction coefficient times the path length the light travels in getting across the container holding the solution, times the concentration of the solution. Rearranging, we see that $\varepsilon = A/lc$. Because absorbance is a unitless value, the extinction coefficient ε has the units $M^{-1} cm^{-1}$. To give you an idea of scale, the enhanced green fluorescent protein has $\varepsilon = 55,000\,M^{-1} cm^{-1}$.

The amount of light that a fluorophore absorbs is one thing, but will the absorbed light be emitted as a photon? A solution of carrot juice will absorb light, but it is not fluorescent. The probability that a fluorophore will emit a photon after a photon has been absorbed is known as the *quantum yield*, Φ.

It is a dimensionless number that is expressed as (# of photons emitted) / (# of photons absorbed). Φ will always be between 0 and 1. For the enhanced green fluorescent protein, $\Phi = 0.6$.

The extinction coefficient and quantum yield are both used to determine the *brightness* of a fluorophore:

$$\text{Brightness} = \varepsilon\Phi$$

This relation should make sense, in that we must consider how many photons will be absorbed by a fluorophore and what the chances are that the fluorophore will spit out a fluorophore once one has been absorbed when we determine how brightly a fluorophore will fluoresce. For the enhanced green fluorescent protein, brightness $= (55,000)\,(0.6) = 33,000\,\text{M}^{-1}\text{cm}^{-1}$.

Intensity, I, is not the same as brightness. While brightness is a descriptor of fluorescence under standard conditions, the intensity that the investigator observes will depend upon the illumination and detection apparatus being used. Intensity is a function of brightness and can be expressed as

$$I = I_0 k \left(\Phi\varepsilon\right)cl,$$

where I_0 is the lamp intensity, k is the machine constant: how well it gathers light, $(\Phi\varepsilon)$ is the brightness, c is the concentration, and l is the path length of light.

Recall from above that $\varepsilon = A/lc$. Substituting this into the equation and canceling out the l and c terms gives us

$$I = I_0 k \Phi A,$$

where A is the molar absorptivity.

8.3   FLUORESCENCE DETECTION

Many times, the selection of the right fluorophore is not made by choosing the most intense fluorophore with the largest Stokes' shift. Fluorophores will often be detected via fluorescence-detection microscopy, fluorescent-plate readers, or even qPCR setups. Each of these utilizes a set of filters to sift out unwanted wavelengths and to verify the wavelength of light being detected.

Light filters come in three main categories. *Long-pass filters* (LP) allow light with wavelengths greater than a certain value to pass through the filter. To allow yellow, orange, and red light all to get through a filter, one might select an LP575 filter (Figure 8.8). *Short-pass filters* (SP) follow the same principle, except that all wavelengths shorter than a given value will get through. An SP530 will allow green, blue, indigo, violet, and ultraviolet light to pass. The third type of filter is the *band-pass filter* (BP). These filters are denoted by two numbers: the width of the band of wavelengths that can pass and the wavelength that sits in the center of the band. For instance, BP500/40 describes a filter centered at 500 nm that lets a band of wavelengths 40 nm wide to get through. Specifically, it will allow photons in the range of (480-520) nm to get through.

FIGURE 8.8   Slightly more than the visible light spectrum, with wavelengths given to get a feel for colors and their associated wavelengths.

Example 8.1

Suppose that you own a microscope setup with a BP 480/40 excitation filter and a LP 530 emission filter. You are going to stain cells with antibodies that have been labeled with a fluorophore. What would you expect to see if you used the following fluorescent markers?

	E_x	E_m
1	440 nm	550 nm
2	550 nm	600 nm
3	490 nm	510 nm
4	488 nm	530 nm
5	496 nm	525 nm

Answer:

Fluorophore 1: Nothing would be seen. The band-pass filter lets the range (460-500) nm light through. With an E_x of 440 nm, it is doubtful that the fluorophore will be excited.

Fluorophore 2: Nothing would be seen. Even though 550 nm is in the range of the second filter, that is the emission filter. The excitation filter will not let the 550 nm light pass, so the fluorophore will not fluoresce.

Fluorophore 3: Nothing would be seen. Even though the fluorophore will be made to fluoresce by the 490 nm light, the fluorescent light of 510 nm will not be able to pass through the emission filter so nothing will be seen.

Fluorophore 4: Cells displaying a beautiful green fluorescence will be seen.

Fluorophore 5: Cells displaying a beautiful green fluorescence will be seen. While it is true that 525 nm light will not be able to pass through the LP 530 emission filter, recall that E_x and E_m are wavelengths corresponding to maximal values on their respective curves (refer back to Figure 8.7). Although the fluorophore is not optimal for the existing filter setup, you should still be able to observe some fluorescence.

8.4   FRET

Consider two fluorophores that are very close to each other in proximity: one has an $E_x = 488$ nm and $E_m = 510$ nm, and the second has an $E_x = 510$ nm and $E_m = 590$ nm. If the first fluorophore is hit with 488 nm light, it will fluoresce at 510 nm. Interestingly, the emitted 510 nm light can be absorbed by the second fluorophore, which will emit light at 590 nm. This phenomenon is known as *fluorescence resonance energy transfer (FRET)*. It has been used to show that two molecules are close to one another, such as inside the cell.

As an example of FRET, consider that an enzyme might have been labeled with the first fluorophore just described, and a certain protein has been labeled with the second. If we then expose a preparation of the enzyme and ligand to 488 nm light and it glows red (actually, if we observe light using a red filter), then the two fluorophores are relatively close to each other, implying that the labeled protein is a substrate for the enzyme. How close is close? That depends upon the sensitivity of the detection equipment, but the efficiency of energy transfer depends on the distance between the donor and acceptor fluorophores raised to the inverse sixth power. Roughly, this translates to distances in the 1-6 nm range.

Conservation of energy applies to FRET as it does to single-fluorophore fluorescence. This means that a FRET pair must have the donor excitation energy > donor emission energy > acceptor excitation energy > acceptor emission energy (i.e., traveling in the violet-to-red direction). This may seem straightforward until a FRET pair is selected for the first time. Figure 8.9 is an illustration of a FRET pair. Note that concessions in efficiency must be made because $E_{m,donor}$ seldom equals $E_{x,acceptor}$.

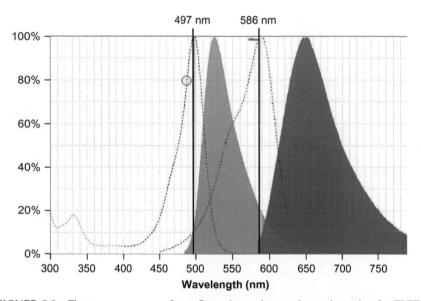

FIGURE 8.9 Fluorescence spectra of two fluorophores that can be used together for FRET. The excitation maximum for each fluorophore is given over each excitation curve (dashed lines). Emission curves are filled in with the color of each fluorophore, and the height of each is based on excitation with the optimal wavelength for each fluorophore. The curve amplitudes are not indicative of what would be achieved in a FRET experiment. If we were to excite with a laser of 488 nm wavelength (yellow circle), the y-values on the green emission curve would be reduced to ~80% of the values shown, which would reduce the amount of excitation of the red fluorophore accordingly. In addition, the excitation energy for the red fluorophore would be dependent upon not only a single wavelength but also the integral of [(the green emission curve (function)) times (the red excitation function)] over the wavelength range spanned by the red excitation curve, plus a similar contribution by the original excitatory wavelength(s) being used. This second term becomes another integral if, instead of a single excitatory wavelength, a range of wavelengths is used as in the case when a mercury bulb is used with a light filter (such as a BP 370/40). The point here is not to teach you how to accurately determine total FRET output, but to appreciate the complexity involved in selecting a FRET pair. Finally, a successful FRET experiment is dependent upon filtering the original excitatory signal and the signal from the first fluorophore out of the second fluorophore emission. In this example, the overlap between the green and red curves must be removed. This can be achieved by selecting an emission filter such as an LP 650 (or LP 700) to remove most of the output seen as emission from fluorophore #1. *Image generated, in part, using BD Fluorescence Spectrum Viewer.*

QUESTIONS

1. Suppose you have a microscope with the following filter sets (you cannot mix and match the filters between sets):

	Excitation	Emission
①	BP 360/40	BP 460/50
②	BP 480/40	BP 535/50
③	BP 535/50	BP 610/75

Which fluorophores could you definitely use with your system?

		E_x	E_m
a.	Pacific Blue	410	455
b.	PotatoGreen 488	488	516
c.	AndiFluor 578	578	530
d.	mmm…Cherry	580	610
e.	mmm…Banana	540	553
f.	mmm…Applesauce	528	577
g.	Red Rain	607	630
h.	At least one of the fluorophores above is obviously made up. Identify it (them) and justify your response		
i.	Considering fluorophores a-g, which pair would work best for FRET analysis?		

2. Why would a small Stokes' shift be undesirable if you wanted to use a fluorophore for microscopy?

3. You go out to a nightclub and are very excited to see that some of the drinks are glowing in the darkened room. You buy one of the drinks and immediately drive to your professor's house to show off this seemingly miraculous wonder. Unfortunately, when you present the drink, it is no longer glowing.
   a. Propose a reason why the drink is no longer glowing.
   b. In desperation to save your skin, you shine a flashlight on the drink. Does this make the drink glow? Would a cigarette lighter or match help? How about waiting until the next day and holding the drink up to sunlight? What would you see in each case?

4. If I have an excitation filter that allows 493-502 nm light through, can it be used to excite a fluorophore with $E_x = 488$ nm? Why or why not?
5. What could be a use of a molecule that absorbs all wavelengths of visible light but does not emit any?
6. Detergent companies used to make the claim that using their product could make your white clothes "whiter than white." There was actually some truth to the claim, and it involved some of the phosphates in the detergent. Come up with a scientific reason why the manufacturers were not lying.

RELATED READING

Förster, T.H., 1948. Intermolecular energy migration and fluorescence. Ann. Physik. 2, 55–75 (Article in German).

McEwen, J. ChemWiki: The Dynamic Chemistry E-textbook. Maintained by UC Davis. http://chemwiki.ucdavis.edu/Physical_Chemistry/Spectroscopy/Electronic_Spectroscopy/Jablonski_diagram (accessed 03/07/2014).

McRae, S.R., Brown, C.L., Bushell, G.R., 2005. Rapid purification of EGFP, EYFP, and ECFP with high yield and purity. Protein Expr. Purif. 41, 121–127.

Stokes, G.G., 1852. On the change of refrangibility of light. Phil. Trans. R. Soc. London 142, 463–562.

Stryer, L., 1978. Fluorescence energy transfer as a spectroscopic ruler. Annu. Rev. Biochem. 47, 819–846.

Chapter 9

Locating Transcriptional Control Regions: Deletion Analysis

In addition to the coding regions (exons) of a gene, there are also control regions such as promoters and enhancers to consider. Here is a miniature application of how we can detect and locate where one of these control regions lies within a stretch of DNA. Having a precise location is very economical for the biotechnologist who delivers genes into cells. If one is constructing a gene under the control of a specific enhancer, and it is known that the enhancer lies within 4000 base pairs of the transcriptional start site, one could use the entire 4000 base pair upstream region in constructing an engineered gene. However, if the enhancer region only spans 500 bases and the precise location of the enhancer is known, it is far more efficient to include only the 500 base pairs in the engineered gene because more copies of the gene could be delivered for a given mass of DNA.

Finding control regions can be accomplished via *deletion analysis*. Suppose that we know the location of the transcriptional start site for a gene. We can take a large region of DNA upstream of start and put it into a circular piece of DNA called a *plasmid*. (Just how this is done will be discussed in Chapter 12) The plasmid will contain, among other things, an exon for a *reporter gene*. A reporter gene produces a product that is easily detectable, such as a fluorescent protein. As a part of the deletion analysis procedure, we will insert the fragment that contains our suspected control region upstream of the exon for the reporter. We will create many copies of this engineered plasmid, and then we will deliver them into cells to see how much of the reporter is expressed. In a parallel set of steps, we will shorten the control region and insert it into another reporter plasmid, create many copies, and deliver those plasmids into cells to again see how much reporter is expressed. The process of shortening the suspected control region is performed several times to generate many plasmids with control regions of decreasing sizes. (It may seem odd right now, but the shortening of the suspected control region to produce many DNA fragments of varying lengths is often performed as a single step.) See Figure 9.1 for a pictorial overview of the process.

An Introduction to Biotechnology. http://dx.doi.org/10.1016/B978-1-907568-28-2.00009-5

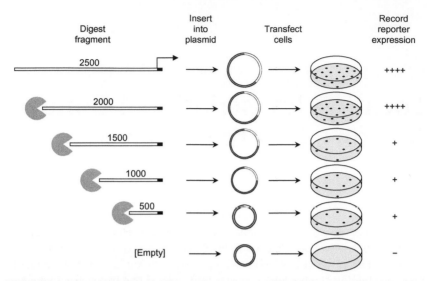

FIGURE 9.1 Overview of the deletion analysis process. DNA fragments are digested with an exonuclease to produce fragments of varying sizes. The fragments are put into plasmids, which are replicated and delivered to cells. Promoter/enhancer/repressor activity is implied by reporter gene expression in the transfected cells. Spots in the culture plates shown imply reporter expression, not the total number of colonies.

9.1 AN EXAMPLE OF DELETION ANALYSIS

Suppose that we generated the following data with our plasmids:

Plasmid	Bases in Control Region	Represented Region	Reporter Expressed (RFU)
1	2500	−2500 to −1	10,473
2	2000	−2000 to −1	10,672
3	1500	−1500 to −1	4873
4	1000	−1000 to −1	4688
5	500	−500 to −1	4907
6	0	NA	—

In analyzing these data, look for where there is a change in reporter expression. Notice that there was a decrease in reporter expression between fragments 2 and 3; this implies that a control region was lost when shortening fragment 2, so the region must lie in the bases that were lost. In this example, it implies that there must exist a control region somewhere between −2000 and −1500 bases upstream of start. Further analysis of the table shows that there is another control region somewhere in the region (−500 to −1) (Figure 9.2).

To solve deletion analysis problems, including the times you might be in the laboratory performing the actual experiment, it might help to make a master map of the region being tested. Every time a control region is elucidated by a change in reporter expression, mark it as illustrated at the bottom of Figure 9.2. As can be seen from the results in the figure, the 2500 bp upstream region has control regions somewhere in the ranges (−2000 to −1500) and in (−500 to −1). Note that the negative numbers indicate that we are in regions upstream of transcriptional start (which is counted as base number +1).

In practice, it is difficult to control exactly how long each fragment will be. An exonuclease, which is an enzyme that chews up a polynucleotide (such as DNA) from the end, is used to digest the DNA region of interest to yield many fragments with a distribution of sizes. Plasmids are made and amplified and the reporter-expression experiments are run to see which fragments produced the greatest amount of reporter expression. DNA sequencing is performed separately, perhaps later, to determine which DNA regions were actually being tested when effects were seen. We will not get the nice, controlled fragment sizes shown in the example, but we will still be able to locate regions of interest in the gene.

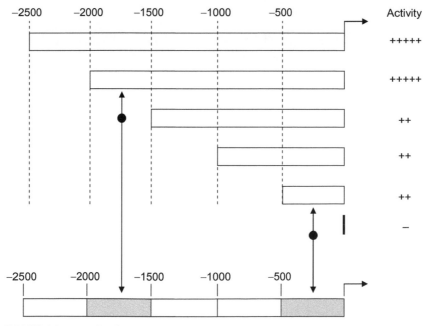

FIGURE 9.2   A straightforward way to map out deletion analysis data. Fragment can be represented as they are here, from largest to smallest. Writing the corresponding reporter expression levels to the right, note significant drops or increases in expression activity. When a significant drop or increase is detected, it is purportedly due to the segment of DNA that is present in the next larger fragment but missing from the present fragment (noted by dotted lines).

QUESTIONS

1. (Refer to Chapter 10 if more information is needed regarding agarose gels.) Suppose we want to locate upstream control regions in a certain gene. We have used an exonuclease to chew the gene from the 5′ end. Below is the result of several runs, where we have let the exonuclease chew for different periods of time.

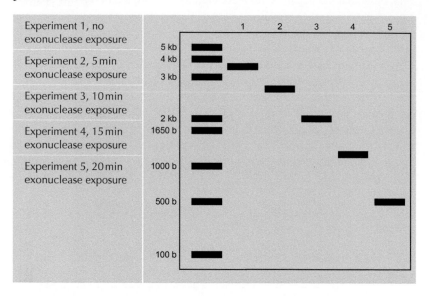

a. This gel has two controls: one for the experiment and one for the gel itself. What are the controls? What does each control show?

b. We then cloned the fragments into a plasmid containing the coding region for a reporter gene and then transfected cells with copies of the plasmid. The following results for reporter expression were obtained:

Relative Fluorescence Units (RFU)	
Plasmid 1	1.000
Plasmid 2	0.993
Plasmid 3	0.658
Plasmid 4	0.648
Plasmid 5	−0.003
(Plasmid *n* was created from fragment *n*, from experiment *n*)	

c. Estimate the location of each control region.

    d. Suppose that RFU=0 for transfections utilizing the five plasmids above. Give two possible explanations for the result. What control could you have run to prevent the ambiguity?

    e. Estimate the nuclease efficiency of the enzyme used in bases/sec.

2. Using deletion analysis, researchers are trying to find the location of transcriptional control elements for a gene. They observe a sharp decrease in polypeptide production corresponding to the (-1092 to -957) range and a lack of any reporter expression corresponding to the (-92 to -1) range. Oddly enough, there was full expression of reporter when the reporter plasmid contained the upstream sequence (-888 to -1). Explain these results, including any identifiable DNA elements.

Chapter 10

Agarose Gels

Gels used for DNA analysis are typically made out of *agarose* (Figure 10.1), a sugar that can be extracted from red algae. In making an agarose gel, the solvent used is typically TAE, which is an acronym for tris(hydroxymethyl)aminomethane (Tris), acetic acid, and ethylenediaminetetraacetic acid (EDTA). Tris is a pH buffer. Acetic acid is used to help titrate the pH of the solution into the proper range. EDTA (Figure 10.2) serves as a chelator for divalent cations, such as Ca^{2+}. (Chelation is where one molecule—technically, a Lewis base—binds a central metal atom simultaneously at two or more separate places.) Agarose will be added to the TAE buffer, and it must be heated in order to bring the agarose into solution. Upon cooling, the agarose will form cross-links to yield a porous gel. The procedure is similar to making Jell-O®, which solidifies when it is cooled after cooking. In fact, casting a gel means pouring it into a mold just like one might do with Jell-O. The mold that we use for agarose gel electrophoresis will produce a rectangular prism. A "comb" will be inserted into the gelling liquid to produce cuboid indentations for insertion of the DNA samples to be electrophoresed (Figure 10.3).

The percentage of agarose used to make the gel will determine its porosity. Higher percentages, such as 1.5-2%, can be used to separate DNA fragments that are relatively small. Using less agarose, such as 0.5%, will yield an agarose network with less cross-linking and therefore larger pores. Larger pores will permit easier separation of large DNA fragments. 0.8-1% is the most common agarose concentration range, allowing good resolution upon separation of fragments in the range of 500-15,000 bp (Table 10.1).

The agarose gel is a good way to separate DNA based upon size, but there must be a way to visualize the DNA. Typically, an intercalating dye is used for this purpose. A very common choice of dye is ethidium bromide (Figure 10.4). Ethidium bromide is a salt, so it will separate in solution to yield the positively charged ethidium ion plus a bromide ion. The planar ethidium ion can insert (intercalate) between the stacked bases of helical dsDNA, interacting with them via van der Waal forces. Interactions between an ethidium cation and the negatively charged phosphates on DNA also occur. While ethidium fluoresces orange when exposed to UV light, it does so much more intensely when intercalated with dsDNA. After it interacts with and becomes concentrated in the gel at the locations of DNA strands, ethidium can be visualized via exposure of the gel to ultraviolet light, generated via a UV-light box, as shown in Figure 10.5.

An Introduction to Biotechnology. http://dx.doi.org/10.1016/B978-1-907568-28-2.00010-1
Copyright © W T Godbey, 2014.

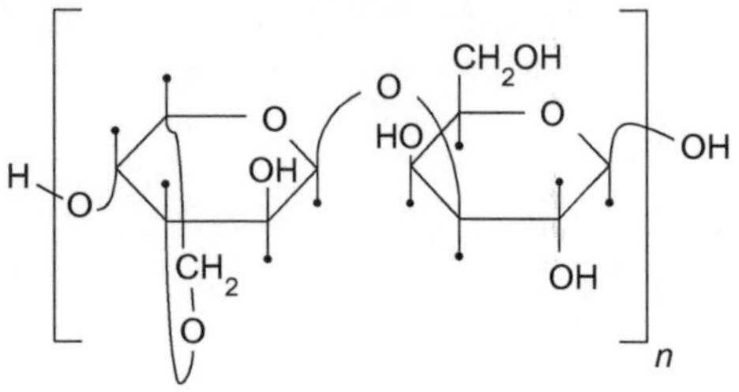

FIGURE 10.1   The structure of agarose.

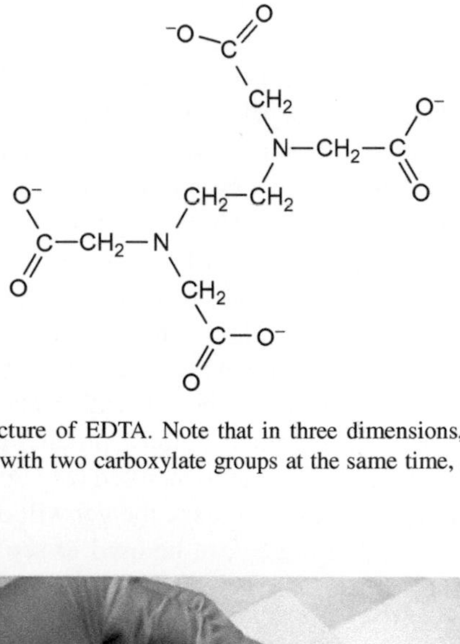

FIGURE 10.2   The structure of EDTA. Note that in three dimensions, a single divalent cation such as Ca^{2+} can interact with two carboxylate groups at the same time, a phenomenon known as chelation.

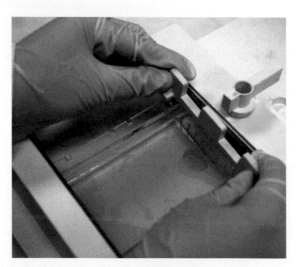

FIGURE 10.3   Removing the comb from a cast agarose gel.

TABLE 10.1 Size Ranges of dsDNA Fragments That can be Resolved Using the Given Percentages of Standard Agarose

% Agarose	Resolvable dsDNA Fragment Sizes (bp)
0.5	700-25,000
0.8	500-15,000
1.0	250-12,000
1.2	150-6000
1.5	80-4000

Other types of agarose exist.

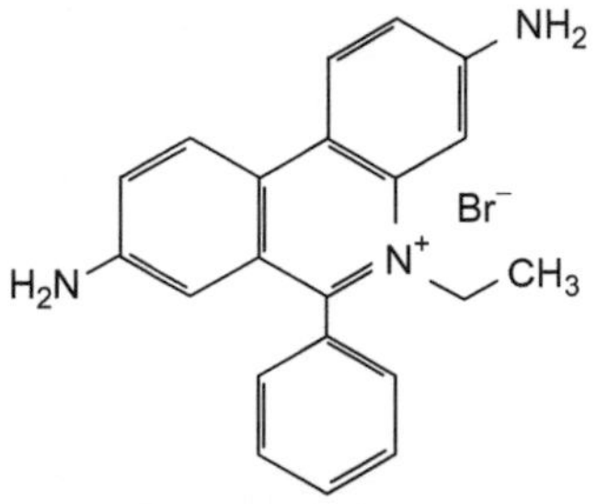

FIGURE 10.4 The structure of ethidium bromide.

Aside

Once intercalated, ethidium can interfere with DNA replication, repair, or transcription, making it carcinogenic. If it gets into your cell nuclei, it's going to intercalate with your own DNA and remain there perhaps for the life of the cell. It can alter DNA folding, alter transcription, and can even interfere with DNA replication and proofreading. While the comparative effects of coming into direct contact with ethidium bromide are not as bad as, say, phenol, the effects of ethidium bromide are cumulative over one's lifetime.

Ultraviolet light can induce mutations in DNA, including DNA fragments separated in an agarose gel. If the fragments are being separated for further use, such as construction of a plasmid, then mutations can be detrimental to the project. An alternative to ethidium is a class of dyes that fluoresce when exposed to blue light. Although these dyes are more expensive than ethidium bromide and they require the purchase of a blue-light box, many researchers prefer these dyes for DNA visualization because they help to preserve DNA sequences in separated strands.

Once the DNA samples have been loaded into the wells of an agarose gel that has been immersed in TAE buffer, an electrical field on the order of 100 mV is

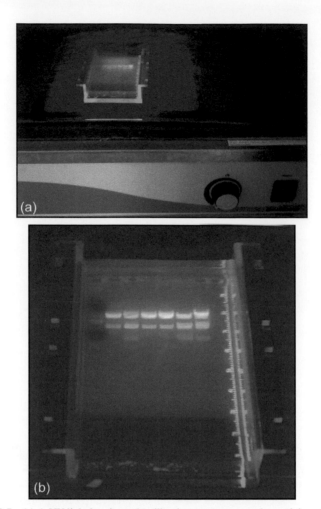

FIGURE 10.5   (a) A UV-light box is used to illuminate an agarose gel containing ethidium bromide. (b) Ethidium that is associated with dsDNA will fluoresce brightly. DNA fragments of equal sizes can be seen as distinct bands in the gel.

applied. DNA, having the same charge as electrons, will migrate in the same direction as electrons: toward the anode. DNA fragments will be separated on the basis of size. However, the separation does not directly result from the charge of the DNA. Consider two DNA fragments: a 100 bp fragment and a 5000 bp fragment. While the larger fragment is 50 times larger, it also has 50 times the amount of total charge, so it will have the same amount of electrical attraction toward the anode per unit length as the smaller fragment. The charge concentration is the same for both of the fragments, namely, one negative charge per repeating unit. Separation of the two fragments is achieved because of their differing sizes. The cross-linked agarose polymer is much like an obstacle course that is more easily traversed by the smaller DNA fragment, which can more easily snake its way through the twists and turns of the agarose. Smaller fragments will travel down the gel more quickly, providing a regular and predictable separation. If one of the lanes of the gel is loaded with a *molecular weight marker*—a commercially available standard comprised of a mixture of DNA fragments of known size—then DNA from sample wells can be compared to the standards and fragment sizes can be estimated.

10.1   APPLICATION OF AGAROSE GELS: GEL SHIFT

Consider two samples of identical DNA fragments. The base sequence of these fragments is such that a certain transcription factor protein will bind to it. If we were to mix one of the samples with the protein and then load both samples into separate lanes of a gel, the DNA in the sample that contained the protein would migrate more slowly than the sample containing naked DNA. This should make sense: running an obstacle course with a monkey on your back will take longer than running the course unfettered. With protein bound to the DNA fragment, it will migrate through the gel more slowly. This is referred to as a *gel shift*. The gel shift experiment is used to show that DNA will interact with an agent such as a new polymer designed for gene delivery or a suspected transcription factor protein. The gel shift only shows general interactions, though. Sequence-specific interactions between DNA and a protein can be examined in greater detail via the DNA footprinting experiment, which can yield a more precise location of where a protein binds to a DNA sequence.

10.2   APPLICATION OF AGAROSE GELS: DNA FOOTPRINTING

DNA footprinting is a way to determine where proteins such as transcription factors will bind on a stretch of DNA. While the goal of DNA footprinting is similar to that of deletion analysis, there are key differences in the procedures that can often trip up the student.

Suppose we have a stretch of DNA that is four nucleotides in length. (Such a short fragment was chosen for simplicity of illustration. In practice, the DNA fragment to be analyzed will be much longer.) The stretch of DNA can be labeled on its 5′ terminus via the radioactive phosphorous isotope ^{32}P. This is done because any fragment that contains the ^{32}P can be detected by exposure to X-ray film.

Next, the labeled DNA fragment will be exposed to an *endonuclease*. An endonuclease will cut a polynucleotide from within the sequence. This is in contrast to an *exonuclease*, which cuts a polynucleotide from one of the terminal (exterior) bases. The endonuclease of choice is often DNase I, which does not have a specific recognition sequence but rather cuts randomly.

Consider for a moment the specific case where the DNA fragment is cut only once, between bases 2 and 3. The resulting fragments would be 1-2 and 3-4, where each digit here represents the position of each nucleotide in the original DNA fragment (1-2-3-4). When we run this out on a gel and then expose X-ray film to the gel, the only fragment that we will perceive via film exposure will be 1-2. The 3-4 fragment will be present in the gel, but because it does not have a radioactive label, it will not be detectable.

If the original DNA fragment is exposed to DNase I for an extended period of time, the result will include only single bases since all of the phosphodiester bonds will be cleaved. However, if the exposure to endonuclease is for a limited period of time, the resulting fragments could be of every possible size: 1, 2, 3, 4, 1-2, 2-3, 3-4, 1-2-3, 2-3-4, and 1-2-3-4 (Figure 10.6a). Note, however, that the only fragments that will be detectable via exposure to X-ray film will be the fragments that contain the 5′ label, that is, all fragments that contain base number one: 1, 1-2, 1-2-3, and 1-2-3-4 (Figure 10.6b). The other fragments will be present on the gel, but the only ones that will be visualized will be the ones that contain base number 1.

Now, consider that there is a protein that will bind to the DNA tetramer in such a way as to obscure the phosphodiester bond between bases 2 and 3 (Figure 10.6c). DNase I would still be able to cut between bases 1 and 2 and between 3 and 4, so the set of all possible fragments would include 1, 4, 2-3, 1-2-3, 2-3-4, and 1-2-3-4. Gel electrophoresis of the set would allow us to visualize only the fragments 1, 1-2-3, and 1-2-3-4 (Figure 10.6d). There would be no visible band that was two nucleotides long. The absence of the two-nucleotide band (1-2) would indicate that the protein obscured the cut between bases 2 and 3. Using this procedure, one can prove that a given protein will not only bind to a specific stretch of DNA but also estimate where the binding occurs.

10.2.1 A More-Detailed Example

In this example, we will once again be looking for where a protein binds on a stretch of DNA. The experiment will entail production/isolation of a DNA fragment (likely using the polymerase chain reaction, which will be discussed later) and placing the DNA into a tube with the protein of interest. The mixture is then incubated to give the protein a chance to bind to the DNA, after which DNase I is added to yield the fragments for gel analysis. It is important that the endonuclease is added after the protein, because if it is added before, the protein will have no chance to protect the DNA from cleavage.

Suppose the protein is an activator and the DNA sequence contains an enhancer that we are trying to locate and that the protein binds as shown in Figure 10.7. When it covers a space between two bases, the endonuclease will be unable to cut at that specific location. The pictured gel shows the pattern of bands that would result from the given coverage. All of the smaller bands would be the same as the control sample (lane 2) that contains no protein. Note that a 16 bp band will be visible because the protein does not cover the phosphodiester bond between bases 15 and 16. However, the protein will obscure the bonds between

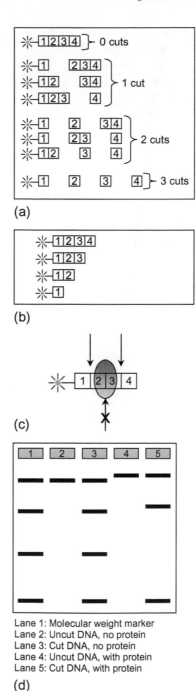

(a)

(b)

(c)

Lane 1: Molecular weight marker
Lane 2: Uncut DNA, no protein
Lane 3: Cut DNA, no protein
Lane 4: Uncut DNA, with protein
Lane 5: Cut DNA, with protein

(d)

FIGURE 10.6   DNA footprinting. Refer to the last paragraph of Section 10.3 in the text for a complete explanation of this figure.

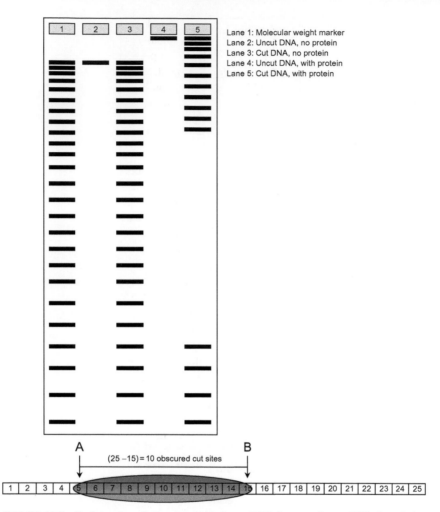

Lane 1: Molecular weight marker
Lane 2: Uncut DNA, no protein
Lane 3: Cut DNA, no protein
Lane 4: Uncut DNA, with protein
Lane 5: Cut DNA, with protein

A   B
(25 –15) = 10 obscured cut sites

| 1 | 2 | 3 | 4 | 5 | 6 | 7 | 8 | 9 | 10 | 11 | 12 | 13 | 14 | 15 | 16 | 17 | 18 | 19 | 20 | 21 | 22 | 23 | 24 | 25 |

FIGURE 10.7   To determine where a protein binds to a DNA fragment from a DNA footprinting gel: (1) Starting from the bottom, count the number of bands in the lowest group in the sample lane (lane 5). (2) Protein binding is located after that many possible cuts (Arrow "A"). Note that it is between cut sites. (3) Subtract the number of bands in the sample lane from the number of bands in the control cut lane (lane 3). (Note that with the protein bound, there is a shift in the apparent weight of the bands with protein bound.) (4) The answer to (3) is the number of potential cut sites protected by the protein and determines the 3′ edge of where the protein binds (Arrow "B").

bases number 5 and 6, 6 and 7, 7 and 8, *etc.* So there will be no corresponding bands in the gel. Also note that there will be a band for the labeled fragment that is 15 base pairs long, but this band will run slightly slower than the 15 mer in the molecular weight marker lane because the activator protein will be bound, an example of a gel shift. Every visible band greater than 15 bp will also be gel shifted for the same reason. (Also note the gel-shifted bands in Figure 10.6d.)

If the protein used binds to the DNA fragment under investigation, the results of the DNA footprinting experiment will show an area on the gel ladder with no bands. This area will correspond to the location of protein binding.

A *control* should be run with every experiment. In the control, every parameter will be set to produce a known, predictable, and repeatable result. Controls are run to verify that an experiment was set up properly and all components are functioning as intended. There are negative controls and positive controls. A negative control is run to show what the result would look like if a certain reaction or response did not take place. A positive control is run to show what the experimental data would look like if the reaction or response did take place. Experimental runs that test for the presence or absence of an effect under varied but controlled conditions are referred to as test samples. In the DNA footprinting experiment, the molecular weight marker functions as two controls. First, it helps to verify that we made a good gel (we did not load it before it cooled completely, we added sufficient and functioning ethidium bromide, our TAE buffer is adequate, we did not run the gel upside down, etc.), and second, it serves as a standard of comparison for the bands that are loaded into the lanes containing test samples. Other controls to be used for this experiment include a lane for which the DNA sample contained no endonuclease or protein, to verify that the starting DNA material was intact (not already cut or sheared), and a lane with endonuclease but no protein (in which every possible DNA fragment size should be visualized), to verify that the endonuclease is functioning and the time of digestion was adequate.

DNA footprinting examples are fairly straightforward when one already knows where the protein binds. In the laboratory, however, one will only see bands on X-ray film. The challenge is to interpret the banding pattern to deduce the location of protein binding. The caption for Figure 10.7 describes an algorithm for determining where a protein binds to a DNA fragment.

Aside: How Can a Bacterium Have Restriction Endonucleases and Not Cut Its Own Genome?

The answer is not that the bacterial genome is devoid of recognition sequences—that would be too restrictive (pun intended). In addition to producing restriction endonucleases, the bacteria will also have enzymes that modify their own genomes. The typical enzyme for such modification is a DNA methyltransferase. The DNA-modifying enzyme will bind to the same recognition sequence that is used by the associated restriction enzyme and will transfer a $-CH_3$ (methyl) group onto one of the bases in the sequence. Since the sequence is a palindrome, a methyl group will be transferred onto each strand. These methyl groups will project into the major groove of the genomic DNA double helix, preventing proper binding or activity of the restriction enzyme. The restriction enzyme and modification enzyme(s) can be separate proteins, or they can be subunits (or even domains) of a larger protein that performs both restriction and modification functions.

10.3   APPLICATION OF AGAROSE GELS: RESTRICTION ANALYSIS

Agarose gels can also be used to analyze results from experiments involving restriction endonucleases. *Restriction endonucleases* are enzymes that cut inside a polynucleotide sequence. They get their name from the fact that they can restrict the proliferation of bacteriophages, which are viruses that attack bacteria. This phenomenon is accomplished by the binding of the restriction endonuclease to a specific sequence in the viral DNA followed by cleavage. Restriction endonucleases are used by bacteria for self-defense against viral invasions. These bacterial enzymes recognize specific sequences that are not contained in the bacterial genome, so they are very effective at preventing the propagation of viruses that do contain the sequences. Once the viral genome has been cleaved, it will not be possible for it to be completely replicated or to be used to produce more viruses. The sequence that is bound by a restriction enzyme is known as the recognition sequence for that enzyme.

We as humans have recognized the existence and mechanism of restriction endonucleases. Specific bacteria that contain a specific endonuclease of interest can be grown and harvested. These bacteria are propagated in the laboratory, after which they are lysed and the endonuclease proteins are extracted and purified. The first restriction endonucleases were discovered at Cold Spring Harbor Laboratory, but they were first commercialized by the company New England Biolabs in 1975. The initial offering by the company provided scientists with a choice of eight enzymes. Today, there are well over 200 restriction enzymes available, and several different companies offer them for purchase. Many thousands of bacteria and archaea have been screened in the search for new enzymes to sell. During this extensive search, it was found that all free-living bacteria and archaea appear to code for restriction endonucleases, suggesting that they serve as a sort of prokaryotic immune system.

The recognition sequence of a restriction enzyme is typically a *palindrome*. In English, a palindrome is a word, phrase, or sentence that reads the same forward or backward. In molecular biology, the definition is similar. Consider the restriction enzyme EcoRI, which has the recognition sequence GAATTC. Although the six letters do not spell the same thing when read backward, if we consider the complementary sequence and impose the rule that we only read in the 5′ to 3′ direction, then the recognition sequence becomes

5′-GAATTC-3′
3′-CTTAAG-5′

which reads the same on either strand in the 5′ to 3′ direction.

There is also a systematic naming scheme for restriction enzymes. The first letter is capitalized and denotes the genus of the producing bacteria, while the next two (uncapitalized) letters represent the species. If there is a fourth letter, it represents the specific strain of bacteria. Following the main name, there may be a number that appears, written in Roman numerals, that indicates that more than one restriction enzyme has been isolated from that particular strain and the order of discovery. For example, EcoRI (pronounced "*Ee'*-koh är one") comes from *Escherichia coli*, strain R̲Y13, and it is the first of multiple restriction enzymes that have been isolated from this strain. EcoRV ("*Ee'*-koh är five") is the fifth restriction enzyme to be characterized from the same strain. HindIII ("Hin-dee three") comes from *Haemophilus influenzae*, strain "d," and is the third characterized restriction enzyme from this strain.

Refer again to the recognition sequence given above for EcoRI. This enzyme will cut the recognition sequence between the G and the A, as indicated by G▾AATTC. Keep in mind that since this is a palindrome, there will be two G▾AATTC cuts at the recognition site, one for each DNA strand. The two fragments will then be separated to yield two fragments that have single-stranded overhangs:

$$\begin{matrix} \text{GAATTC} \\ \text{CTTAAG} \end{matrix} \rightarrow \begin{matrix} \text{G}|\text{AATTC} \\ \text{CTTAA}|\text{G} \end{matrix} \rightarrow \boxed{\begin{matrix}\text{G} \\ \text{CTTAA}\end{matrix}} + \boxed{\begin{matrix}\text{AATTC}\\ \text{G}\end{matrix}}$$

The fragments above have 5′ overhangs. There do exist restriction enzymes that will cut to yield 3′ overhangs. If the DNA fragments have overhangs, they are said to have *sticky ends*, because these ends are made up of complementary bases that can hydrogen bond. Not all restriction enzymes will yield sticky ends. If a restriction endonuclease cuts in the middle of its recognition sequence, as is the case for EcoRV (with the recognition sequence and cut site GAT▾ATC), then the product is said to have *blunt ends*.

$$\begin{matrix} \text{GATATC} \\ \text{CTATAG} \end{matrix} \rightarrow \begin{matrix} \text{GAT}|\text{ATC} \\ \text{CTA}|\text{TAG} \end{matrix} \rightarrow \boxed{\begin{matrix}\text{GAT} \\ \text{CTA}\end{matrix}} + \boxed{\begin{matrix}\text{ATC}\\ \text{TAG}\end{matrix}}$$

The production of sticky ends is more than a random phenomenon presented merely to trip you up on a test; it is a valuable tool that serves to make the lives of many genetic engineers much easier. It is often the case that one might want to remove a section of DNA from one source and transfer it into a plasmid vector. If the source and the vector are cut by the same set of sticky end-producing restriction enzymes, then the desired fragment can be transferred into the plasmid with relative ease because hydrogen bonding will hold it in place. This application of restriction enzymes is presented in greater detail in Chapter 12.

QUESTIONS

1. Given that a restriction enzyme has a hexameric, palindromic recognition sequence beginning with GCA, write the entire sequence. If the enzyme cuts after the G, will it produce blunt ends, a 5′ overhang, or a 3′ overhang?

2.
   a. Consider the gel below *carefully*. Lane 1 is the result of exposing a set of identical DNA fragments to a 5′ exonuclease for just enough time to

cut each individual fragment one time. Lane 2 is the result of the same treatment, plus the addition of a protein suspected to bind somewhere on the uncut fragment. On the sketch of the uncut DNA fragment beside the gel, indicate with an arrow where the protein might bind.

b. Suppose that the protein binds to three consecutive DNA bases. In lane 3 of the gel, draw the bands that you would expect if you were to run a DNA footprinting procedure on the original set of uncut DNA fragments.

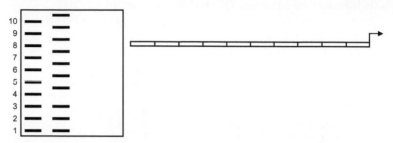

3. Suppose we have isolated a protein that we suspect binds to a specific regulatory region for a certain gene. We perform a DNA footprinting experiment and obtain the following results:

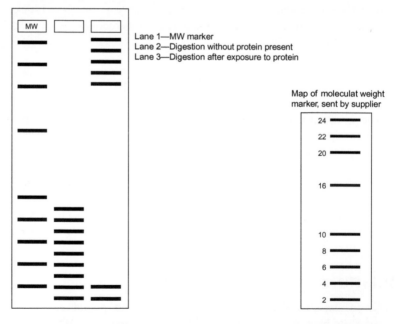

Lane 1—MW marker
Lane 2—Digestion without protein present
Lane 3—Digestion after exposure to protein

Map of moleculat weight marker, sent by supplier

a. On a number line like the one below, draw where you think the protein binds.

919   920   921   922   923   924   925   926   927

b. When the protein binds, it causes an upregulation in the transcription of the gene. Give the one-word term that describes the region of DNA to which the transcriptional factor binds and the one-word term that describes the transcriptional factor.

c. Is this region of DNA upstream or downstream of start?

4. Suppose we have isolated three proteins that we suspect bind to a specific regulatory region at −100 to −120 for a certain gene. We perform a DNA footprinting experiment and obtain the following results:

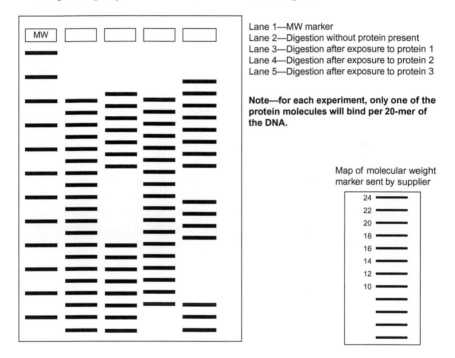

Lane 1—MW marker
Lane 2—Digestion without protein present
Lane 3—Digestion after exposure to protein 1
Lane 4—Digestion after exposure to protein 2
Lane 5—Digestion after exposure to protein 3

Note—for each experiment, only one of the protein molecules will bind per 20-mer of the DNA.

Map of molecular weight marker sent by supplier

a. Draw where you think each protein binds on number lines similar to the one below.

−120 −119 −118 −117 −116 −115 −114 113 −112 −111 −110 −109 −108 −107 −106 −105 −104 −103 −102 −101

b. Name the molecule that permits us to visualize our DNA fragments in this experiment. On a molecular scale, where is it located on the DNA fragments?

5. Agarose gels might also be used as part of the deletion analysis protocol. (Refer to Chapter 9 for more information on deletion analysis. This question is intentionally duplicated from Chapter 9.) Suppose we want to locate upstream control regions in a certain gene. We have used an exonuclease to chew the gene from the 5′ end. Below is the result of several runs, where we have let the exonuclease chew for different periods of time.

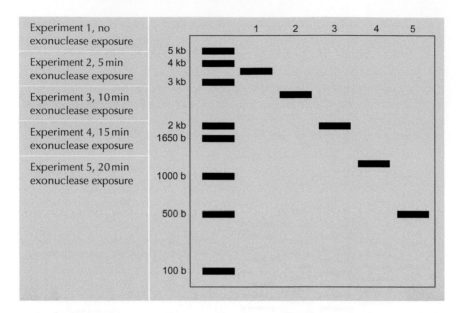

a. This gel has two controls: one for the experiment and one for the gel itself. What are the controls? What does each control show?

b. We then cloned the fragments into a plasmid containing the coding region for a reporter gene and then transfected cells with copies of the plasmid. The following results for reporter expression were obtained:

Relative Fluorescence Units (RFU)	
Plasmid 1	1.000
Plasmid 2	0.993
Plasmid 3	0.658
Plasmid 4	0.648
Plasmid 5	−0.003

(Plasmid *n* was created from fragment *n*, from experiment *n*)

c. Estimate the location of each control region.

d. Suppose that RFU=0 for transfections utilizing the five plasmids above. Give two possible explanations for the result. What control could you have run to prevent the ambiguity?

e. Estimate the nuclease efficiency of the enzyme used, in bases/second.

RELATED READING

New England Biolabs. https://www.neb.com/sitecore/content/nebsg/home/faqs/cutsmart-restriction-endonucleases/restriction-modification-systems (accessed 01/2014).

Chapter 11

The Polymerase Chain Reaction

DNA polymerase, which we have already discussed in terms of cellular repli-cation, can be used in the laboratory to make multiple copies of a DNA frag-ment. While the cell will use DNA polymerase to make a single copy of its genome before division, in the laboratory, we can use the polymerase repeatedly to amplify a region of DNA (called an *amplicon*) at an exponential rate. In the first round of amplification, the amplicon will be replicated once. In the second round, the amplicons of the original DNA and the copies made during the first round will be replicated. In the third round, the amplicon and all of the previ-ously constructed copies (a total of four dsDNA molecules) will be replicated, leaving us with a total of eight dsDNA fragments at the end of the round. These rounds of replication will continue, ultimately yielding (by one theory) 2^n cop-ies of the desired DNA fragment. We will discuss the expected number of copies in more detail later. For now, note that these repeated cycles of replication are known as the polymerase chain reaction (PCR).

The core of the PCR procedure takes place in three steps: melt, anneal, and extend.

11.1   MELT

Suppose that we have some dsDNA that we want to amplify. (We will discuss the reasons why one might want to amplify this DNA a little later.) Knowing that we're going to use Taq DNA polymerase for the amplification and that DNA polymerase requires a single-stranded piece of DNA as a template, we must begin by separating the dsDNA into two ssDNA pieces. This is accom-plished through heat. Raising the temperature will disrupt the hydrogen bonds between paired bases (A-T and G-C base pairs). We choose 95 °C because it is a high temperature that is still below the boiling point of water.

It is not advisable to go above 100° for the melting step because water is a necessary part of B-form DNA structure. (B-form is the structure of DNA in an aqueous environment, as opposed to A-form DNA, which is waterless, as in the case of a DNA crystal. B-form dsDNA will have a double-helix structure that includes a major and minor groove. Water molecules intercalate between two of the strands of the double helix and pull them together to create a minor groove. In pulling these strands together, two other sections of the double helix will

An Introduction to Biotechnology. http://dx.doi.org/10.1016/B978-1-907568-28-2.00011-3

be pulled slightly apart, resulting in a major groove in the DNA double helix. Think of the strands as being the same distance apart, but when water gets in there, because of polarity, hydrogen bonding, and electronegativity, two parts of the strands will be pulled closer. When starting with evenly spaced strands, pulling two of them closer means they will be further away from one another on adjacent turns, hence the rise of the major and minor grooves.) Going above 100° will serve to cook the DNA, much like going above 100° serves to alter the structure of proteins in an egg white when we cook it. At 95°, the dsDNA will be melted into two ssDNA strands without permanently altering the DNA structure.

11.2   ANNEAL

The next step, anneal, refers to the binding of short DNA primers to each of the ssDNA strands. PCR primers are short DNA sequences that are complementary to the 3′ ends of the amplicon. Keep in mind that there are two 3′ ends in the amplicon, which refers to dsDNA. The primers are designed so that they will bind to sites that flank the region that you want to amplify (Figure 11.1).

After the third round of replication, shown in Figure 11.2, five of the six different dsDNA species involved with the PCR process will be present, including the first appearance of dsDNA consisting solely of the amplicon. If our goal is to have many copies of only the amplicon, should we be concerned about the presence of the other five extraneous forms of DNA shown in the figure? The answer is "no," because we will not have *many* of these extra forms. Figure 11.3 accounts for all of the different species of dsDNA that occur during PCR. One of the extraneous forms will disappear, two of them will remain constant at two copies each, while the fourth and fifth extraneous forms will increase by one copy with each cycle. However, copies of dsDNA representing only the amplicon will increase exponentially over time. This means that after 30 cycles of PCR, we will have 60 molecules of dsDNA that contain extra nucleotides but will also have 1,073,741,764 molecules of dsDNA that strictly represent the amplicon.

In the three figures just presented, the amplicon is represented by darker colors. The original dsDNA is in blue, with the desired amplicon in dark blue and flanking regions in light blue. Primers have been designed to bind at the extreme ends of the amplicon. We designate the difference between these primers by calling one the "forward primer" and one the "reverse primer." The forward primer will have the same sequence as the first few (18-24) nucleotides of the sense strand. After the dsDNA has been melted and slightly cooled, the forward primer will bind to the 3′ end of the antisense strand (the left side as shown in Figure 11.3). The reverse primer will have the same sequence as the first few nucleotides of the antisense strand when reading it from the 5′ to 3′ direction (right to left in Figure 11.3). After the dsDNA has been melted slightly and cooled, the reverse primer will bind to the 3′ end of the sense strand (the right

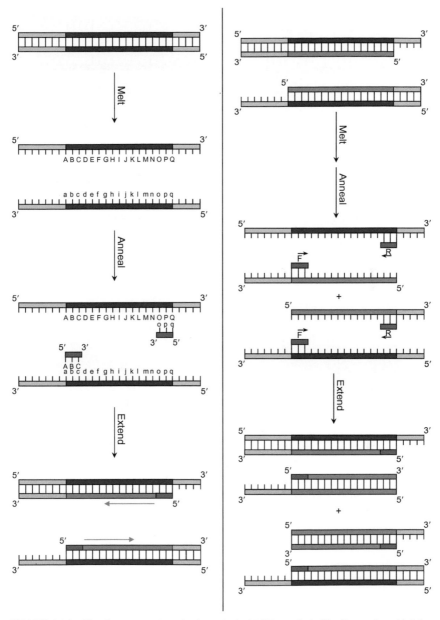

FIGURE 11.1 The first two cycles of a hypothetical PCR set. *Left:* The first cycle, which begins with the original dsDNA shown in blue. Complementary base pairs are reflected by matching upper- and lower-case letters. Note that the forward primer has the same sequence as part of the coding strand, but it will bind to the complementary strand. Primers are shown in red, and newly synthesized DNA is shown in orange. *Right:* The second cycle begins with two pieces of dsDNA, each containing one of the original (blue) strands. Sections that will not be replicated are shown in lighter colors.

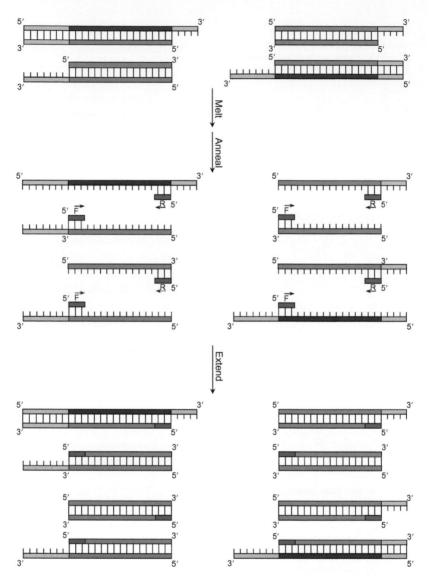

FIGURE 11.2 The third cycle of a hypothetical PCR set (continued from previous figure). Note that, although there are several species of dsDNA after this round of replication, dsDNA consisting of only the amplicon first appears after this round. This species will be amplified exponentially in subsequent rounds.

side as shown in Figure 11.3). It is very important to keep track of the strand and direction when referring to DNA sequences. For instance, if you were to order primers from a company, the company will expect primer sequences defined in the 5′ to 3′ direction. That is a straightforward and easy to remember standard, but when designing your own primers, always remember to which strand and

Species   **Number of copies**

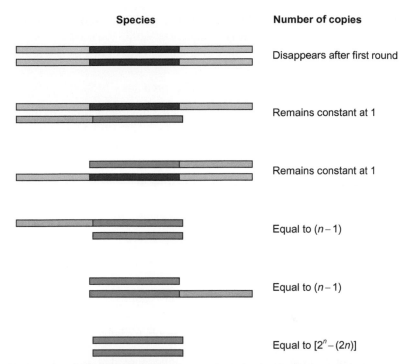

Disappears after first round

Remains constant at 1

Remains constant at 1

Equal to $(n-1)$

Equal to $(n-1)$

Equal to $[2^n - (2n)]$

FIGURE 11.3   Different dsDNA species to consider during the PCR process. The number of copies of each species after n rounds of PCR is shown to the right of each species ($n > 0$).

which direction you were referring (top versus bottom, sense versus antisense, and left to right versus right to left).

While 95° is used to melt the dsDNA, a temperature that high will also prevent the binding of primers to ssDNA. The temperature must be lowered. *Annealing* is the term given to the process whereby the primers attach to the ssDNA. The annealing temperature depends on the specific DNA sequence of both the forward and reverse primers. Consider two 20-nucleotide primers, one made entirely of As and Ts and the other comprised solely of Gs and Cs. Recalling that G-C base pairs involve three hydrogen bonds while A-T base pairs only involve two, it should be apparent that more energy would be required to separate the GC primer after it has bound to ssDNA. In other words, if we were to slowly heat up a solution of the two primers that are bound to their targets, the GC primer would separate last. Looking from the other direction, if we had a hot (95°) solution of ssDNA plus the two primers, which were to be cooled gradually, the GC-rich primers would bind to their targets first. This principle explains why the annealing temperature for a given primer depends upon its specific base content. In designing a primer set (which contains both forward and reverse primers), it is desirable to keep the annealing temperatures of the two primers as close to equivalent as possible. A range of 58°-62° is desirable. As the solution is cooled from the 95° melting step, eventually, enough energy will have been removed from the system to finally allow the binding of the primers with their complementary sequences within the ssDNA.

The optimal temperature for annealing can be determined by the following formula:

$$T_a = 0.3T_m\left(\text{primer}\right) + 0.7T_m\left(\text{product}\right) - 14.9$$

where T_a = annealing temperature and T_m = melting temperature of the primers or product, defined as the temperature at which 50% of the oligonucleotides and the associated complimentary strand are in duplex. T_m can be calculated by the following:

i. For short sequences (≤20 bases), such as primers,
$T_m = 2(A+T) + 4(G+C)$, where A, T, G, and C are the number of adenine, thymine, guanine, and cytosine bases in the sequence, respectively.

ii. For longer sequences, such as amplicons,

$$T_m = 81.5 + 16.6\left(\log M\right) + 0.41\left(\%G + \%C\right) - 0.62\left(\%\text{formamide}\right) - \left(\frac{500}{n}\right),$$

where M is the concentration of monovalent cations (such as Na^+ and K^+); %G and %C are the mole fractions of guanine and cytosine, respectively; %formamide is the percentage of formamide in the solution; and n is the number of nucleotides in the sequence.

It is desirable to have an annealing temperature that is relatively high, but it must be cooler than the temperature that will be used for the extension step, which will be discussed in the next section. The reason why a high annealing temperature is desired has to do with preventing *nonspecific binding*. If we have designed a 20-base primer, we expect it to be complementary to its target sequence in all 20 bases. However, if there were a sequence of DNA that was complementary to 17 of the bases in our primer, there is a possibility that our primer could bind to this region of DNA even though it's not specifically complementary to every single base in the primer. Just like the analogy with a primer made solely of As and Ts versus a primer made solely of Gs and Cs, it would take more heat energy to remove the primer that was bound to 20 bases as opposed to the same primer bound to only 17 bases. Likewise, in cooling the solution containing primers and ssDNA, the primers would preferentially bind to the DNA region for which 20 nucleotides could interact with the primer as opposed to a less specific region. This means that at the restrictively high temperature (T_m) of annealing, nonspecific binding will be prevented.

11.3   EXTEND

The extension step of traditional PCR is typically carried out using a DNA polymerase that was isolated from *Thermus aquaticus* (Taq), a heat-loving (thermophilic) bacterium that was first discovered in a hot spring associated with a geyser in Yellowstone National Park. This enzyme, aptly named Taq DNA polymerase, performs optimally at 73 °C and can polymerize at a rate of approximately 1000 nucleotides per 60 s. Supposing our amplicon is 300 base pairs, it will take approximately $(300/1000)(60\,s) = 18\,s$ for Taq DNA polymerase to extend the ds-DNA from our primer over the entire length of the single-stranded amplicon. It is okay to increase the time allotted for the extension step to make sure that the enzyme will have enough time to completely copy the amplicon; extraneous bases will not be a factor in the final solution, as was shown in Figure 11.3.

At this point, we have gone from one dsDNA sequence to two dsDNA sequences. We can repeat the entire process of melt, anneal, and extend to yet again double the amount of dsDNA in our solution. We can continue to repeat this cycle for a theoretical maximum of 28 times. (This theoretical maximum is limited by the concentration of dNTP that can be present in the starting solution.) In reality, the maximum number of repeated cycles will be higher because the efficiency of every ssDNA binding with a primer is <100%.

Assuming 100% primer efficiency, after one repeat of the cycle, we will have two copies of dsDNA that contain the amplicon. After two cycles, we will have four copies (2^2), and after n cycles, we could hope to have 2^n copies.

11.4 PCR LOOPS

The PCR reactions take place in a very small (200 µl), thin-walled tube. The total volume of the reaction can be as little as 25 µl. To obtain the changes in temperature, the tube will be placed into a thermocycler, which is basically a metal block that is heated and cooled precisely and relatively quickly. The thermocycler is automated, allowing for the programming of temperatures and times. To get the most efficient transfer of heat into the PCR solution, it is recommended that the walls of the tube be very thin.

The core purpose of the thermocycler is to provide the correct temperatures for each of the three steps of PCR in a repeated fashion. It will heat the tubes to 95° (melt), cool them off to ~58°-62° (anneal), and heat them up to 73° (extend), then to 95°, then to 62°, then to 73°, etc. A typical run might look something like the following:

	95°	10 min
Repeat 34 times	95°	15 s (melt)
	59.2°	30 s (anneal)
	73°	20 s (extend)
	73°	5 min
	4°	Indefinitely

The initial heating to 95° is held for several minutes to allow the sample to equilibrate. The repeated core process then follows. Recall that the annealing temperature was determined during the primer design process. The time and temperature for the extension step are determined by the specific DNA polymerase and amplicon size that are being used. The supplier for the DNA polymerase will provide the optimal temperature for the enzyme, and the time for extension is based on the size of the amplicon divided by rate of polymerization for the enzyme. The final 4° step is simply to hold the sample until the operator returns to collect it.

When we started out, the sample tube contained some dsDNA, a DNA polymerase, deoxyribonucleic acid triphosphates (dNTPs), a buffer, water, and primers. Pay attention to the dNTPs. During the first time the core steps of the reaction are performed, they should work very well. The same goes for the second time. However, by the 28th time, we will have used a large number of dNTPs. To run 71 cycles of PCR to obtain 2^{71} copies of a 300 bp amplicon would require that we start with over a mole of dNTP. This is unrealistic. Even running 56 cycles would require that we use over 1 M dNTP in the reaction tube. The point is we only have so many dNTP molecules in the tube, so we cannot amplify for an infinite number of cycles. The reaction will eventually be limited by the number of dNTPs that are available. During the later repeats in the PCR loop, the DNA polymerase will operate more slowly because time will be spent waiting for the polymerase to bind with the correct dNTP.

If we were to graph the dsDNA concentration versus *cycle number* (how many times the core reactions have been repeated), the data would take on a sigmoid-like shape as shown in Figure 11.4. During the early cycles, the number of dsDNA strands would increase exponentially. However, as the concentration of dNTPs decreases, the rate of increase in the number of dsDNA strands will also begin to decrease. The rate of dsDNA increase would decelerate, eventually coming to a plateau, meaning a constant amount of dsDNA would be present even if further cycles were performed. The exponentially increasing portion of

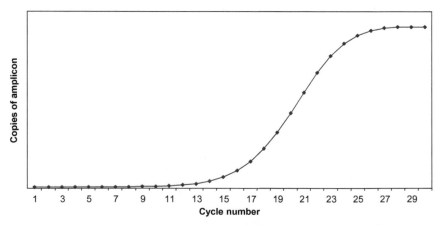

FIGURE 11.4   Although the number of amplicons increases exponentially during the early cycles of PCR, the overall shape of the curve of amplicon copies versus cycle number is sigmoid.

the curve represents where the number of new double-stranded DNA molecules per PCR cycle is approximately doubled during each cycle.

11.5 AN APPLICATION OF TRADITIONAL PCR

Consider the following experiment: we hypothesize that increased glucose levels will cause an increase in the transcription of the *endothelin 1* gene in microvascular endothelial cells. If the hypothesis were correct, then *endothelin 1* would be implicated in diabetic vascular disease. To test our hypothesis, an experiment will be run whereby two cell cultures of microvascular endothelial cells will be maintained: one in normal culture conditions and one in identical conditions except for an elevated glucose concentration in the medium. We have designed a primer set for *endothelin 1* mRNA and will use PCR to determine the relative amount of expression between the control cells and the cells cultured with elevated glucose levels.

Recall that PCR requires dsDNA, but our hypothesis is concerned with the amount of *endothelin 1* transcription, and the product of transcription is mRNA. The approach that is commonly taken is to isolate all of the mRNA from the cells in each culture and to generate cDNA from the mRNA molecules using reverse transcriptase. This process, combined with PCR, is referred to as RT-PCR (reverse transcription polymerase chain reaction).

Cells will be grown in cultures under the conditions just mentioned. The cells will be lysed and the mRNA collected, followed by reverse transcription to yield a cDNA library. The complete set of genes that is being transcribed in a cell at a particular moment is referred to as the *transcriptome*. Ideally, the cDNA library will be a representation of all of the mRNA molecules in the cells from a given culture at the point of cell lysis, and the cDNA molecules corresponding to individual genes will be at the same proportions as the corresponding mRNAs in the transcriptome.

Even though there will be many cDNAs in the library, we will only be interested in amplifying the cDNA that corresponds to one specific mRNA transcript (*endothelin 1* in this example). We can amplify the cDNA from only *endothelin 1* cDNA by using primers, one forward and one reverse, that are specific for sequences that are only found in *endothelin 1* cDNA. (Note that a cDNA sequence is not necessarily the same as the corresponding genomic DNA sequence, which may include introns.) The primers must be long enough and specific enough so that they will not bind to any of the other cDNAs in the transcriptome library. Verification of specificity is accomplished through database searches of known gene sequences. A popular tool for performing the search is BLAST, which stands for Basic Local Alignment Search Tool. This tool is provided by the National Center for Biotechnology Information in the United States and can be found online at www.ncbi.nlm.nih.gov/BLAST. The need for specificity should explain why we use primers that are 18-24 nucleotides long as opposed to being only 4-6 nucleotides in length.

Again, our tube of cDNAs will contain many sequences for which we have no interest. However, after PCR, we will have amplified our sequence of interest so that it will comprise the vast majority of all dsDNA molecules (Figure 11.5). We can take a small sample from the tube following PCR and load it into a gel. Following electrophoresis, we should be able to visualize a band that represents the amplification product of PCR using primers directed at a specific cDNA sequence—a sequence that corresponds to a specific mRNA sequence, which represents the amount of transcription of a given gene at a given time under given circumstances.

Back to our example, we had two cell cultures, one normal and one incubated with increased glucose. Our hypothesis was that the cells exposed to an increased glucose concentration would increase their transcription of the *endothelin 1* gene. While PCR will be performed on the transcriptomes of the two samples, if our hypothesis is correct, then there will be more copies of the *endothelin 1* cDNA in the tube that corresponds to the cells that were exposed to the increased glucose concentration. Suppose there are seven times more cDNA molecules corresponding to *endothelin 1* in the treated sample. If we started with 7 versus 1 *endothelin 1* cDNA molecules in the treated versus untreated samples, after one doubling, we would have 14 versus 2 molecules, after two doublings, we would have 28 versus 4 molecules, then 56 versus 8, 112 versus 16, etc. Even though the number of copies of the *endothelin 1* cDNA in each sample is doubled each time, there will always be a sevenfold difference in the number of *endothelin 1* amplicons between the two tubes.

We extracted a set of mRNA for each culture and used them to produce two cDNA libraries. We performed PCR on supposedly equal portions of each of the two libraries, using the exact same conditions and number of repeated cycles because the two samples were loaded together into the thermocycler block. Equal portions of the resulting solutions were loaded into a gel and electrophoresed, and the resulting bands were then visualized under UV light. The result might look like the following:

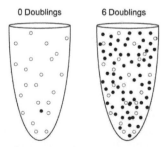

FIGURE 11.5   Even though the cDNA of interest is only a minor constituent of an initial cDNA library, after PCR amplification, it will comprise the majority of the dsDNA in the tube. Notice the prevalence of the black dots after only six doublings.

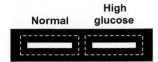

These bands (in white) appear to be of equal thickness and brightness, and there do exist computer programs that can return a quantitative number to describe the overall brightness in a defined region. These programs operate off of the intensity, using a gray scale value between 0 and 255, for every pixel in the defined region (shown by dotted lines). Taking a ratio of the two returned values will give an indication of which sample had more *endothelin 1* mRNA in its transcriptome and by how much (fold intensity).

One must always be able to account for mistakes. Perhaps, when loading the gel, slightly unequal amounts of each sample were placed into the respective wells, which is entirely possible considering that we are dealing with microliter amounts of solution. If we loaded 10 µl but each measurement was off by just 1 µl in each direction, then the difference in band intensities could be off by 22% (1-11/9). Similarly, if there was a pipetting error made during the collection of RNA, during the generation of cDNA, or while loading the tubes for PCR, significant errors could result in our final band intensities. Even errors in the number of cells loaded onto each plate at the very beginning of the experiment will make a difference in the final PCR result. What is needed is an *internal control* to allow the researcher to tell how much sample was initially loaded. The ideal internal control would capitalize on a gene for which expression is constant regardless of whether the cells were treated or untreated in our experiment. While there is no perfect internal control, there are several candidates that are used in the laboratory. The best candidates are housekeeping genes because they are needed by the cell at all times for survival. For instance, in cells that perform glycolysis as their main producer of energy, the genes that code for enzymes to take part in this vital cascade must always be expressed. One such gene codes for glyceraldehyde 3-phosphate dehydrogenase (*GAPdH*). Other common genes used as internal controls are *18S ribosomal RNA*, used because this molecule is a part of the functioning eukaryotic ribosome and cells translate the transcriptome at a roughly constant rate, and β-*actin*, which is always being expressed because it is a part of the cytoskeleton. (Do not confuse this with α-actin, which is part of the muscle contraction machinery.)

(If you get way down into the rigorous details of this science, there's no perfect internal control. For every system that one studies, several internal controls must be evaluated to find the one that remains most constantly expressed during the experimental treatment regimen.)

Going back to our experiment, after we isolate the RNA from our two cell cultures and produce two cDNA libraries, we can perform PCR as before with the exception that we will be amplifying two genes in each sample: *endothelin 1* and, say, *18 S rRNA*. The resulting gel might now look something like

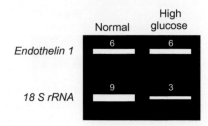

Numbers have been placed over the bands to indicate their relative strengths. Notice that the bands for the 18 S rRNA are unequal. Based upon our assumption that the amount of 18 S rRNA expression should remain constant regardless of experimental treatment, the above data indicate a pipetting or loading error. This error is easily corrected by normalizing the strength of the bands for the gene of interest to the strength of the bands for the housekeeping gene. In other words, since it appears from the 18 S rRNA bands that three times more of the normal sample was loaded into the gel, the strength of the *endothelin 1* band in the glucose-treated sample should be normalized up by a factor of three. Whereas our previous experiment that lacked an internal control seemed to indicate that there was no effect of glucose concentration upon *endothelin 1* transcription, this properly controlled experiment indicates that an increasing glucose concentration in the culture medium will result in greater *endothelin 1* transcription. The amount of upregulation in this example appears to be (6/3)/(6/9) = 3-fold. A general formula for this relation is

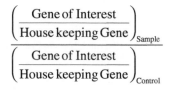

Keep in mind the PCR method described thus far is not recommended to quantitate gene expression. The above formula is only given to demonstrate the concept behind normalizing to an internal control. For traditional PCR, it is safer to make any resulting claims to the effect of "an upregulation or downregulation was observed," rather than "a threefold upregulation was observed." One must normalize to the amount of DNA that was initially loaded into the gel. Even though we may *intend* to load the same amount for every sample, volumes of 1 μl lend themselves very easily to measurement errors. Even temperature and pressure differences between tubes and ambient conditions can affect how much fluid is pulled up with a pipettor, as will pipettor error (for which 4% or less is considered acceptable) and operator error. Even in perfect conditions, the number of cells in each original sample will affect the ultimate results. The use of an internal reference will help to take into account unseen factors such as these.

11.6   TRADITIONAL VERSUS REAL-TIME PCR

Soon, we will discuss real-time PCR, also known as quantitative PCR (qPCR). An advantage of traditional PCR over qPCR is that it is very straightforward, it is technically simple to perform, and one does not need an expensive piece of equipment other than a thermocycler, which can cost under $1000. Real-time PCR involves taking a measurement of fluorescence once per cycle, which means expensive fluorescent molecules and an even more expensive fluorescence detector must be used. In addition, the detector must be linked to a computer to store and process the fluorescence data. Real-time PCR is a lot more accurate, as we shall soon learn, but it's a lot more expensive, costing several tens of thousands of dollars for the equipment alone.

One source of error produced by traditional PCR that is not present in real-time PCR stems from the use of postproduction data. In traditional PCR, the amplifications are carried out to a predetermined number of cycles before the DNA is examined, and further manipulation is required beyond the PCR process (i.e., a gel must be run). On the other hand, real-time PCR takes a measurement during each cycle. Another problem with traditional PCR is illustrated by the amplification curves shown in Figure 11.6. There is a maximum amount of amplification that can occur before differences between samples are obscured (a plateau phase is reached). If the optimum number of cycles for visualizing the difference between two samples is, say, 24 cycles, but we performed a traditional PCR reaction for 40 cycles, the amplification curves for both of the samples would be in the plateau phase, meaning a maximum discernible brightness

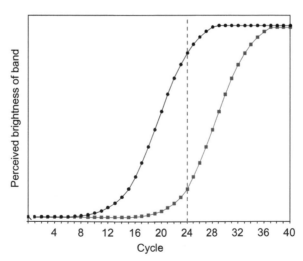

FIGURE 11.6   In traditional PCR, as the number of PCR cycles increases, so does the perceived brightness of the resulting bands, up to a point. Since results are only observed after the total number of PCR cycles have been run, erroneous results can occur if too many amplification cycles have been employed. Compare the differences between the two curves at 24 versus 40 cycles.

had been produced from the initial cDNA samples, which would result in a gel that revealed bands of equal strength. The maximum might be due strictly to perception of brightness, or it might be due to depletion of dNTPs as more and more copies of the amplicon are produced. Either way, running the PCR series for too many cycles will lead to saturation of all samples. This might, in turn, lead the investigator to conclude that there is no difference in the expression of a given gene between the samples being analyzed.

11.6.1   Problems Specific to Traditional PCR

Other problems associated with traditional PCR include odd or unexpected DNA patterns in the gel. Figure 11.7 demonstrates common gel results that indicate the thermocycler program must be refined. The second lane of the gel shown in the figure depicts a large smear instead of a well-defined band. If the annealing temperature is too low, then we have a less restrictive environment for primers to bind to the DNA. In such a case, not only will the desired amplicon be produced, but also DNA fragments can be amplified from multiple starting positions in the DNA sequences. This can result in undesired amplicons that are both too large and too small relative to what the experiment was designed to amplify. The remedy to this situation is to raise the annealing temperature, which will restrict primer binding to sites in the DNA that are completely complementary to the primer sequence (no mismatches).

The third lane of the gel shows another smear, but it reflects DNA sequences that are less than or equal to the desired amplicon size. This might indicate that the extension time is too short. Even though calculations might indicate that

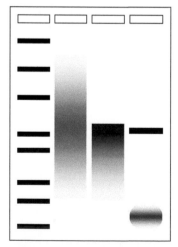

FIGURE 11.7   Common problems with traditional PCR are shown in the above gel. Lane 1: MW marker. Lane 2: annealing temperature too low. Lane 3: extension time too short. Lane 4: primer dimers.

18 s is all that is needed for amplification of the desired amplicon, conditions such as suboptimal salt concentrations, partial enzyme damage, improperly controlled temperature, and other environmental factors may reduce the rate of DNA polymerization. This situation can often be corrected by increasing the extension time. As shown in our discussion of PCR theory, having extension times that are longer than needed should not produce an abundance of amplicons that are larger than the DNA segment that is bounded by the forward and reverse primers.

The fourth lane of the gel shows that, although we get the desired amplicon, a second and much smaller band has appeared. This reflects a common problem known as *primer dimers*, where parts of the forward and reverse primer sequences are complementary (especially at their 3′ ends), providing an additional double-stranded DNA region to which DNA polymerase can bind. As shown in Figure 11.8, the amplicon produced from polymerization off of a primer dimer can be replicated in future PCR cycles. While not a significant problem in traditional PCR, other than the fact that dNTP resources will be used to produce amplicons that are meaningless, primer dimers can be a significant issue in real-time PCR (explained below).

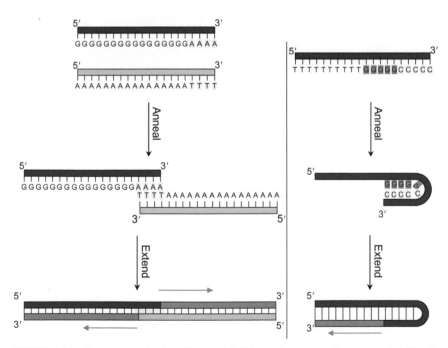

FIGURE 11.8 Two sources of primer dimers. *Left*: Primers that are complimentary at their 3′ ends can pair to allow polymerization in the 5′ to 3′ direction. *Right*: Individual primers with 3′ ends complimentary to an internal sequence can form hairpins that serve as a double-stranded primer for DNA polymerase. Polymerization is shown in orange.

In traditional PCR, even after multiple experiments have been formed to determine the optimal conditions for the reaction—the optimal number of cycles, the optimal annealing temperature, the optimal extension time, etc.—the best result that we can hope to expect will be a gel that reflects PCR results that have undergone further processing even after the PCR cycles have been concluded. The operator will have to pipette a specific volume of each PCR product into the wells of a gel. Again, there is possible pipettor error, plus loss of differing amounts of sample from each well by diffusion or by the turbulence created when the pipette tip is removed after placing the sample into the gel (which is submerged in solution). Every step of processing is associated with some degree of error, so the less manipulation a sample must undergo following a reaction, the closer the resulting data will be to reflecting the true state of the reaction.

11.7   REAL-TIME PCR

Real-time PCR (qPCR) is a method that removes the problem of postprocessing, plus it yields quantitative data. This method utilizes the same thermal cycling steps as those discussed for traditional PCR, but the amount of dsDNA in the sample is inferred through fluorescence measurements taken directly of each sample during each cycle. There are two strategies for determining the amount of dsDNA with qPCR. The first uses a compound known as SYBR green, and the second utilizes double-tagged probes.

11.7.1   SYBR Green

If one were to take a known quantity of SYBR green, load it into a fluorescent plate reader, and expose it to blue light, it will fluoresce a small amount. This is known as *background fluorescence*. However, when SYBR green is intercalated with dsDNA, the dye will fluoresce a great deal when exposed to blue light. If adding one unit of dsDNA to a sample of SYBR green yields one unit of fluorescence (after subtracting background fluorescence and assuming an excess amount of SYBR green), adding two units of DNA will yield twice as much fluorescence. Think of it as you've got these molecules of SYBR green floating around in a test tube or sample well, and we will observe little fluorescence in the absence of dsDNA. We try to rig the experiments so that there will always be more SYBR green than dsDNA, so that as DNA becomes more and more plentiful, there will be more and more sites for the SYBR green to bind to become fluorescent. The fluorescence will increase proportionally with the amount of dsDNA in the sample. Suppose that an experiment commenced with one piece of dsDNA, and a certain amount of SYBR green fluorescence associated with it (Figure 11.9). When we perform the melt step to obtain two ssDNA molecules, there will only be a background level of fluorescence in the sample. After the anneal and extension steps have been completed, there will be twice as much dsDNA as there was at the beginning of the cycle, which will provide

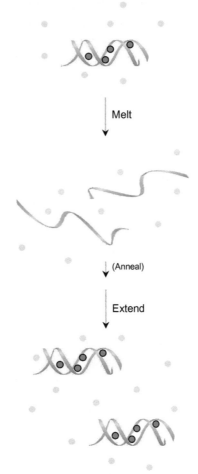

Melt

(Anneal)

Extend

FIGURE 11.9 SYBR green fluoresces brightly when intercalated in the minor groove of dsDNA. When there is no dsDNA (i.e., after the PCR Melt step), there is only a background amount of fluorescence. After the extension step, twice as much dsDNA will yield twice as much fluorescence.

twice as many binding sites for SYBR green molecules, thereby yielding twice as much fluorescence for detection. Over time, we will be able to track the amount of fluorescence in the sample. Whereas earlier, a graph was presented in Figure 11.4 to reflect the theoretical amount of dsDNA in a sample as PCR cycles were performed, now, a graph is presented in Figure 11.10 to reflect the total amount of fluorescence in our sample—fluorescence that is directly related to the amount of dsDNA in the sample.

Returning to our thought experiment, we now have two units of fluorescence in our sample. In comparison with the vast number of unused SYBR green molecules in solution, there will not be much difference between one and two

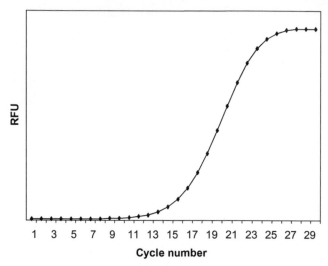

FIGURE 11.10   The general shape of the fluorescence curve as the number of cycles is increased in a real-time PCR experiment. RFU, "relative fluorescence units."

fluorescent units after subtracting background fluorescence. The same is true for the next several cycles, where we will have 4, 8, 16,... units of fluorescence in the sample. Eventually, however, there will be enough fluorescence for the apparatus to reliably detect its presence. The minimum amount of fluorescence that can be reliably detected by the qPCR apparatus is known as its threshold value. The cycle number at which this threshold value is crossed is known as the *cycle threshold* (C_t).

Suppose that the threshold limit for the machine in our thought experiment is 256 fluorescent units. In the sample we have been discussing, it would take eight PCR cycles (2^8) to yield enough fluorescence to meet or exceed the threshold value of 256 fluorescent units. Now, consider a second sample of DNA that contains 32 dsDNA molecules at the start of the experiment. It would only take three PCR cycles to generate enough SYBR green fluorescence for reliable detection ($32 \times 2^3 = 256$). The C_t values for samples 1 and 2 are 8 and 3, respectively: $C_{t_1} = 8$; $C_{t_2} = 3$.

Now, suppose that we run qPCR on the same two samples and observe the same C_t values as before ($C_{t_1} = 8$; $C_{t_2} = 3$), but this time, the question is "by how much does the starting amount of dsDNA differ between the two samples?" The answer can be obtained mathematically from the C_t values: $2^8/2^3 = 256/8 = 32$. In other words, there was a 32-fold difference in the amount of dsDNA in sample two versus sample one. This is consistent with the data presented in the previous paragraph.

The way to use C_t values to determine the fold difference between the amounts of starting dsDNA between two samples can be generalized as follows:

$$\text{Fold difference} = \frac{2^{C_{t_2}}}{2^{C_{t_1}}} = 2^{C_{t_2} - C_{t_1}}$$

C_t values are returned directly as the output of the qPCR instrument, but these values are determined from data such as those given in Figure 11.10. The fluorescence curves for all samples will first be generated, and afterward, the computer will determine where reliable fluorescence increases are detected—the threshold value. (Technically, this value is the number of relative fluorescence units (RFUs) after the exponential phases of the amplification curve have all progressed to 10% of what they will in the y-direction.) The computer notes the cycle number at which each fluorescence curve crosses the threshold value and reports it as C_t. (Note that fluorescence readings are taken in a discrete way, once per cycle. It's not always the case that the threshold value will be exactly reached at the end of a given cycle, it may be exceeded during that cycle. Noninteger C_t values are therefore possible and are arrived at by interpolation.)

Consider another example: We hypothesize that drug A acts by causing an upregulation in the transcription of a certain gene. To test this, we grow two cultures of cells, leave one culture untreated while treating the other with drug A, isolate RNA from each of the two cell cultures, produce two cDNA libraries, and perform qPCR on the two libraries with primers specific for our gene of interest. Suppose the C_t value for the untreated sample is 18, while the C_t value for the sample treated with drug A is 11.

Question: Which sample transcribed more of the gene of interest, and by how much?

Answer: By inspection, it should be evident that treatment with the drug causes an *up*regulation in the transcription of the gene of interest. A *lower* C_t value implies a *higher* amount of starting material, because it takes fewer doublings to reach the detection threshold of the machine. This is a point of confusion for many students. As long as you keep in mind that C_t values report how many doublings were required for detection, the concept of lower C_t indicating higher starting concentrations should present no problem. Using the formula presented earlier, we can roughly ascertain that drug A caused a $2^{(18-12)} = 2^6 = 64$-fold increase in the mRNA levels associated with the gene of interest.

A few words about semantics: Just because the treated cells contained 64 times more mRNA associated with the gene of interest, that doesn't mean that these cells produced 64 times more of the corresponding protein. The polypeptide that is associated with this gene may require other polypeptide subunits, from other genes, before a functional protein is produced. It is also possible that the mRNA that is produced in the presence of the drug will have an altered half-life, meaning it is degraded at a different rate in the presence of the drug. The particular polypeptide may have more than one function, appearing as a subunit in more than one protein or perhaps serving to regulate gene expression in another capacity. It is important to be aware that PCR measures the amount of RNA associated with a particular gene; it doesn't measure the amount of gene expression. Gene expression is the amount of protein that is produced from a given gene (by way of the central dogma). While transcription is a part of this process, simply transcribing a gene is not expressing the gene. PCR only measures RNA levels and therefore can only be used to make inferences regarding gene transcription, not gene (protein) expression.

Returning to the thought example: As with every good experiment, one must not forget the controls. We will not cover every possible control here, such as running the reaction without primers. However, one necessary control is the same as it was for traditional PCR: the internal control. Although we initially ascertained that the transcription rate for the gene of interest is increased by 64-fold in the presence of drug A, we cannot rely on this value because we have not accounted for loading errors. The internal control, which involves a gene for which transcription rates are relatively constant under the experimental conditions, will take care of such errors. The C_t values for the gene of interest must be normalized to the C_t values of the internal control. The example in Figure 11.11 is given to demonstrate that we must take into account the internal controls, which are represented by dotted lines in the figure. We will basically normalize each reading to its internal control. In other words, we are interested in the distance between the two curves—gene of interest and internal control—and will

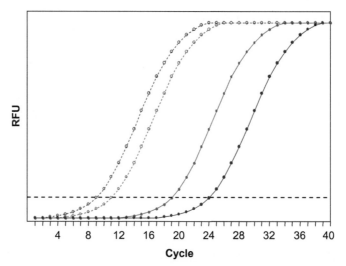

FIGURE 11.11   For qPCR, as with traditional PCR, an internal control must be run to account for loading or sample concentration errors. The above graph reflects readings obtained for a sample of cells treated with a drug (red lines) and an untreated control sample (blue). The gene of interest is shown by the solid curves, and the internal control is represented by the dashed curves. By inspection, one can determine that the distance between the red curves is less than the distance between the blue curves. This implies that the treatment of the cells with the drug causes an increase in the amount of transcription of the gene of interest. Using the formula presented in the text, we can determine that the amount of upregulation is equal to 2^7, or 128-fold.

subtract the distance obtained from a treated sample from the distance obtained for a negative (untreated) control. Simply put, this is a double normalization.

The formula for determining the effect of the treatment of one transcription of the gene of interest now becomes

$$\text{Fold difference} = \left[\dfrac{\dfrac{2^{C_{t\,\text{Gene of Interest, Negative Control}}}}{2^{C_{t\,\text{Reference Gene, Negative Control}}}}}{\dfrac{2^{C_{t\,\text{Gene of Interest Sample}}}}{2^{C_{t\,\text{Reference Gene, Sample}}}}} \right] = \dfrac{2^{\left(C_{t\,\text{Gene of Interest, Negative Control}} - C_{t\,\text{Reference Gene, Negative Control}}\right)}}{2^{\left(C_{t\,\text{Gene of Interest Sample}} - C_{t\,\text{Reference Gene, Sample}}\right)}}$$

$$= \left[2^{\left(C_{t\,\text{Gene of Interest, Negative Control}} - C_{t\,\text{Gene of Interest Sample}}\right)} \right]\left[2^{C_{t\,\text{Reference Gene, Sample}} - C_{t\,\text{Reference Gene, Negative Control}}} \right]$$

$$= 2^{\left(C_{t\,\text{Gene of Interest, Negative Control}} - C_{t\,\text{Gene of Interest, Sample}}\right) - \left(C_{t\,\text{Reference Gene, Negative Control}} - C_{t\,\text{Reference Gene, Sample}}\right)}$$

This last version of the formula is often called $2^{-\Delta\Delta}$, where each delta represents the change in C_t value between the untreated control and the treated sample for each respective gene.

The number for fold difference ($2^{-\Delta\Delta}$) will always be greater than zero. If it's greater than one, then there has been an upregulation in transcription, and if it is less than one, we have downregulation.

If we had no loading error at all, the C_t values for the internal control gene (let us suppose it was GAPdH) would be the same for both the untreated sample and the sample treated with drug A. Suppose this C_t value was 9. Plugging in the numbers from our example into the above formula gives us

$$\text{Fold difference} = 2^{\left[(18-12)-(9-9)\right]} = 2^{(6-0)} = 2^6 = 64,$$

just like before.

Just to check the concept, consider the same example, with the same C_t values, with the exception that for some reason, only half as much cDNA from the treated sample was loaded into the PCR tube. Because half as much was used, we would expect the signal for the reference gene to require one additional doubling to reach the detection threshold, meaning the C_t value would be one higher than that observed for the internal reference gene in the untreated control sample. The C_t values returned from the experiment would be

	$C_{t\,\text{Negative Control}}$ (untreated)	$C_{t\,\text{Sample}}$ (received drug)
Gene of interest	18	12
GAPdH	9	10

$$\text{Fold difference} = 2^{\left[(18-12)-(9-10)\right]} = 2^{(6+1)} = 2^7 = 128.$$

Make sure that this example makes sense to you. If we only loaded half as much treated sample as we should have, whatever fold difference we would have seen without correcting for loading error would be off by a factor of two, meaning the values for fold difference we were getting before would only be half of the actual difference. Likewise, notice that the value of 64 that we obtained without utilizing the internal reference was one-half of the actual value: 128.

11.7.1.1 The Fold Difference: What it Means Versus What it Implies

The results of PCR, whether traditional or real-time, infer how much *cDNA* there was that corresponded to a given RNA sequence. It is often assumed that the cDNA sequences being amplified correspond to *m*RNA and are unique to a specific gene (both of which are not always true). In the end, if we got to a detectable level of fluorescence in sample 1 sooner than we did for sample 2, it generally means we started with more cDNA for the gene of interest in sample 1 than we did in sample 2. This implies we had more copies of RNA (assumed to be mRNA) produced off of the gene of interest in sample 1 before we ever performed reverse transcription to generate the cDNA. This implies that more transcription for the gene of interest took place in the cells used to generate sample 1 (or, far less likely, that for some reason, that mRNA was being degraded more slowly in sample 1). There are cases where some mRNAs are more amenable to reverse transcription than others, but that phenomenon is beyond the scope of this book and will not be covered further.

Using the previous formula, we may make a conclusion such as, "there is an *x*-fold up- or downregulation of the transcription of our gene." The logic tree, progressing backward from output C_t values, is as follows: the C_t values indicate relative levels of cDNA in the material after several PCR runs, which indicate the relative levels of cDNA in the starting materials before the first PCR cycle had been run, which imply the relative amounts of specific mRNAs before reverse transcription was performed, which imply the relative amounts of specific mRNAs in the original cell extracts, which imply the relative amounts of transcription of the gene of interest in our treated or untreated cells. This tree is a chain where each statement depends on the validity of the previous one.

If the experimental conditions and data from these experiments were absolutely perfect, then the choice of threshold value would be arbitrary. In Figure 11.12, notice that the distance between C_t values is the same between the two curves no matter which of the two threshold values is chosen, meaning the distance between the two curves is the same for a given *y*-value (assuming the values are both taken before the plateau phase). Unfortunately, experiments are never perfect, and for that reason, we want to choose the lowest threshold value possible. Suppose we had a single sample of cDNA, and we took three aliquots from that one sample for a PCR experiment. The three amplification curves generated would not be identical. Figure 11.13 is an illustration of the three different curves, although in practice, they would not differ quite as much as what

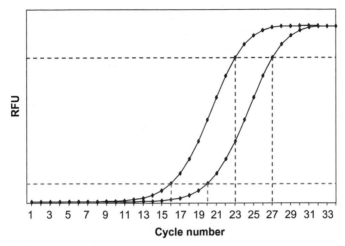

FIGURE 11.12 In analyzing qPCR data, the threshold value chosen is, in theory, arbitrary. In this example, notice that it takes four additional cycles for the PCR curve on the right to cross the threshold value (horizontal, dashed line) whether the blue or red threshold is used.

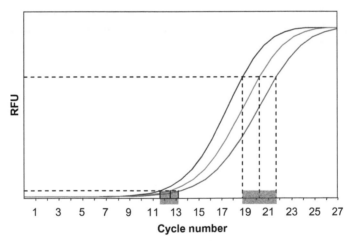

FIGURE 11.13 In practice, there is a difference in results obtained using different thresholds. The width of the gray boxes on the x-axis illustrates the error in reported C_t value that could result from splitting a single sample into three aliquots and running qPCR simultaneously on each. Having a threshold set at a higher value results in a greater range of C_t values. This principle is different from that illustrated in Figure 11.12. Here, the rate of increase for the three curves is different, due to the different amounts of time taken for individual dNTPs to bind to DNA polymerase and the exact efficiency of each individual DNA polymerase molecule in each sample tube. To minimize the discrepancy between theory and practice, the lowest reliable threshold value is used.

is shown. Notice that, at the higher threshold, the C_t values are quite different. However, for the lower threshold, the three C_t values converge toward the same point. This is not the same as what is presented in Figure 11.12! Figure 11.12 deals with ideal, theoretical conditions. Figure 11.13 shows a difference in ΔC_t at higher versus lower thresholds; what is presented in Figure 11.13 is differ-ent from theoretical values because the amount of time for an individual dNTP to bind to DNA polymerase will differ ever so slightly because of probabili-ties involving molecule orientations, diffusion constraints, etc. The exact rate of polymerization for each DNA polymerase molecule is not a mathematical constant, either. As a qPCR experiment proceeds to higher cycle numbers, any such minor differences will be amplified exponentially. To minimize the vari-ance seen in real-world conditions, data for all samples and genes are collected at a low threshold.

The ability to collect reliable data at low cycle numbers is an advantage of real-time PCR over traditional PCR. For traditional PCR, one must allow the reaction to proceed to high numbers of cycles in order to be able to see with our eyes the thickness of bands on an agarose gel. For real-time PCR, data are taken at low cycle numbers to minimize error. Another advantage to taking data earlier in the amplification process is that there are no issues with limited re-sources, which can skew results. There should be sufficient dNTPs so that DNA polymerase will not be prevented from polymerizing as rapidly as possible. At higher cycle numbers such as those used by traditional PCR, dNTP availability can become an issue as these resources are depleted, which may result in data inaccuracies.

11.7.1.2 Primer Efficiency

So far in our discussion of PCR, we have assumed that for every cycle, we dou-ble the amount of dsDNA. That is a fine assumption when learning the theory behind PCR, but in real life, it is seldom the case that the number of dsDNA strands is exactly doubled per round. One explanation for this has to do with the efficiency of the primers that are designed and used. You've got ~20 nucleotides per primer that must find and bind to a sense or antisense strand in the ~30 s that we give it during the annealing step, at the annealing temperature that we give it, to allow the DNA polymerase to then find this newly formed 20-mer double-stranded patch for subsequent polymerization. These events do not occur 100% of the time, especially when primer dimers, secondary structures, or damaged RNAs come into play. In these (and other) cases, *primer efficiency* will not be 100%. Different laboratories may have different ranges of primer efficiencies that are considered acceptable, but 90-105% is generally used. (Primer efficien-cies above 100% often indicate that there was an error in experimental design when determining the efficiencies, commonly because dilutions were too high.) Before, when we used terms such as $2^{C_{t_2}-C_{t_1}}$, the 2 was based upon perfect dou-bling stemming from our assumption of 100% primer efficiency. If a primer set were determined to yield only 95% efficiency, the 2 would be changed to 1.95, to reflect the quantity $(1 + \text{primer efficiency})$. (Note that when using a useless primer, meaning 0% efficiency, the amount of dsDNA would remain constant, which is why 1 is added to the primer efficiency.) A more robust way of expressing $2^{C_{t_2} - C_{t_1}}$ is now $(E + 1)^{C_{t_2} - C_{t_1}}$, where $E = \text{primer efficiency}$, because the amount of dsDNA may not be fully doubled completely per cycle.

Our formula for fold difference now becomes

$$\text{Fold difference} = \frac{\left(E_{\text{Gene of Interest}} + 1\right)^{\left(C_{t_{\text{Gene of Interest, Negative Control}}} - C_{t_{\text{Gene of Interest, Sample}}}\right)}}{\left(E_{\text{Reference Gene}} + 1\right)^{\left(C_{t_{\text{Reference Gene, Negative Control}}} - C_{t_{\text{Reference Gene, Sample}}}\right)}},$$

where $E =$ the efficiency of the primer set for the gene of interest or the reference gene, as indicated.

11.7.2 Probes

In the qPCR experiments described up to this point, SYBR green was used to indicate the amount of dsDNA in each sample. Another method to achieve dsDNA quantitation involves the use of double-labeled probes. For quantitation, the mathematics will be the same as already described and the same formulas can be used. Both the $2^{-\Delta\Delta}$ and the efficiency-corrected methods are valid. The difference between the use of probes as opposed to SYBR green lies in the fundamental use of fluorophores in the test tube.

When TaqMan® DNA polymerase encounters the double-stranded segment created when the probe binds to the ssDNA segment, the intrinsic 5′-3′ exonuclease activity of the polymerase will chop up the probe, thus permitting the fluorophore and quencher to diffuse away from one another. The increased distance will render the quencher unable to absorb the fluorophore's emitted light, thus allowing the fluorescence of the fluorophore to be detectable. Increased fluorescence implies more amplicons have been replicated.

Consider that we have a single fragment of dsDNA (cDNA). Just like before, there will be a melt step, the annealing of forward and reverse primers, and extension via DNA polymerase. When probe sets are used, the fluorophore that is present will not be an intercalating dye, but rather a double-labeled fragment of DNA that is complementary to an interior portion of the amplicon (Figure 11.14). This fragment, labeled with a fluorophore on one end and a quencher on the other, is the *probe*. It is not necessary to use probes that are complementary to each fragment (both sense and antisense) in the dsDNA amplicon; only one fragment need be probed. The probe will contain two light-sensitive molecules that serve as labels: a fluorophore and a quencher. From our discussion in Chapter 8, we already know that the fluorophore will accept light of one wavelength and emit it at a longer wavelength. The quencher works much like the secondary fluorophore of a FRET experiment: its excitation wavelength range should include the emission wavelength of the fluorophore. However,

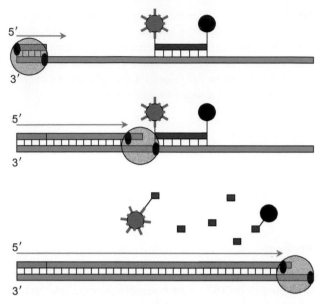

FIGURE 11.14   Primer/probe sets utilize a fragment of DNA that is complementary to an interior portion of one strand of the amplicon. A fluorophore and a quencher are covalently bound to the probe fragment. The quencher will be close enough to the fluorophore that a large percentage of emitted photons will be absorbed, meaning the fluorescence will be quenched down to background levels. When TaqMan® DNA polymerase encounters the double-stranded segment created when the probe binds to the ssDNA segment, the intrinsic 5′-3′ exonuclease activity of the polymerase will chop up the probe, thus permitting the fluorophore and quencher to diffuse away from one another. The increased distance will render the quencher unable to absorb the fluorophore's emitted light, thus allowing the fluorescence of the fluorophore to be detectable. Increased fluorescence implies more amplicons have been replicated.

unlike with FRET, the quencher molecule will not reemit the absorbed photon in the visible light spectrum.

Before a probe molecule has attached to a complementary DNA sequence, the quencher molecule will be close enough to the fluorophore that the quencher will absorb any light emitted by the fluorophore. The fluorophore and quencher molecules are attached to the ends of an oligonucleotide that is complementary to a sequence of DNA in the interior of the amplicon. The probe will anneal to ssDNA at the same time as the primers. When the DNA polymerase attaches to the primed, double-stranded portion of the DNA fragment, it will polymerize as before in the 5′ to 3′ direction, but when it encounters the probe (which creates a small, double-stranded patch of DNA), its 5′ to 3′ exonuclease activity will chew up the oligonucleotide probe, releasing the nucleotides that contain the fluorophore and quencher molecules to be free to move about the reaction solution via diffusion. Since the fluorophore and quencher are no longer tied to the same molecule, they will move apart by diffusion, and the quencher will no longer be able to absorb the photons emitted from the fluorophore, thus allowing for detection of fluorescence.

With SYBR green, we were able to detect how much more dsDNA was in the reaction vessel by the production of more potential spots for intercalation. With probes, no fluorescence is detected unless a probe is chopped up by DNA polymerase. Consider a new thought experiment, which starts with only a single dsDNA molecule. It can bind our two primers and one probe, and only one probe molecule will be chopped up during the first cycle. During the second cycle, two more fluorophores will be separated from quenchers (one for each new dsDNA molecule that is being produced), and the amount of fluorescence in the reaction tube will be doubled (assuming 100% primer efficiency). Fluorescence will be detected during each cycle, and it should increase exponentially in a 1:1 relation with the amount of new dsDNA molecules that have been produced. The curve showing detected fluorescence versus cycle number should be similar to a curve that could have been generated using SYBR green.

The advantage of using a primer/probe system is that several primer/probe sets can be used in the same reaction vessel, which will allow for the monitoring of multiple genes at the same time in the same tube, which will reduce variability. In qPCR instruments that can detect four colors simultaneously, four different primer/probe sets can be designed, each probe with its own unique fluorophore/quencher system, which will allow for the monitoring of amplification of four different genes at the same time.

QUESTIONS

1. If the region of DNA you wish to replicate with Taq polymerase is 750 nucleotides long, and Taq works at 100 nucleotides per minute, how long should you allow for extension during PCR?
2. Is this a proper PCR sequence? Why or why not? Assume there are about 400 bases in the amplicon.

Melt 95°	45 s
Anneal 42°	30 s
Extend 73°	24 s

3. Why is the annealing temperature so precise in PCR? What happens if the temperature is too high or too low? Why is the annealing temperature usually different from experiment to experiment?
4. Give two reasons why qPCR is more accurate than traditional PCR.
5. Compare qPCR using SYBR green versus a primer/probe set.
   a. What is literally being measured with each method?
   b. How does each method imply mRNA level?

6. An entire transcriptome is converted into cDNA by reverse transcriptase, and qPCR analysis is performed using SYBR green and primers determined to be specific for a certain transcript. How can one verify that the results of the PCR are not the result of the primers binding to more than one site on the DNA?

7. Why is cDNA not the same as genomic DNA?

8. What does PCR analysis tell us about a gene?

9. What protein is used to turn mRNA into cDNA? For PCR, why is this step performed?

10. Why do we use Taq DNA polymerase and not human DNA polymerase for PCR?

11. Dr. Frank runs a PCR experiment to demonstrate that his new drug *increases* the activity of a gene that is 10,000 base pairs in length. For his test sample, his gel reflects a band at 300 bp for the gene of interest, and he is happy about it! Is he crazy, did he stay up too late studying, or is there some other reason why he is not upset by the lack of a band at 10,000 bp? If you pick the third option, give the reason. (Hint: He is not crazy and he went to bed early.)

12. Your instructor will give you a bonus point if you point out this question to him/her. Do not contact the author of this book on this matter; it is between you and your instructor.

13. Why do we use Taq DNA polymerase and not human DNA polymerase for PCR?

14.

    a. How many cycles must pass in a PCR reaction for the *majority* of your DNA strands to be encompassed by your target sites for primer annealing?

    b. Will the strands produced in your first round of PCR be longer, shorter, or the same length as your target amplicon?

15. The following graph reflects readings obtained for a sample of cells treated with a drug (red lines) and an untreated control sample (blue). The gene of interest is shown by the solid curves, and the internal control is represented by the dashed curves.

    Black lines: untreated controls
    Red lines: cells treated with the drug
    Solid lines: gene of interest
    Dashed lines: GAPdH

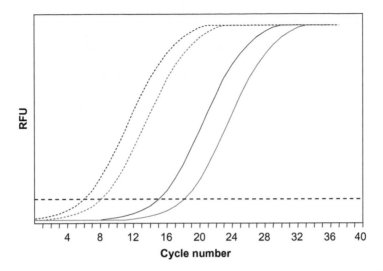

16. Why is the number of repeats used in a PCR experiment limited? If re-sources were not a problem, could one use 40 cycles for everything? Why or why not?

17. Suppose you are going to perform PCR in a buffer solution that contains 150 mM KCl and 100 mM MgCl (and no formamide). Give the T_m of the following primers:
    5'-CCT TGT CAA GT-3'
    5'-CCA CCT CTG CGA TGC TCT TAC-3'

18. Suppose you wanted to generate a large number of poly(A) polynucleotides (at least 100 A residues per strand) via PCR. You happen to already have a template with the sequence of 5'- $(T)_{100}$ AAA AAA AAA-3', so you design a primer that has a DraI cut site at the end. (The recognition sequence of DraI is TTT|AAA.)
    a. If you use 200 nM of the primer with 1×10^6 copies of the template, how many copies of the template would you have after 30 cycles of PCR?
    b. Write the sequence of the primer you will use.
    c. Give two reasons why using a 6 mer as your primer is not a good idea.
    d. Determine the optimal annealing temperature for your PCR.
    e. Write out a modification to the template that would substantially in-crease the amount of poly(A) you would obtain with the PCR primer that you designed.

19. Suppose that an entire transcriptome has been converted into cDNA by re-verse transcriptase, and RT-PCR analysis is performed using SYBR green. How can you prove that the results of the PCR are not the result of the prim-ers binding to more than one site on the cDNA?

RELATED READING

Marmur, J., Doty, P., 1962. Determination of the base composition of deoxyribonucleic acid from its thermal denaturation temperature. J. Mol. Biol. 5, 109–118.

Meinkoth, J., Wahl, G., 1984. Hybridization of nucleic acids immobilized on solid supports. Anal. Biochem. 138, 267–284.

Rychlik, W., Spencer, W.J., Rhoads, R.E., 1990. Optimization of the annealing temperature for DNA amplification in vitro. Nucleic Acids Res. 18, 6409–6411.

Chapter 12

Genetic Engineering

In this chapter, we will see how to engineer a gene and obtain enough copies of it to be used in experiments or, potentially, clinical treatments. The amplification process will consist of transferring our engineered genes into bacterial cells and letting them do the work of making many copies of the gene for us. But first, we have to make the gene.

12.1 PLASMID ARCHITECTURE

A *plasmid* is a circular piece of DNA. Much of genetic engineering utilizes plasmids, which can be constructed relatively easily in the laboratory through the judicious use of natural enzymes. A robust plasmid has several elements that are of great importance. First, there is the gene that the scientist wishes to deliver into target cells. The gene will contain at least a promoter and an exon. The promoter is used to help initiate transcription of the encoded gene. The exon codes for a predetermined polypeptide, which will be expressed using the steps already mentioned as the central dogma. Another feature of the plasmid is an *antibiotic resistance marker* (denoted by α^r in general), which is a bacterial gene that will be useful as a selection marker when many copies of the plasmid are produced in a process called *amplification*. An antibiotic resistance marker codes for a protein that will give resistance to a specific antibiotic, such as ampicillin or kanamycin, to bacteria that express the protein. Another necessary feature of the plasmid is a bacterial *origin of replication*, which will allow the researcher to harness the power of *E. coli* for amplification, using the bacteria as biological factories to churn out many copies of the plasmid. A prokaryote is used for amplification because it lacks a membrane-bound nucleus, thus making it a relatively simple task to get the plasmid into the presence of DNA polymerase, the enzyme responsible for DNA replication. (In eukaryotic cells, DNA polymerase resides in the nucleus.) DNA polymerase does not distinguish between genomic and foreign DNA. The cellular replication machinery will assemble and begin replication upon exposure to an origin of replication (*ori*) in the genome or on the plasmid.

Figure 12.1a is a map of a common reporter plasmid, *pEGFP-N1*. Plasmid names usually begin with a lowercase "p" to denote the DNA is a plasmid. In this example, "EGFP" is an acronym for enhanced green fluorescent

An Introduction to Biotechnology. http://dx.doi.org/10.1016/B978-1-907568-28-2.00012-5
Copyright © W T Godbey, 2014.

pEGFP-N1 Vector Information

GenBank Accession #U55762

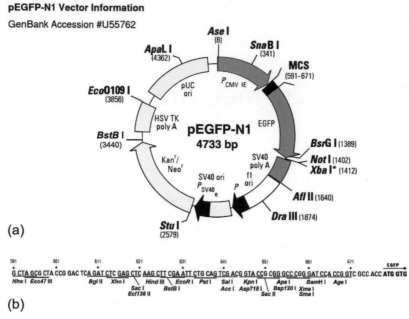

(a)

(b)

FIGURE 12.1 (a) Map of the plasmid pEGFP-N1. Locations of certain restriction sites are noted by enzyme name, with the precise location of each cut in parentheses. Segments of the plasmid are explained in the text. (b) Specific sequence of the multiple cloning site, with specific recognition sequences notated and underlined.

protein, and "N1" indicates that this plasmid can be used to produce a fusion protein, where the protein portion of interest in our final product will be on the N-side (N-terminus side) of the complete protein. Later, a thought experiment will be presented that will lead to the insertion of an additional exon into the plasmid. Notice that some of the features in the figure are emphasized with green and some are emphasized with yellow. This is to illustrate that the plasmid performs different functions in prokaryotic versus eukaryotic cells. The portions of the plasmid denoted in yellow represent sequences relevant to bacteria:

- $^PSV40_e$ is a promoter from simian virus 40 that will dictate the transcriptional start site for the bacterial gene *Kan^r*.
- *Kan^r* is a bacterial gene encoding a protein that will confer antibiotic resistance to kanamycin.
- *HSV TK poly A* encodes a signal (obtained from the herpes simplex virus gene for thymidine kinase) that will cause a poly(A) tail to be put onto the 3′ end of the bacterial gene during transcription.
- *pUC* is a DNA sequence that serves as a starting point, or *origin*, for bacterial replication.

The portions shown in green represent sequences relevant to eukaryotic (in this case, mammalian) cells:

- PCMV IE is a promoter for "immediate early" genes in the cytomegalovirus genome. This is a *strong promoter* in mammalian cells, meaning it supports a large amount of transcription initiation.
- *EGFP* encodes the gene for an enhanced green fluorescent protein.
- *SV40 poly A* encodes a signal for polyadenylation of mRNA transcripts.

(For completeness, the region shown in white encodes the *f1 ori*, which serves as an origin for single-stranded DNA production. Single-stranded DNA is used in some viral applications.)

One other feature of commercially available plasmids, including the plasmid shown in Figure 12.1, is the *multiple cloning site* (MCS). The MCS is a stretch of DNA that is rich in restriction enzyme recognition sequences. It is placed in the DNA sequence on purpose, to allow researchers the ability to insert DNA strands of their own making into the plasmid at a useful location with relative ease. The specific bases of the pEGFP-N1 MCS and the names of the enzymes that can be used to cut it are given in Figure 12.1b.

There are many plasmids that are commercially available to aid the biotechnologist in constructing genes. They may be purchased with or without a promoter, with or without a reporter exon, and with or without an enhancer region. However, without the bacterial features of an origin for replication and an antibiotic resistance gene, amplification of the plasmid inside bacteria in selective media would not be feasible. Commercially available plasmids (vectors) should also have an MCS to allow for insertion of promoters and/or exons with relative ease.

12.2   MOLECULAR CLONING

It was mentioned earlier that the "N1" in the reporter plasmid name denoted that the vector could be used to produce a fusion protein. This is due, in part, to the location of the MCS. A *fusion protein* is a single protein that is produced by the expression of two formerly separate genes that have been combined into a single gene. By inserting an exon in the MCS of *pEGFP-N1*, expression of the resulting gene will be a protein that has a fluorescent region. The exon could code for almost any protein, but for the sake of example, we will use an exon that codes for bone morphogenetic protein 5 (BMP5). In this example, we are creating a fluorescent protein consisting of $H_3{}^+N-BMP5-EGFP-COO^-$.

There are different ways to obtain the DNA sequence for insertion into a vector. One could create a sequence from scratch or use the polymerase chain reaction (Chapter 11) to amplify a specific sequence directly from a genomic preparation. In this example, the *bmp5* sequence will reside in a plasmid that was sent to us from a collaborator's laboratory. Our goal is to make a fluorescent version of BMP5 so we can verify its expression in target cells. (As a matter of semantics, note that genes are written in lowercase italics, and protein names are written in unitalicized uppercase.)

When a plasmid is sent from one laboratory to another, it would typically be included with a map to help the recipient determine the sequence of bases upstream and downstream from any regions of interest. If the exact DNA sequence is not given, at the very least, a map showing enzyme names and cut locations should be supplied. To remove an exon from the plasmid for insertion into our own vector, sometimes, the task is easy and the DNA region we are interested in will be flanked by two enzymes that are also found in the MCS of our vector and in the same order. Other times, the problem is slightly more complex, and a pair of enzymes must be selected from a list of candidates determined by examination of the supplied maps of the donated plasmid and the vector. In this case, it is important to ensure that the enzymes selected are compatible in the same digest buffer and that the cut sites appear in the same order in the donated plasmid and the vector. The order is important because the exon of interest will almost certainly not be a palindrome, meaning that if it were to be inserted backward, then the DNA message would be gibberish.

12.2.1 Cutting (and Ligating) Sticky Ends

Returning to our example, suppose the vector is pEGFP-N1, with an MCS that is given by Figure 12.1b. Also suppose that the plasmid we obtained, call it *pBMP5*, contains the cut sites shown in Figure 12.2. Which two enzymes would work best for cutting the *bmp5* exon out of pBMP5 for insertion into the vector? XhoI and NheI might at first appear to be the perfect pair, but this selection would be a poor choice because the *bmp5* exon would be inserted in the wrong orientation. The sticky ends that will be created in the vector and insert are very important because they will dictate the orientation of the insert in the final plasmid. XhoI and NheI will both produce sticky ends, but the sequence of bases in the overhangs will be different so each sticky end will only match up

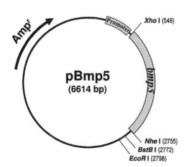

FIGURE 12.2 Partial restriction map of *pBmp5*, a fictitious plasmid used as an example in the text.

with the end of the vector that was produced by the same enzyme. If the exon is inserted backward, the (former) transcriptional and translational start sites will not be on the end closest to the promoter. RNA polymerase would utilize a different transcriptional start site, and the ribosome would use a different AUG translational start (if there was one at all after the flip), and the roles of the coding strand and the template strand would be reversed, yielding an mRNA that would most likely produce a useless protein if any protein were to be produced. The use of the BstBI enzyme would not be the best choice because it cuts twice in the vector (although there is a low-yield method to get around this). XhoI and BstBI could be used to cut the exon out of the BMP5 plasmid, and this enzyme pair would yield sticky ends in the correct orientation inside the MCS of the vector, but an additional cut would be generated, which would render the vector essentially useless. Next, consider XhoI and EcoRI. This enzyme pair would effectively remove the exon from the insert, would yield sticky ends in the correct orientation in the vector, will not generate extraneous cuts in either the vector or insert, and are active in the same buffer, so the pair would be a good choice for the task at hand.

When *pBMP5* is cut with XhoI and EcoRI, the result will contain two fragments as illustrated by Figure 12.3. The exon for *bmp5* will have sticky ends on either side, and they will match up perfectly with the sticky ends created by cutting the vector plasmid with the same enzymes. Notice that, while a large portion of the plasmid containing *bmp5* will be lost, a very small portion of the vector plasmid—the few bases between XhoI and EcoRI—will also be lost. But how will they be lost? They will be thrown away. Following exposure of each plasmid to the restriction enzymes, each resulting DNA sample will be run out in separate lanes of an agarose gel (Figure 12.4). Unless the two enzymes cut at sites that are exactly opposite of one another in a plasmid, a condition that won't happen because our choice of enzymes will prevent such a possibility, the

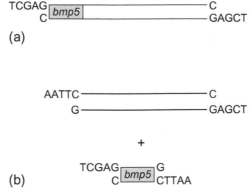

FIGURE 12.3   (a) When the plasmid shown in Figure 12.2 is cut with Xho I, a single, linear, 6614 bp fragment with sticky ends is generated. (b) Cutting the plasmid with Xho I and EcoR I gives fragments that are 4364 and 2250 bp in length.

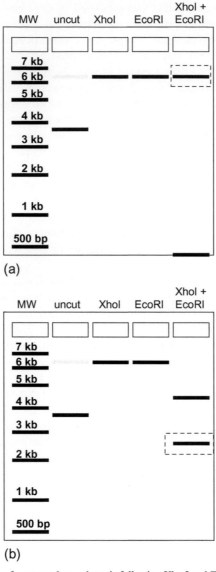

FIGURE 12.4   Results of agarose electrophoresis following Xho I and EcoR I restriction cuts on the vector (a) and insert (b) plasmids described in the text. Fragments to be excised are denoted with dotted lines. Note that the 16 bp fragment in (a) may run off of the gel.

two fragments will migrate different distances through the gel. Because we will have a map of each plasmid, we will be able to predict the fragment sizes that will be created from the restriction cuts. We will simply visualize the fragments with UV light (remember that a fluorescent DNA marker such as ethidium bromide will be used in the gel) and cut out the one(s) that we want to keep using a scalpel or razor blade. It is a simple matter to melt away the agarose to then isolate the DNA via an anion exchange resin.

After melting the agarose, the solution will be put into a column that contains a matrix that carries a positive charge. The solution will be forced through this matrix via centrifugation, and the DNA will adhere to the matrix by virtue of the negative charges it carries. After washing the matrix to remove residual compounds such as agarose and salts, the DNA will be eluted under slightly alkaline conditions (pH of 7-9) and low salt concentrations (<10 mM).

The restriction cuts will be performed in a small tube, which will contain several things: water, various salts, a buffer, possibly an additive such as bovine serum albumin (BSA) or S-adenosyl methionine (SAM), the DNA sample, and the restriction enzymes. These will typically be combined to yield a total volume of 10 µl as follows:

Water*	$(10 - x)$ µl
10x buffer	1.0 µl
10x extras	(BSA or SAM. 1.0 µl is used if req. by one enzyme)
DNA	y µl (enough for 0.1-1.0 µg)
Enzyme 1	0.5 µl
Enzyme 2	0.5 µl
Total	10.0 µl

*Water will be added to bring the total volume up to 10 µl. As written, x=the sum of the volumes of all other additives, including "extras" and y.

The above formula, while robust, can be confusing at first. If our vector DNA is kept at a concentration of 1 µg/µl, and we are using the enzymes XhoI and EcoRI, then the recipe becomes

Water	6.0 µl
10x "buffer 4"	1.0 µl
10x BSA	1.0 µl
DNA	1.0 µl
Xho I	0.5 µl
EcoR I	0.5 µl
Total	10.0 µl

"10x" refers to a concentration that is 10 times what is needed for the final working concentration. If the final volume of a different reaction was to be 100 ml, then we would use 100/10 = 10 ml of a 10x solution in constructing the solution, or 1 ml of a 100x solution, or 0.1 ml of a 1000x solution.

1000× concentrations are often used in the laboratory because the math requires only a prefix change: If the final solution is to be 2 l, then we would use 2 ml of a 1000× solution. If the final volume is to be 58.3 ml, then we would use 58.3 μl of a 1000× solution.

Regarding the amount of enzyme, 0.5 μl is often used for practical reasons rather than because of the strict amount of enzyme that is needed for the given reaction. Restriction enzymes are often measured in *"units"*(U), where one unit is equal to the amount of enzyme needed to completely cut 1 μg of phage lambda DNA in one hour at optimal temperature. The definition includes phage lambda DNA only for standardization purposes; one can expect one unit of enzyme to cut 1 μg of most plasmids in one hour. Most restriction enzymes have optimal activity at 37 °C. In terms of our example, suppose that EcoR I was sold at a concentration of 20,000 U/ml. That implies that 1 μl of HindIII will contain 20 U. In our example, we are cutting 1 μg of plasmid, so we only need 1 U or 1/20 μl of HindIII. It's usually impractical, if not infeasible, to measure 0.05 μl, so this number will be adjusted up to 0.5 μl based upon the accuracy of the smallest pipette available for such an experiment in a typical laboratory. (While in theory, the enzyme could be diluted, in practice, this will impair enzyme activity.)

Bovine serum albumin (BSA) is a common additive to the recipe for restriction digests. This protein has been found to stabilize the folding of some restriction enzymes, and it balances the potential negative effects of enzyme interactions with pipette tips, reaction tubes, and/or the air-liquid interface. The positive effects of BSA had been found to increase the activity of restriction enzymes by as much as twofold. This result is so encouraging that some companies now include BSA in every restriction enzyme sample that they sell. Some enzymes (such as Xho I) are said to require BSA to work properly, while others (such as EcoR I) are not listed as having the requirement. If only one enzyme in a chosen pair requires BSA, the protein should be included in the reaction mix. Some laboratories use BSA with every enzyme regardless of whether it is listed as a requirement.

As with any good experiment, the restriction cut must include the appropriate controls. Again referring to our example, we should include a control where only Xho I is used to cut the DNA, and another control where only EcoR I is used, adjusting the volume of water for each reaction accordingly. These controls will help to ensure that each enzyme is working properly and cuts the plasmid as the maps predict. One should also include a control for plasmid that has not been exposed to a restriction enzyme. This control should demonstrate where the uncut, supercoiled plasmid will run. More often than not, however, this particular lane will show two bands: one for supercoiled plasmid and one for the linear form that results when a plasmid is cut once. As a plasmid sample undergoes several cycles of freeze/thaw and repeated pipettings, the plasmids in solution will occasionally be torn open into a linear form. The ratio of brightness between the bands representing supercoiled and linear will help one to measure of the integrity of the plasmid sample. For this part of the restriction digest experiment, it is also important to visualize where uncut DNA will run on the gel to help verify that each of the enzymes is indeed cutting the DNA in the other control lanes. There will be a third use for the uncut control that will be discussed soon.

At this point in the experiment, there will be two tubes, one that contains vector DNA that is now linear with sticky ends and one that contains the *bmp5* insert, also with sticky ends. In theory, it is best to quantitate the amount of DNA in each tube so that, again in theory, equimolar amounts of the vector and insert can be combined. (In practice, it is not uncommon to use a greater number of insert molecules than vector molecules.) The way to determine the mass of DNA will be discussed later, but this step is often skipped because the DNA yield from gel extractions in this particular procedure is often very low.

After the vector and insert fragments have been combined, they will pair up because of their sticky ends. The overhangs from the sticky ends are complementary so hydrogen bonding will keep the fragments associated. However, the bonding is not covalent and there will still be gaps at the four ends between the insert and vector, meaning there are two separate phosphoribose backbones for each component. If an enzyme such as DNA polymerase were to attach to this pseudoplasmid, it would not be able to progress beyond the gaps. Many copies of the plasmid are to be made in E. coli by virtue of DNA polymerase in an upcoming step, so these gaps must be sealed. The sealing is accomplished by the enzyme *DNA ligase*, which will create phosphodiester bonds between the vector and the insert fragments. The covalent linkages will yield a new plasmid.

12.2.2   Blunt-End Ligation

The above has been a very straightforward example of removing a segment of DNA from one plasmid and inserting it into another. Sometimes, however, the insert will be flanked by enzymes that are not contained in the MCS of the vector. In such a case, the biotechnologist might resort to *blunt-end ligation*. The insert can still be removed as before, but an additional step must be performed to remove the sticky ends. Consider the same example as before, except that the plasmid that contains the insert has the *bmp5* exon flanked by Mlu I and Pvu I, which will not cut the vector (Figure 12.5). Following excision with the two enzymes, the enzymes will be *heat-killed*, which means they will be inactivated by exposure to temperatures in excess of 70 °C for 10 min. Next, a blunting enzyme will be added to the solution. Two common enzymes used for this purpose are the *Klenow fragment* and *T4 DNA polymerase*. The Klenow fragment is the large subunit of DNA polymerase I, isolated from *E. coli*. In the wild, T4 is a bacteriophage that can infect *E. coli*, and one of its genes codes for a DNA polymerase. Both of these DNA polymerases have $5' \rightarrow 3'$ DNA polymerase activity and also $3' \rightarrow 5'$ exonuclease activity (Figure 12.6). While the two enzymes have very similar activities, T4 DNA polymerase is usually the choice

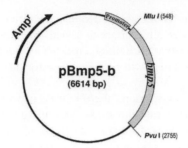

FIGURE 12.5   Partial restriction map of *pBmp5-b*, a fictitious plasmid used as an example in the text.

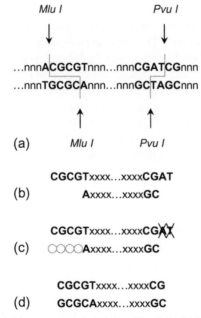

FIGURE 12.6   (a) Recognition sequences for Mlu I and Pvu I. (b) Sticky ends produced by the two enzymes. Note that Mlu I leaves a 5′ overhang and Pvu I leaves a 3′ overhang. (c) The Klenow fragment and T4 DNA polymerase both have 5′ → 3′ DNA polymerase activity and 3′ → 5′ exonuclease activity, which work on the given overhangs as shown. (d) The final result is a linear fragment with blunt ends. Note that both recognition sequences have been lost.

for blunting for two reasons. First, the 3′ → 5′exonuclease activity of T4 DNA polymerase is roughly 200 times that of the Klenow fragment. Second, T4 DNA polymerase does not displace downstream oligonucleotides as it polymerizes (the Klenow fragment does).

The end result from either of the above two DNA polymerase enzymes will be a blunt-ended fragment. The same process can be carried out on the vector.

In this case, the vector would only need to be cut by one enzyme instead of a pair, since a single restriction cut would yield a linear DNA fragment. After blunting with the Klenow fragment or T4 DNA polymerase, the vector and insert can be combined into a single tube for blunt-end ligation. DNA ligase will still be used to carry out this task, but much more time will be needed since the respective DNA fragments will not be held in place by hydrogen bonding.

Besides generating a lower yield because of the lack of hydrogen bonding from sticky ends, another problem with blunt-end ligation is that there is no guarantee that the insert will be introduced in the proper direction. To check for this possibility, one might try cutting the engineered plasmid with an enzyme pair where one of the enzymes cuts inside the insert and one cuts outside (Figure 12.7). The sizes of the resulting fragments would be used to verify the direction of the insert.

With blunt-end ligation, it is also quite possible that the vector will seal upon itself without an insert to yield what is sometimes referred to as an *empty vector*.

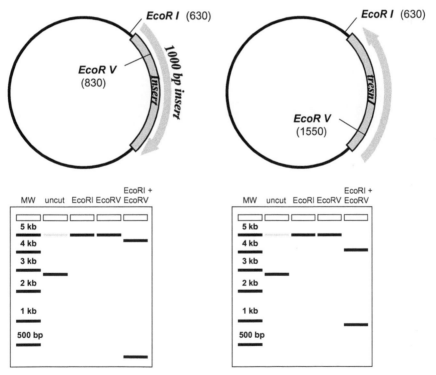

FIGURE 12.7 Detection of insert in reverse orientation. Suppose the entire plasmid, after ligation, is 5000 bp, the insert is 1000 bp, and the insertion occurs 10 bases after the EcoR I site. If EcoR V cuts 190 bases from the 5′ end of the true coding strand of the insert, then the result of an EcoR I/EcoR V double digest would produce fragments that are $(10+190)=200$ and $(5000-200)=4800$ bp (left). If the insert were ligated in the reverse orientation (right), the fragments would be $(10+1000-190)=820$ and $(5000-920)=4180$ bp.

This phenomenon can be avoided with two different methods. In the first, a molar excess of insert—perhaps five- to eightfold—can be used to help increase the probability of an insert fragment being lined up with the vector for ligation. The second method involves an *alkaline phosphatase* such as calf intestinal phosphatase, shrimp alkaline phosphatase, or Antarctic phosphatase. These products will remove the 5′ phosphates from DNA (or RNA) samples. The alkaline phosphatase will be added to the (blunted) vector and later heat-killed or removed by a purification step. Since 5′ phosphates are required for DNA ligase to work (Figure 12.8), self-ligation of the plasmid will be prevented. By this reasoning, the only circular plasmids that will be produced will contain inserts, thus theoretically eliminating colonies containing empty plasmids after plasmids have been delivered. There will still be two nicks in the plasmid, giving the plasmid an architecture known as *open circle* (Figure 12.9). While DNA deliveries using this architecture are not as efficient as when closed circle plasmids are used, yields are substantially higher using closed circle plasmids versus linear fragments.

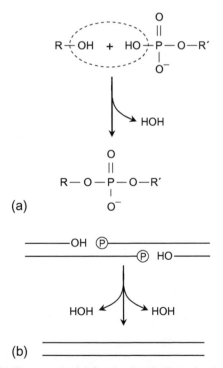

(a)

(b)

FIGURE 12.8   (a) DNA ligase acts by joining the phosphodiester bond of the 5′ terminal phosphate with the 3′ hydroxyl to create a phosphodiester bond. (b) This mechanism is used to join DNA fragments. Sticky ends are pictured, although the same action is used to join blunt ends. It is important to note the requirement of a phosphate group in the reaction.

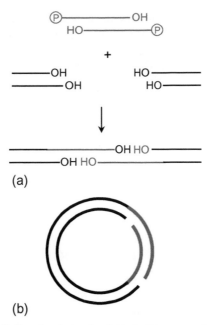

(a)

(b)

FIGURE 12.9   (a) An alkaline phosphatase has been used to remove the 5′ phosphates from the DNA vector, shown in black. Without these phosphates, DNA ligase will be prevented from sealing the ends of the vector together. Only the insert will contain the phosphate groups needed for ligation. (b) Because the insert only contains two phosphates, only two of the four gaps will be sealed. This plasmid architecture is known as open circle.

12.2.3   Direct Extraction of a Gene from the Genome

Up to now, we have been customizing our plasmids by utilizing sequences obtained from other existing plasmids. While it is common to utilize commercial vectors that already contain an MCS, a selection marker (such as antibiotic resistance), and a prokaryotic origin of replication, scientific research will often involve promoters, enhancers, or exons that are not commercially available. Cutting-edge research will often use DNA sequences that are not even available from other scientists. So where do these sequences come from? Many times, we can take them from the genome itself, and we have already covered the techniques used for such a feat! Genomic DNA can be extracted from an existing culture of cells, and a region of interest can be amplified from the extracted DNA via PCR. Through thorough investigation of the literature, including published genome sequences (including human), one can design primers that are unique to the region of interest and amplify it to concentrations that will allow isolation and insertion into a plasmid. Insertion will occur, as we have already covered, via restriction enzymes and DNA ligase, usually via the use of sticky ends. PCR will generate blunt-ended amplicons, but there will be restriction sites within the amplicons that allow for the creation of sticky ends.

It is certainly reasonable to wonder how often a genomic region of interest, flanked by *unique* sequences to allow for primer binding, will also contain restriction sites for the generation of sticky ends. From a genomic perspective, the answer is "almost never." However, primers can be designed in such a way that additional bases can be inserted into the amplicon (Figure 12.10). These additional bases will include a restriction site (a different one for the forward versus reverse primer), plus a few extra bases (usually around six) to allow the restriction enzyme to bind and operate when the time comes following PCR. Modification of the amplicon via restriction digest will yield linear, sticky-ended fragments ready for insertion into the plasmid being constructed via the methods already described. This method allows the biotechnologist to engineer plasmids containing genomic DNA sequences, whether the sequence is coding, regulatory, wild-type, or mutated. It is a very powerful tool.

12.3   A SINGLE PLASMID IS NOT ENOUGH

After the new plasmid has been constructed, it is not immediately useful. Suppose that only one good copy of the plasmid has been created. If it were created for gene delivery, only one cell could receive the one plasmid. If an organism were to be treated with gene delivery, transfecting only one cell would have little if any effect on the organism. To get an idea of scale, when 1 µg of a 5000 bp plasmid is loaded into a gel, there are ~1.95×10^{11} copies of the plasmid. These examples are meant to illustrate how relatively useless a single plasmid is. What is needed is a way to produce many copies of the engineered plasmid. This can be accomplished via *amplification*. Amplification is a straightforward technique whereby a plasmid is inserted into an *E. coli* cell, which then creates more copies of the plasmid using its own DNA polymerase. As the *E. coli* cell grows and divides, the number of plasmids that are being copied per unit time also increases. The amount of plasmid being produced increases exponentially as long as the *E. coli* have ample room to divide and enough food to eat. The process will take 12-16 h for optimal results.

Before the *E. coli* can begin to amplify the plasmid, the plasmid must first be inserted into one or many of the bacteria. This can be accomplished via *transformation* or *electroporation*. In transformation, *E. coli* cells must first be made *competent*, or able to take up exogenous DNA. This can be accomplished via membrane disruption during exposure to high concentrations of Ca^{2+} or Mg^{2+} and a temperature shock. The *E. coli*, stored frozen but allowed to thaw very slowly on ice in the presence of the transforming plasmids, will be exposed to 42 °C for 30-45 s and then returned to the near-0° environment used for thawing. Ca^{2+} has been shown to be essential for the process because it helps to protect the plasmid DNA from nucleases. Most of the transforming DNA will enter the cells during the incubation period following the heat shock.

Electroporation involves the delivery of an electrical current across the cells. Pores will spontaneously open in the cell membranes, and DNA will flow in the same direction as the electrons. When the current is removed, the pores

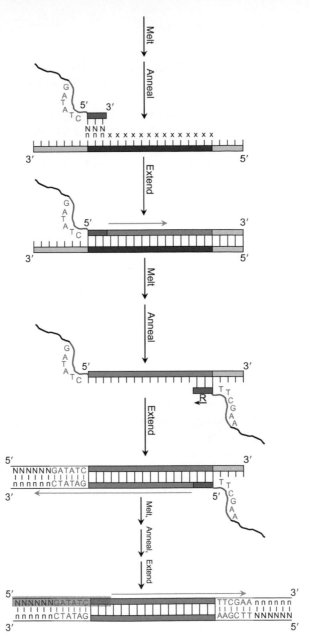

FIGURE 12.10 In adding a restriction site to an amplicon, the PCR steps are the same as those covered in Chapter 11 (see Figure 11.1). However, extra bases have been added to the 5′ end of each primer. These bases (shown as red bases and a nondescript spacer) will not pair with the genomic DNA. "N" stands for any nucleotide, and "n" is the base that is complimentary to "N." The genomic fragment is in blue. The first cycle of PCR will yield dsDNA fragments with the added cut site and spacer still unpaired. The cycle is shown for only one of the two original ssDNA strands. The second cycle of PCR will yield a completely paired dsDNA fragment on the side where the primer from the previous cycle is located. However, the primer used for this step has an unpaired tail with a second restriction site and spacer. The third cycle of PCR will yield the first completely paired dsDNA fragments. The fragment shown at the bottom of the figure was created by a forward primer, shaded in red, binding to the bottom strand of the dsDNA pair shown in the previous step.

will spontaneously close and any DNA in the cytoplasm (or nucleus when this technique is used for gene delivery to eukaryotic cells) will be trapped inside the cell.

Once transformed, the *E. coli* will be grown in suspension for 20 or more minutes (the length of time of one *E. coli* generation) in a batch reactor. The suspension will consist of cell medium that contains all of the nutrients necessary for bacterial growth. The medium is typically Luria-Bertani broth (LB), which consists of tryptone, yeast extract, and NaCl at isotonic concentrations. Note that no antibiotic is present at this time.

Following this incubation, a sample of the cell suspension will be streaked onto plates that contain LB agar plus one more very important component that is added to the agar to permit selection for our transformed cells: an antibiotic. Recall that our plasmids have an antibiotic resistance marker. Suppose this marker codes for resistance to kanamycin. Typical *E. coli* cells will die in the presence of kanamycin. However, if one of the cells has been transformed by our engineered plasmid, it will be expressing the antibiotic resistance gene and be able to survive the kanamycin-laced agar medium. The reason that no antibiotic is present during the first 20-40 min following transformation (previous paragraph) is to allow enough time for the transformed cells to express (transcribe and translate) the bacterial antibiotic resistance gene. After the kanamycin-laced agar medium has been streaked with the transformed culture, it is allowed to incubate overnight to allow each of the transformed *E. coli* to proliferate to form individual colonies. Each colony arises from a single parental bacterium, so all cells of the colony will contain many copies of the same plasmid.

After colonies have grown on the LB-antibiotic-agar, small samples from individual colonies can be grown in a small suspension (~3 ml) of LB plus antibiotic for extraction of plasmid DNA for further analysis. After one or more of the colonies have been verified to contain the correct plasmid, cells from the colony will be permitted to grow in greater volumes of medium (100 ml or more, perhaps liters in the research laboratory, or perhaps thousands of liters on the commercial scale) for a mass amplification of the engineered plasmid. Plasmid amplification is the subject of the next section.

12.3.1   Plasmid Amplification

When we want to obtain many copies of a plasmid, we harness the power of *E. coli* to perform the work for us. Because the plasmids we engineer will contain a bacterial origin of replication, after we transform the plasmids into these bacteria, the bacteria will make copies of the plasmids just as they make copies of their own genomes (which also have origins of replication). As the bacteria proliferate, more copies of our plasmids will also be created. This is the process of amplification.

Following amplification, how do we extract our plasmids from the *E. coli* cells? That will be presented by a classic extraction protocol. Although there now exist kits that skip a couple of the following steps and combine a couple of others, a classic protocol is being presented here because it has many good chemical principles worthy of consideration. Many consider the plasmid preparation protocol a biological technique, but it can also be considered a separation process from a strictly chemical standpoint. We will have to extract our plasmids from the bacteria, and we will do so by lysing the microorganisms. After the *E. coli* cells have been split open, we will first be faced with the problem of separating something hydrophilic from something hydrophobic. We will later be faced with the separation of proteins from nucleotides, as well as the separation of plasmid DNA from RNAs and genomic DNA.

We begin the process with a Petri dish loaded up with bacterial food (LB broth, discussed later) mixed with agar. (Agar is very similar to Jell-O® in consistency.) We smear the plate with our bacterial sample, which after the transformation process includes both transformed and untransformed *E. coli*. We then place the plates in a 37 °C oven (the same as normal human body temperature) for an overnight incubation. During this incubation, the cells will divide approximately every 20 min. As the cells continue to divide, they must stay in contact with the food source (the LB agar), so they will populate an ever-expanding circular region around the initial cell. This cell-containing disk is referred to as a *colony*, which will be perhaps 1-2 mm in diameter after the overnight incubation (Figure 12.11). As long as the colonies do not overlap on the plate, it is assumed that they each originated from a separate individual cell. The idea is that each colony is made up of identical bacteria because they all originated from a single microorganism and, hopefully (although not guaranteed with 100% statistical certainty), the DNA that was taken up by the original bacterium during transformation was a single plasmid.

Faced with several colonies upon a single agar plate, we must determine which ones contain our plasmid in the form in which we are interested. One way to do this is to take a small sample of bacteria from an individual colony and analyze the DNA on a very small scale. The sample can be taken by "picking"

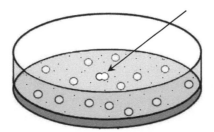

FIGURE 12.11   LB agar with bacterial colonies growing. The black dots represent untransformed *E. coli*, which are dead because they did not have resistance to the antibiotic that was mixed in with the medium. Note the doublet (indicated by an arrow), which is the overlapping of two colonies. With the purpose of this plate being to separate individual bacteria for further study, doublets should be avoided because the origin of each individual bacterium cannot be guaranteed.

the colony, which entails inserting a sterile toothpick or pipette tip into the center of the colony. Upon removal, some of the bacterial cells will adhere to the toothpick. By swishing the toothpick around in a tube containing LB medium, some of the bacteria will be dislodged and a new culture, this time in suspension, will have been created. After an additional incubation of this new culture for several hours at 37 °C, we will end up with a relatively dense suspension of bacteria that are roughly identical.

So, how do we know the bacteria in the culture have the right plasmid? The first method to ensure this is to use a *selective medium*. A selective medium is a medium that everything will die in except the cells of interest. In the present experiment, this will entail the use of an antibiotic such as ampicillin. The LB medium will contain it, as will the LB agar. The only bacterial cells that will be able to survive will be the cells that are expressing the bacterial antibiotic resistance gene that happens to reside in our engineered plasmid. Cell selection is not restricted to bacterial experiments, nor is it restricted to antibiotics.

Suppose there is a cell that did not take up a plasmid during the transformation process. It will die because it doesn't have any resistance to ampicillin. Its contents, including its DNA, will still be on the agar plate after we streak the plate, but when it is time to pick the colonies, we will not see this bacterium because it has not proliferated (because it is dead).

The right thing to do is to check many of the colonies that grow on the plate because you might get colonies that contain a variation to the plasmid we've constructed. If you get colonies that touch each other, then they should not be used because there is no guarantee that, in picking one side of the doublet, we will not take some of the bacteria from the other colony, which would mean we had nonidentical microbes in our next culture. After 12 h, the colonies will still be rather small, and having colonies that touch will not be a prevalent problem, unless we have so many colonies that they must touch each other for spatial reasons. If that ever happens, then the researcher may have reason to be quite happy because the transformation and amplification procedure have gone better than expected. In such case, a new plate will be streaked with a lower density of bacteria. The common yield for this experiment will only be 10 to 100 colonies.

Won't all of the colonies have identical plasmids since they were able to survive the selective medium? Most of them will, but the mutation rate is fairly high in *E. coli*. If a mutation occurred fairly soon after plating, then this new mutation would be propagated into all of the progeny cells, yielding a colony that contains a very high percentage of mutated plasmids. It is also possible in some situations (especially when blunt-end ligation is used) that the vector reseals on itself without an insert. Such an "empty plasmid" could still survive and multiply in the selective medium because the antibiotic resistance gene typically resides in the vector. Also in blunt-end ligations, it is possible that the insert is oriented backward. Sticky-ended ligations have the possibility of housing multiple inserts. For these reasons, it is necessary to pick several colonies and examine the plasmids from each. One of the common small-scale procedures for doing this—known as plasmid miniprep—is relatively quick and straight-forward. The plasmid prep procedure that will be discussed in detail in the next section is a larger scale of the miniprep procedure.

Each colony that is picked will be grown in its own tube that contains ~3 ml of LB plus antibiotic (Figure 12.12). Several hours later, the bacterial density will be high enough to visualize the microorganisms *en masse*. Each tube can then be split into two samples: one that will be frozen (with a total of 40-50% glycerol to prevent cell lysis from the expansion of water during freezing) and one from which plasmid DNA will be extracted and analyzed. While we keep the first tube in the freezer, let us turn to the contents of the second tube. We are going to perform a restriction digest in the way we have already discussed. We are doing this to help verify that the plasmid we think we have is indeed the plasmid we have isolated. The selection of restriction enzymes will be key in this step. It is important to not only verify the size of the isolated plasmids but also verify some predicted identity.

Earlier, we discussed some of the controls that are used for the restriction cut experiment. Now that blunt-end ligations have been discussed, we should revisit the subject of controls for the restriction digest experiment. There are special problems that can occur with blunt-end ligations: the insert can be put in back-ward, the vector can be sealed with no insert at all, or multiple vectors or vector/ inserts can be ligated into a single large plasmid. If we were to simply rely upon the control that cut the plasmid once inside the vector, we would not be able to discern the case where two vector/inserts are ligated into a single plasmid (Figure 12.13). However, running our ligation products with the uncut control will reveal the case where multiple vector/inserts have been ligated, which will be visualized by the supercoiled band running more slowly than usual. Since we will be picking multiple colonies for DNA analysis, we will be able to detect a slow running supercoiled band because this is a very rare occurrence and most (if not all) of the plasmids from the other colonies will be of the appropriate size. To detect the direction of the insert, we simply run a digest with an enzyme that cuts inside the insert and another enzyme that cuts outside of it, preferably close to one of the ends (this was presented earlier in Figure 12.7).

Sticky-ended ligations can also yield plasmids with multiple copies of the vector/insert because sticky ends made by the same enzyme will line up (Figure 12.14a and b). While this is not a highly probable event, it must be guarded against. Notice that a restriction cut to verify proper insertion for the desired plasmid will yield the same results as the same restriction digest per-formed upon a plasmid with multiple vector/insert copies (Figure 12.14c and d left). This is where the control sample containing uncut plasmid sample may come in handy. Migration distance of the supercoiled plasmid will indicate how many vector/insert copies are in the given plasmid. However, if insert sizes are small, cutting the plasmid with only one enzyme might be a better approach (Figure 12.14d, right). Note that the commonly used molecular weight markers only indicate the sizes of linear fragments. They are irrelevant with regard to

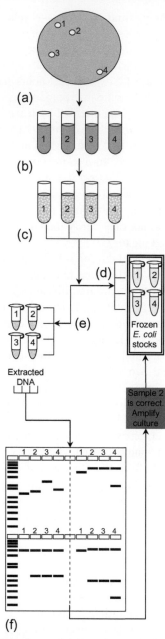

FIGURE 12.12   Logical progression to deciding which colony is appropriate for further amplification. (a) Colonies of transformed bacteria are picked and (b) placed in individual tubes with
culture medium. (c) After incubation, (d) a small amount of each culture will be frozen and (e) the
rest will undergo DNA extraction. (f) Extracted DNA will be analyzed via restriction digests and gel
electrophoresis. If and when a suitable DNA sample has been identified, the corresponding frozen
culture (d) will be thawed and amplified for a larger-scale extraction.

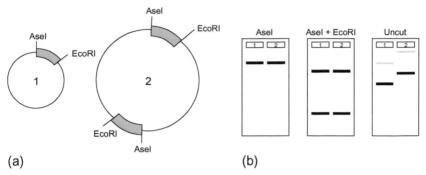

(a)   (b)

FIGURE 12.13   (a) Plasmid 1 represents a typical vector with a single insert (blue). Plasmid 2 represents a ligation of two (vector + insert) units. The inserts are bounded by recognition sequences for AseI and EcoRI, respectively. (b) Routine restriction analysis can be misleading. The first gel shows the results of cutting plasmids 1 and 2 with the enzyme AseI, a procedure typically used to produce a linear version of each vector/insert. This technique will not reveal the full picture, though. Cutting the inserts out of each plasmid (second gel) will likewise obscure the truth because the inserts, and the sequences between the inserts, are identical for both plasmids. In this special case, only running uncut plasmid samples will give a clue as to what is going on. (Note: Sometimes in uncut samples, there will be linear DNA sequences due to plasmids being opened by shear forces. This effect is shown by faint gray bands in the third gel.)

supercoiled dsDNA sizes. We can still obtain information, however, by comparing supercoiled dsDNA plasmids from different samples to each other.

Our focus on implying plasmid identity via restriction analysis begs the question "How do we know that the vector or insert has not gained a mutation?" While restriction analysis gives us a rough idea as to the identity of a plasmid, in reality, it does little more than verify the existence of recognition sequences and the relative distance between such sequences. Other methods must therefore be employed if a more rigorous analysis of plasmid identity is desired. DNA sequencing is a fairly reliable way to verify plasmid identity, base by base. This technique will be employed after the final amplification in our process. We will cover DNA sequencing in greater detail later.

At this point, we know how to test for empty vectors, multiple inserts, and insert orientation. We should be able to identify which of our picked colonies is/are most likely to contain the plasmid we originally designed. Refer back to Figure 12.12, especially part *f*. The identity of each plasmid represented in the gel will not be revealed here—that is the subject of one of the problems at the end of this chapter—but suffice it to say that sample #2 is the correct one. Each DNA sample is paired with a specific *E. coli* sample that we have stored in glycerol in the freezer. It is now a simple matter to expand the correct culture (#2). We can take the frozen cultures that contain undesirable forms of our plasmid, kill the bacteria (with bleach), and discard them to preserve freezer space.

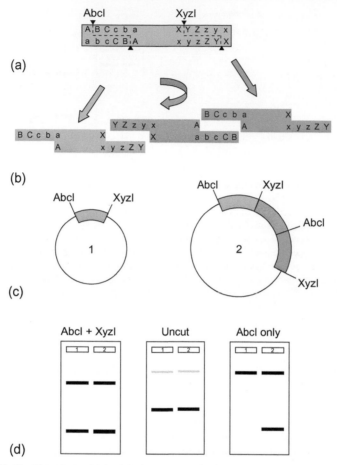

FIGURE 12.14   Detecting multiple sticky inserts. Complementary base pairs are reflected by matching upper- and lowercase letters. (a) Consider a stretch of DNA with recognition sequences for the two hypothetical enzymes shown. (b) Cutting several sequences will yield several sticky-ended fragments. Note that the middle (orange) fragment is the same as the other two, only rotated 180°. (c) While most open vectors will receive a single sticky-ended insert (plasmid 1), there is a possibility that some vectors will receive multiple inserts (such as what is depicted in plasmid 2). (d) It is not always easy to identify a case of multiple inserts. Cutting out the fragment as usual, using enzymes that recognize the bounding restriction sites, will yield similar results, as shown by the gel on the left. Having comparing uncut versions of the plasmids will, in theory, show the difference in insert sizes, but if the inserts are small, the difference may go unnoticed (middle gel). Cutting at only one of the bounding restriction sites will make the difference between the plasmids clear, as shown by the gel on the right.

To expand the once-frozen, desirable culture, we will take a sample of it, measured in microliters, and place it into a flask containing LB plus antibiotic with a 1000-fold greater volume. For example, a 300 μl aliquot of the *E. coli* could be transferred into 300 ml of LB + antibiotic. In the laboratory, the volume of LB might be anywhere from 50 to 2000 ml. On the industrial scale, volumes can be much larger.

12.3.2 The Plasmid Prep Procedure, or, the 12-Step Program for Plasmid Recovery

The small-scale procedure just described is an application of the plasmid prep procedure. The second larger-scale procedure will involve the same basic steps. We will now examine a classic version of the procedure in greater detail:

1. *Establish a culture*

This may involve going back to a frozen stock of bacteria (as in Figure 12.12) or repicking a colony that has been shown to contain the correct version of the engineered plasmid. The culture will be expanded in a suitable volume of medium, depending on the application. Following the establishment of an *E. coli* culture that contains our engineered plasmid, several separations must be performed to isolate the plasmid. Consider that, at this point, the *E. coli* culture contains cells, cell metabolites, and medium (LB broth).

"LB" is an acronym given to Luria-Bertani broth, a bacterial culture medium whose recipe is attributed to Salvador Luria and his research associate Giuseppe Bertani. LB broth contains water, sodium chloride, tryptone, and yeast extract. *Water* is present as a diluent because it is the solvent of life. *Sodium chloride* is used to create an isotonic solution. *Tryptone*, derived from a digest of casein (a protein in cow's milk), provides amino acids, which serve as building blocks for protein synthesis as well as a carbon source for the bacteria. *Yeast extract* provides nutrients and growth factors to enable cell growth.

Aside: Osmolarity

"Isotonic" is a descriptive term that refers to the osmolarity of a solution relative to a cell. *Isotonic* indicates that there is no osmotic gradient between the inside of the cell and the solution in which it resides. A *hypertonic* solution is a solution that has a higher osmolarity than the interior of the cell. Cells in a hypertonic solution will lose water and crenate (shrink) as a thermodynamic result of the osmotic gradient being driven toward zero. In a *hypotonic* solution, there is lower osmolarity relative to the interior of the cell. This time, water will rush into the cell, causing it to swell and even lyse (burst) in extreme conditions.

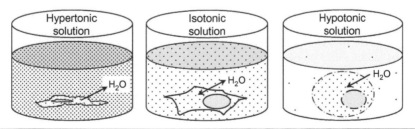

The separation of our engineered plasmid from everything else in the flask is performed stepwise to address distinct components in the flask. The process will begin with a modified alkaline lysis, followed by plasmid DNA being separated via an anion exchange column, followed by elution in a high salt buffer, and end with salt removal via alcohol.

2. *Remove materials that surround the cells*

The first step is straightforward: There are no plasmids outside the cells, so separation of the plasmids will include the removal of all salts, proteins, waste products, etc. from outside the cells. One might consider evaporation for this step, but this would only remove the water from outside (and subsequently inside) the cells. The salts, proteins, and other molecules from the medium would still remain around the cells. One might consider letting the bacteria grow until the media is exhausted, but then, the cells would be surrounded by waste and a separation of the cells from what surrounds them would still be required. One could filter the cells. The medium would flow through the pores of the filter while the bacteria would be left behind, awaiting washing and removal from the filter. This is a feasible possibility, but it is relatively expensive. What is normally employed is centrifugation, which capitalizes upon the difference in density between the cells and the medium.

Relative centrifugal force (RCF) is described by

$$\text{RCF} = r\frac{\omega^2}{G} = r\frac{(2\pi N)^2}{G} = 1.119 \times 10^{-5} \, rN^2$$

where ω is the angular velocity, G is the gravitational constant of Earth $(9.8 \, \text{m/s}^2 = 3.528 \times 10^6 \, \text{cm/min}^2)$, r is the radius of rotation (distance, in cm, from the center of the rotor to the sample), and N is the number of revolutions per minute (RPM).

Practically, what we need to decide upon is the RPM. If a protocol requires a centrifugal force of, say, $6000 \times G$, we would need to know how fast to set the centrifuge. This can be determined by

$$N = \sqrt{\frac{8.937 \times 10^4 \, \text{RCF}}{r}}$$

After centrifugation, we will still have a bacterial soup, but all the chicken and noodles will be in one wad at the bottom of the bowl and all the liquid will be everywhere else. We can then simply pour off the broth, or in laboratory terms, we discard the supernatant. With this simple procedure, we will have accomplished the removal of the salts, proteins, and wastes from the outside of the cells and be left with just cells.

3. *Resuspend the cells*

We now have cells that contain our engineered plasmid, plus a plasma membrane, cell wall, and a cytoplasm that contains proteins (including enzymes and ribosomes) and messenger RNA. There is no nucleus because *E. coli* are prokaryotes, but they do have circular genomic DNA that is attached to the plasma membrane at one or several points (Figure 12.15). While separating the plasmid from most other cellular constituents is relatively easy because of their differing

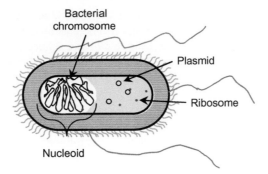

FIGURE 12.15   Location of DNA in the bacterial cell. The chromosome, which is attached to the inner cell membrane, is located in a non-membrane-bound area of the cytoplasm known as the nucleoid. Translation will be carried out by 70S ribosomes.

sizes, charges, and hydrophilicities, some conceptual finesse is required to separate plasmid DNA from genomic DNA and the polynucleotides of mRNA. If we are going to be working with the guts of the cell, we need to get to the guts of the cell, which means we must lyse the cell. This is the modified alkaline lysis portion of the protocol. The cells will be split open with a combination of detergent and high pH. First, the cell pellet will be resuspended. Whereas before, we had a cell soup containing salts, proteins, wastes, etc., at this point, we will have a controlled suspension that contains tris(hydroxymethyl)aminomethane (Tris), EDTA, and RNase A. In discussing this protocol, the compositions of several buffers or additives will be discussed. The point is not to have the reader memorize the compositions of solutions but rather understand why certain constituents are used as the goal. Tris is a pH buffer. EDTA, introduced in Chapter 10, is a chelator of divalent cations such as Ca^{2+}. RNase A is an enzyme that catalyzes the hydrolysis of RNA. It serves a very important role because it will degrade the RNA (including mRNA) present in the cell lysate.

Perhaps the most difficult components to separate in the cell lysate are genomic DNA, plasmid DNA, and RNA, because they are all polynucleotides. The first to be removed will be the RNA, which can be accomplished because it is chemically distinct from both plasmid and genomic DNA in two key aspects: RNA is single-stranded, and it contains a hydroxyl on the number two carbon of each ribose in the backbone. RNase A acts upon single-stranded RNA by a two-step mechanism that hydrolyzes at phosphodiester bonds by temporarily having the phosphate attached to both the 3′ carbon and the 2′ carbon (using the oxygen atom on the 2′ hydroxyl) simultaneously.

4. *Lyse the cells*

The cells will be lysed with 200 mM NaOH and sodium dodecyl sulfate (SDS). The sodium hydroxide serves to raise the pH significantly, which will result in the denaturation of proteins. The SDS, being an ionic detergent, will both disrupt membranes and bind to the denatured proteins. The proteins will be coated with SDS to give them a consistent, anionic charge concentration, thus rendering them soluble in aqueous solution. The plasma membrane will be disrupted by SDS in the manner discussed in Chapter 2. Notice that the genomic DNA that is shown attached to the plasma membrane in Figure 12.15 will still be attached to the plasma membrane after cell lysis. At this point, if we remove the plasma membrane from our solution, we will also remove the genomic DNA by association. This brings us to the next step in the process.

5. *Precipitate the plasma membrane (and genomic DNA)*

The high-pH cell lysate solution will be neutralized with 3 M potassium acetate (KAc). KAc will completely dissociate in aqueous solution into K^+ and Ac^-, yielding 3 M of potassium ions.

Recall from the previous step that we used sodium dodecyl sulfate, which dissociates into Na^+ and dodecyl sulfate (DS^-) ions. When combined with an excess of potassium ions, which is assured by the high concentration (3 M) of KAc added, potassium dodecyl sulfate (KDS) will be formed. KDS will precipitate out of solution, effectively clearing the solution of detergent.

$$SDS \rightleftharpoons Na^+ + DS^-$$
$$KAc \rightleftharpoons Ac^- + K^+$$
$$\rightleftharpoons NaAc + KDS \quad \rightarrow \text{a precipitate}$$

During the precipitation, the above reactions will be drawn to the right, and cellular components that were associated with SDS will be dragged along as the KDS precipitates. These components will include the plasma membrane, which is composed mainly of phospholipids, and proteins. Because of the density of the phospholipids, the precipitate will fall *up* out of solution. Also recall that in prokaryotes, which do not have cell nuclei, organization of the circular genome is maintained through its attachment to the plasma membrane (Figure 12.15). As such, when the plasma membrane is precipitated out of solution, genomic DNA and membrane proteins will also be pulled into this layer of precipitate.

When the cells were lysed, the incubation was only allowed to proceed for 5 min. It was not permitted to proceed for 5 h because some of chromosomal DNA would have been hydrolyzed, which would have prevented removal of the entire genome during the present step. (In addition, some of the plasmid DNA would have also been hydrolyzed, which would have defeated the purpose of this entire procedure.)

6. *Removal of membranes, proteins, and genomic DNA*

At this point, the three types of nucleotides have been either hydrolyzed or separated. We degraded RNAs with RNase A, and we separated out the genomic DNA via KDS precipitation. The bulk solution now contains a mixture of our plasmid, degraded RNA, small proteins, and some salts. The tube contains a slightly cloudy solution topped by a fraction that has the appearance of little chunks of Crisco®. The two fractions can be separated by pushing the bottom fraction of the solution through a filter into a collection tube. The filtrate will contain our plasmid, degraded RNA, small proteins, and some salts.

7. *Endotoxin removal E. coli* are gram-negative bacteria

As such, they contain a membrane on the exterior side of the cell wall. This outer membrane contains phospholipids, proteins, and lipoproteins but is mainly composed of *lipopolysaccharides* (LPS). Another name for LPS is *endotoxin*. Endotoxins can elicit an inflammatory response in animals, and they can activate the alternate pathway of the complement cascade (not covered). LPS molecules will almost certainly have been washed through the filter during the previous step of this protocol. If the ultimate purpose of the engineered plasmids involves an animal or clinical application, it is important to remove the endotoxins from the DNA solution to help prevent an adverse response to delivery of our DNA.

There exist several different routes to lower endotoxin levels in pharmaceutical preparations. However, the fact that so many different methods are in use indicates that there is still a problem with endotoxin removal. While endotoxins will most likely not be completely removed from our solution, endotoxin levels can be lowered to nonbioactive levels. A method commonly used for DNA purification procedures involves the use of a *nonionic* surfactant such as Triton X-114. *Surfactants* are "surface-active" agents, with both hydrophilic and hydrophobic regions in each molecule. As such, they can be used to form micelles around endotoxin molecules for removal from our bulk solution.

8. *Isolation of the plasmid via anion exchange*

Imagine that we have a column, and the column is packed with beads that carry positive charges. If we surround the beads with a solution that contains water and a salt such as NaCl, the salt will split into Na^+ and Cl^- in the aqueous solution. Because of the positive charges on the beads, Cl^- ions from the solution will adhere to them. Now suppose that we pour a solution that contains plasmid DNA into the column. The large polyanionic DNA molecules will displace many of the Cl^- ions from the beads. In other words, the Cl^- *anions* will be *exchanged* for DNA in the *resin*, hence the name "anion exchange resin." (There do also exist cationic exchange resins.) In many of the popular plasmid preparation kits, an anion exchange column contains a filter-like material that carries multiple positive charges.

The product from step five, containing salts, small proteins (including RNase A), degraded RNA, and micelles housing endotoxins, will be allowed to flow through the column. Most of the proteins, cations from salt dissociation, and endotoxin-containing micelles will flow completely through the column because they do not contain the correct charge for adhering to the anion exchange surface. (It should now be clear why a nonionic detergent was used for the removal of endotoxins.) Plasmid DNA will bind to the column preferentially over small molecules such as individual nucleotides or Cl^- because of its size (high number of negative charges) and charge density.

9. *Wash the column*

The column now contains plasmid DNA plus small molecules that are not desired in the final isolate. To remove many of these unwanted molecules, we will wash the column with two column volumes of a solution that contains a pH buffer, salt (NaCl), and isopropyl alcohol. The buffer is important because it will help to maintain the negative charge on the plasmid DNA molecules, which will serve to retain the DNA adherence to the column. The salt, at 1 M, can serve to help wash out any residual proteins. The isopropanol, at a concentration of 15%, is used to condense the DNA. It serves to prevent nonspecific binding of molecules to the cationic surface of the column membrane. The overall result of this step is that the salt and protein concentrations in the adhered, DNA-containing fraction will be greatly reduced.

10. *Elute the DNA*

At this point, the column fraction contains our plasmid DNA and little else. We must now collect the DNA by getting it to separate from the beads, a process called elution. This will be accomplished with a high-salt solution. Just like we were able to displace Cl^- with plasmid DNA, we will be able to displace the plasmid DNA with a high concentration of Cl^-. The elution solution will contain a buffer (at pH = 8.5, up from pH 7.0 in the previous wash step), 1.25-1.60 M NaCl (considered a high-salt concentration, up from 1.0 M in the previous step), and 15% isopropanol.

11. *Remove the salts and isopropanol*

The solution at this point primarily contains plasmid DNA, NaCl, and isopropanol in aqueous buffer. Desalting is relatively easy, since the DNA will be condensed in the presence of isopropanol and can therefore be separated by centrifugation. The RCF required for separating plasmid DNA from isopropanol is much higher than what we used in the beginning of this protocol: $15,000 \times G$ (or more). The pellet will contain primarily plasmid DNA with some residual salt. The majority of salt and isopropanol can simply be decanted from the tube after centrifugation.

The pellet will then be washed with 70% ethanol. The DNA will still be condensed in the ethanol and can therefore be pelleted once again via centrifugation. The purpose of this step is to remove the remainder of the isopropanol and salt. Once again, the supernatant can be decanted, leaving the pellet of plasmid DNA, a minute quantity of ethanol and water, and virtually no isopropanol or salt. The advantage to washing with ethanol here as opposed isopropanol is that ethanol is more volatile and thus can be removed more easily via evaporation.

Prior to the evaporation, the pellet rests in a very small amount of 70% ethanol in 30% water. Because under normal conditions the ethanol is more volatile than water, the percentage of ethanol around the pellet will decrease during the evaporation step, eventually nearing or reaching 0%. There should be some residual water left behind. It's important to never let the DNA dry out completely, as it is relatively difficult to resuspend it without inducing physical damage.

12. *Resuspend the DNA*

While one could resuspend the plasmids in water, the task will be easier if a buffer such as Tris (pH=8) is used because the slightly alkaline pH will ensure the DNA carries multiple negative charges, making it more soluble in aqueous solution.

12.4 SPECTROPHOTOMETRY

At this point, we have created a plasmid and amplified it in *E. coli* and have purified it via the multistep chemical separations procedure just presented. We now have a tube with our engineered plasmid in it. Great! How much DNA, in micrograms, do we have? What volume will we need per experiment? Before we can use the DNA for its intended purpose, such as delivering it into cells, it must first be quantified. This can be accomplished through spectrophotometry.

12.4.1 Beer's Law

Double-stranded DNA, like many substances in solution, will absorb light energy if the photons are of the correct wavelength. Different substances have different optima for absorbing light. For dsDNA, the optimal absorption wavelength is 260 nm (Figure 12.16). This implies that if we were to put dsDNA of an unknown concentration into a chamber and then deliver a known amount of 260 nm light, we should be able to determine the concentration of the dsDNA

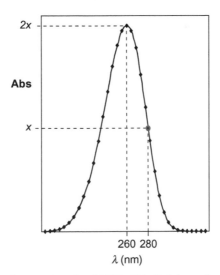

FIGURE 12.16 Absorption spectrum for dsDNA. Note that, for a given concentration of pure dsDNA, the amount of 260 nm light absorbed is approximately twice the amount of 280 nm light that is absorbed.

solution in the chamber. This calculation of *absorbance* (also known as *optical density*) can be performed using Beer's law:

$$A = alc,$$
$$\text{Absorbance} = (\text{molar absorptivity})(\text{path length})(\text{concentration}).$$

(Note that molar absorptivity is the same as the extinction coefficient, ε, discussed in Chapter 8.)

Suppose we have a known amount of light of a specific wavelength that shines in a defined path toward a transparent cuvette. The cuvette will contain a solution of solvent plus a substance of interest. In the present case, consider that the solution is elution buffer (from the plasmid prep procedure) plus ds-DNA (plasmids). The light will pass through the solution-containing cuvette and come out the other side, where it will then strike a photodiode detector. This detector measures the amount of light that hits it. In the equation of Beer's law, molar absorptivity is a constant for the specific solute at standard concentration. Sometimes referred to as the molar extinction coefficient, it can be found in reference books expressed in either $cm^2/mole$ or M 1 cm^{-1}, which are both valid because of the unit conversion $1\,ml = 1\,cm^3$. Path length refers to how far the light must travel to get through the sample (in centimeters) and is represented by the width of the cuvette. Note that values obtained for absorbance are unitless.

Consider that the molar absorptivity of cupric sulfate is around 20, which means a 1 M solution of cupric sulfate, an easily seen blue liquid, will yield an absorbance of 20. For comparison, a 1 M solution of beta carotene (which is responsible for the orange color in carrots) has an absorbance of ~100,000. The reason for the large difference is that beta carotene is a huge molecule in comparison with cupric sulfate. One mole of these huge molecules in a liter of water will be a comparatively opaque solution, as opposed to a 1 M solution of the much smaller cupric sulfate molecules, which will allow much more light to pass between the molecules of solute. We can describe the intensity of light that makes it through the cuvette when the sample molecule is present (I_s) as a function of the intensity of light that makes it through the cuvette when no sample molecule is contained in the solvent (I_0) using Beer's law:

$$I_s = I_0 \times 10^{-alc}$$
$$\Rightarrow \frac{I_s}{I_0} = 10^{-alc} \equiv T$$

T is the *transmittance* of the solution at a given solute concentration.

The transmittance is the ratio of light getting through the solution in the presence *versus* absence of solute. Since intensity cannot be negative and energy cannot be created, the value for transmittance will always be between 0 and 1. Multiplying T by 100 gives the % transmission, or the percentage of light that gets through the sample in the presence of solute.

Using the above equation, we can solve for c to get

$$c = \frac{-\log(T)}{al} = \frac{\log\left(\frac{1}{T}\right)}{al},$$

which shows that the concentration of solute has a negative log relationship with the transmittance of the solution.

If we rearrange the Beer's law equation to get $c = A/al$, it should be easy to see why *absorbance* (also known as *optical density*) is typically used to determine solute concentration instead of transmittance. (Keep in mind that A is expressed on a logarithmic scale.) Spectrophotometers use light that is monochromatic, meaning the light is of single color (has a single wavelength). Values thus obtained for absorbance or optical density are written as A_λ or O.D.(λ), where λ is the wavelength in nanometers.

12.4.2 Determination of DNA Concentration

The above is a general representation of Beer's law. In applications that involve dsDNA, certain facts become important. First, the maximal absorbance of ds-DNA in solution occurs at a light wavelength of 260 nm. Second, the molar absorptivity of dsDNA (assuming an equal presence of the bases A, G, C, and T) is such that a 50 µg/ml solution has an $A_{260} = 1.0$. Note that this value involves unit conversions to switch from moles to µg/ml. The use of the 50 µg/ml value implies that, assuming path length = 1 cm,

a reading of 0.500 for A_{260} indicates a dsDNA concentration of 25 µg/ml,
a reading of 0.400 for A_{260} indicates a dsDNA concentration of 20 µg/ml.

Note that the above calculations determine the concentration of dsDNA *in the cuvette*. In almost every situation, however, the cuvette will contain a dilution of your DNA stock solution. Suppose that you have just finished isolating some plasmid DNA that was amplified in *E. Coli*. Loading the entire volume of product into the spectrophotometer cuvette would not make sense because there would be no DNA left over for your intended experiments! It makes much more sense to load a small aliquot (on the order of microliters) into the cuvette. The equation for dsDNA concentration in the cuvette must be multiplied by a dilution factor to yield the concentration of DNA in your initial stock:

$$c = \left(50 A_{260}\right)\frac{V_{tot}}{V_{sample}},$$

where V_{tot} = total volume in the cuvette (in ml) and V_{sample} = volume of sample taken from stock solution (use µl).

While V_{sample} *can* have units of milliliters, microliter volumes are typically used. The use of µl for V_{sample} in the calculation will yield unit concentrations of µg/µl, which are equivalent to mg/ml.

Most spectrophotometer cuvettes are 1 cm wide, which allows for the dropping of l from Beer's law calculations. Keep in mind that not all cuvettes necessarily hold 1 ml. Smaller volumes allow the user to use a smaller amount of precious stock solution for the analysis. Decreased sample volumes are achieved by using cuvettes with thicker walls and/or shorter heights. These specialized cuvettes will not alter the equation used to determine concentrations.

The above equation is for dsDNA. When RNA is being measured, change the 50-40 μg/ml (because $A_{260} = 1.0$ for RNA is achieved at a concentration of 40 μg/ml). When ssDNA is being measured, the value of the constant is debatable; 34.5 μg/ml is generally acceptable.

Example

We have a cuvette that holds 600 μl. We load 598 μl of buffer into the cuvette and take a reading on the spectrophotometer at 260 nm. The reading comes out =0.000. We then load 2 μl of our DNA stock (stored in the same buffer) into the cuvette, mix, and obtain a value of $A_{260} = 0.231$. What is the concentration of our DNA stock?

Answer: Using

$$c = \left(50A_{260}\right)\frac{V_{tot}}{V_{sample}},$$

we have

$$c = \left(50\mu g/ml * 0.231\right)\left(0.600ml / 2.000\mu l\right)$$
$$= 3.465\mu g/\mu l.$$

12.4.3   Determination of DNA Purity

Knowing the concentration of DNA in our stock solution is important, but another important piece of information that can be obtained from absorbance data is the purity of the DNA solution. Proteins represent the major contaminant in DNA preparations. It is important to be aware of how much protein is in a given plasmid solution because proteins can foul downstream procedures when we try to use our DNA. Similar to DNA, proteins affect light transmission through solutions. By virtue of the aromatic amino acids, especially tryptophan, proteins will maximally absorb light with a wavelength of 280 nm. The ratio of plasmid to protein can be determined using the ratios of maximal absorptions: A_{260}/A_{280}. For dsDNA, ratios between 1.8 and 2.0 are desirable. For RNA, a ratio above 2.0 is wanted. There's an upper limit to the A_{260}/A_{280} ratio because the absorbance curve for pure DNA (or RNA) will have a nonzero value at $\lambda = 280$ (Figure 12.16). Let's call that value x. (For pure dsDNA, $A_{280} = x \approx A_{260}/2$.) Therefore, as the concentration of protein in the plasmid solution goes to zero, the A_{280} value obtained for that solution will go to x. (For dsDNA, A_{260}/A_{280} will go to 2.) Refer to Figure 12.17.

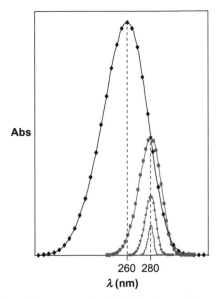

FIGURE 12.17   The absorption spectrum for DNA (black curve) is centered at 260 nm, and for protein (red curves), it is centered at 280 nm. To determine DNA purity, A_{260}/A_{280} is calculated. Both DNA and proteins will contribute to the A_{280} value. As protein content goes to zero (pure DNA), A_{260}/A_{280} goes to 2. This is represented by the progressively lower amplitudes of the red curves.

For plasmid preparations, typical values range from 1.6 to 2.0. For quality control reasons, it is best not to use plasmid DNA preparations for which A_{260}/A_{280} is below 1.8. In such cases, the DNA solution can be further purified through a technique called phenol extraction or by using commercially available cleanup kits. Once purity has been verified and DNA concentration has been determined, it is common to dilute the stock solution to a convenient concentration such as 1 μg/μl.

12.5   WHAT WE HAVE LEARNED SO FAR

We have discussed the engineering, amplification, and purification of plasmids that will be used for biotechnical applications such as gene delivery. The point will be to deliver the genes into a cell or organism so that the encoded proteins can be expressed. These techniques are mainstays to the biotechnologist who wants to harness cells for the production of specific proteins. The goal of the gene delivery application may be to have the proteins act on the expressing cell itself, or the goal may be the generation of a specific protein product on a large scale for commercial use. An example of the former case is having a stem cell express a growth factor that will aid in differentiation. The latter case is exemplified by the production of recombinant insulin. Recombinant insulin can be routinely produced in such large quantities now that it's available for home and clinical use. This technology has been great news for type I diabetics, who require insulin injections every day. While it may seem somewhat tedious that we have covered proteins, cells, restriction enzymes, DNA ligase, bacterial transformations, and chemical separations, it should be becoming clear that the biotechnologist might want to be able to construct a specific plasmid and put it into a chosen cell or cell type so that a desired product will be expressed. The product could be a drug intended for later distribution, or it could act as part of a treatment for the host cell or organism itself.

QUESTIONS

1.

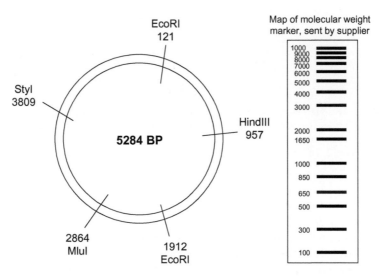

a. Copy the gel on the right, and draw in lines to predict the results of the indicated enzyme cuts. Indicate the length of each band you draw.
b. The last lane shows what you actually saw after running the gel, which differs from your prediction. Explain a reason for the difference.
c. What control could you have run to prove your hypothesis from part *b*?

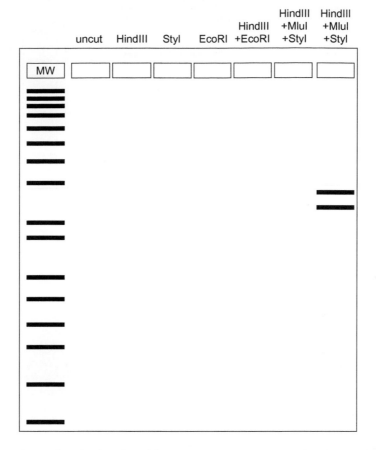

	uncut	HindIII	StyI	EcoRI	HindIII +EcoRI	HindIII +MluI +StyI	HindIII +MluI +StyI

2. At what points in the plasmid prep procedure are genomic DNA and messenger RNA separated from the plasmid DNA?

3. When is RNase A separated from the plasmid-containing fraction during the plasmid prep procedure?

4. Refer to figure of a gel below. Suppose that your design was to insert a 1000 bp fragment into a 4000 bp vector using EcoRI and BamHI sticky ends:

$$5'\text{-EcoRI-}[\text{Insert}]\text{-BamHI-}3'.$$

Also suppose that there is an AseI cut site 800 bp into the insert.

The four quadrants of the gel are

(upper left) uncut plasmids,

(upper right) plasmids cut with EcoRI,

(lower left) plasmids cut with EcoRI + BamHI,

(lower right) plasmids cut with EcoRI + AseI.

a. Identify the compositions of samples 1-3, given the banding patterns shown in the figure.

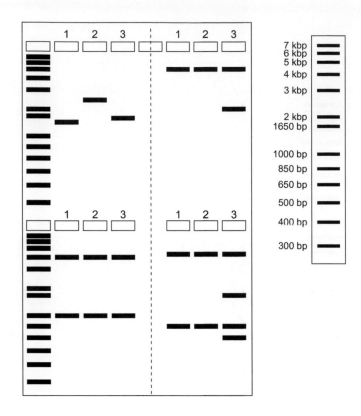

b. Is it possible that an unpicked colony had an insert that was ligated in the reverse orientation? Why or why not?

c. Exactly how many bases make up the bands for all lanes containing sample 3?

5. Refer to figure of a gel below. Suppose that your design was to insert a 1000 bp fragment into a 4000 bp vector using blunt-end ligation:
   Also suppose that there is an AseI cut site 800 bp into the insert.
   The four quadrants of the gel are
   (upper left) uncut plasmids,
   (upper right) plasmids cut with EcoRI,
   (lower left) plasmids cut with EcoRI + BamHI,
   (lower right) plasmids cut with EcoRI + AseI.

a. Identify the compositions of samples 1-3, given the banding patterns shown in the figure.

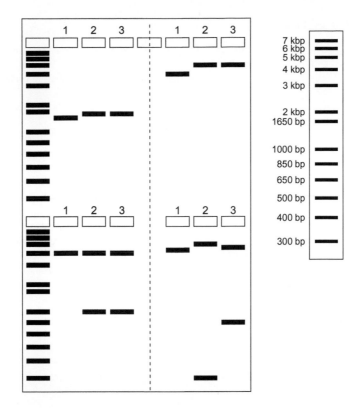

b. Is it possible that an unpicked colony had an insert that was ligated in the reverse orientation? Why or why not?

c. Exactly how many bases make up the bands for all lanes containing sample 2?

6. Given the plasmid above with a 1000 bp sequence we want to cut out to use for a blunt-end ligation, how large will the insert be after it is subjected to the with insert, and how large will the insert be after cutting it out with ClaI and AatII, followed by blunting with the Klenow fragment? (The recognition sequences of the two enzymes are shown below.)

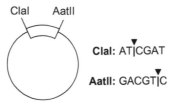

ClaI   AatII

ClaI: AT|CGAT

AatII: GACGT|C

RELATED READING

Bergmans, H.E., van Die, I.M., Hoekstra, W.P., 1981. Transformation in *Escherichia coli*: stages in the process. J. Bacteriol. 146, 564–570.

Jacob, F., Brenner, S., Cuzin, F., 1963. On the regulation of DNA replication in bacteria. Cold Spring Harb. Symp. Quant. Biol. 28, 329–348.

Lark, K.G., 1966. Regulation of chromosome replication and segregation in bacteria. Bacteriol. Rev. 30, 3–32.

Magalhães, P.O., Lopes, A.M., Mazzola, P.G., Rangel-Yagui, C., Penna, T.C., Pessoa Jr., A., 2007. Methods of endotoxin removal from biological preparations: a review. J. Pharm. Pharmaceut. Sci. 10, 388–404.

Rosenberg, B.H., Cavalieri, L.F., 1968. Shear sensitivity of the *E. coli* genome: multiple membrane attachment points of the E. coli DNA. Cold Spring Harb. Symp. Quant. Biol. 33, 65–72.

Sambrook, J., Russel, D.W., Sambrook, J., Russel, D.W., 2001. Molecular Cloning, a Laboratory Manual. third ed. Cold Spring Harbor Laboratory Press, Cold Spring Harbor, New York.

Sezonov, G., Joseleau-Petit, D., D'Ari, R., 2007. *Escherichia coli* physiology in Luria-Bertani broth. J. Bacteriol. 189, 8746–8749.

Stec, B., Holtz, K.M., Kantrowitz, E.R., 2000. A revised mechanism for the alkaline phosphatase reaction involving three metal ions. J. Mol. Biol. 299, 1303–1311.

Sueoka, N., Quinn, W.G., 1968. Membrane attachment of the chromosome replication origin in *Bacillus subtilis*. Cold Spring Harb. Symp. Quant. Biol. 33, 695–705.

Chapter 13

Gene Delivery

At this point, we have discussed how to engineer a plasmid and how to amplify it so that there will be enough plasmid to use for experiments, processes, or treatments. We will now address the topic of gene delivery, which is one way the biotechnologist may choose to use the engineered plasmids. Another way one could use the DNA constructs is to put them into bacterial cells with the goal of having the bacteria produce a product. One example of this is the production of recombinant insulin for human diabetic patients by *E. coli* or yeast. *Recombinant* proteins result from the expression of recombinant DNA. Recombinant DNA is the result of the processes we have already discussed that involve recombining series of bases from different sources. All of the effort that we have devoted to a discussion of genes is not solely for applications of gene therapy. The production of proteins via bioreactors, the production of genetically modified plants or animals, the guidance of stem cell differentiation, and even the production of beer are all applications that should require knowledge of genes and their functions to be performed well. However, gene delivery is behind many different aspects of biotechnology.

Gene therapy is the delivery of genetic material into cells for the purpose of altering cellular function. This seemingly straightforward definition encompasses a variety of situations that can at times seem unrelated. The delivered genetic material can be composed of DNA or RNA. The alteration in cellular function can be an increase or decrease in the amount of a native protein that is produced or the production of a protein that is foreign. The delivery of the genetic material can occur directly, as is the case with microinjection, or involve carriers that interact with cell membranes or membrane-bound proteins as a part of cellular entry. Polynucleotides can be single- or double-stranded and can code for a message, or not (as is the case for antisense gene delivery or miRNA). Even the location of cells at the time of gene delivery is not restricted. Cells can be part of a living organism, can exist as a culture on a plate, or can be removed from an organism, transfected, and replaced into the same or a different organism at a later time.

Gene therapy came into being after the development of recombinant DNA technology and initially moved forward using viruses to target sites *in vitro*. With the understanding of retroviruses and their possible uses as vectors for delivery, gene therapy progressed to applications involving mammalian organisms during

An Introduction to Biotechnology. http://dx.doi.org/10.1016/B978-1-907568-28-2.00013-7
Copyright © W T Godbey, 2014.

the 1980s. Since then, new techniques for gene delivery have been developed that utilize both viral and nonviral carriers. These techniques have been successful enough that proposed disease treatments have evolved to the clinical trial stage with encouraging success. Although many cellular processing mechanisms remain unclear and the search for the ideal gene delivery vector remains ongoing, the tools and methods for gene delivery have provided a strong base from which to build.

13.1   GENE DELIVERY VEHICLES: AN OVERVIEW

Before considering what cells do with gene delivery complexes, one should consider the makeup of gene delivery vehicles. We will begin by organizing gene delivery vehicles and methods into groups. Figure 13.1 is a basic classification of gene delivery vehicles with some examples. Although there exist cases in the literature where classifications have been mixed, as in, for instance, viral proteins being used in conjunction with synthesized polymers, the presented chart covers the rudimentary ideas for vehicles and approaches that have been used to deliver genes up to now.

If you were to talk to a gene therapist about a gene delivery vehicle, the first question that might come up is whether the vehicle is a virus or not. This is the primary delineation for gene delivery vehicles. The methods of gene delivery can be further divided into the three branches of natural science: biology, chemistry, and physics. The viral vectors and the biological vectors are nearly synonymous (despite viruses not being alive) because viruses were produced by

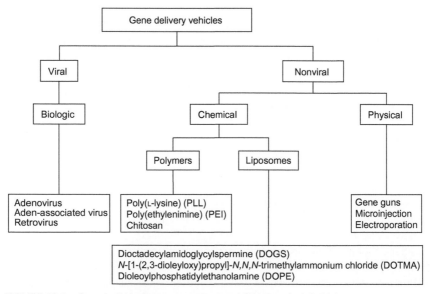

FIGURE 13.1   Organization of some common gene delivery methods/vehicles.

nature as gene delivery vehicles—scientists didn't have to do anything to create the first viral gene delivery vehicles. Viruses seem to exist in nature only to propagate their own DNA in biological systems.

Moving down one level in Figure 13.1, focusing upon the biological methods, some of the viruses that are used for gene delivery include adenovirus, adeno-associated virus, herpesvirus, and lentivirus. Lentivirus is a type of virus known as a retrovirus, which delivers RNA into cells instead of DNA.

Nonviral gene delivery methods can be divided into chemical and physical approaches. The chemical methods include certain polymers and lipids, while the physical methods utilize physical properties and forces to transport genetic material into cells. For instance, you can coat some shot (similar to the shot in a shotgun) with DNA, load it into a special gene gun, and blast away at cells or tissues. The shot goes into or through the cell and the genetic material is left behind. As absurd as this description may seem, it represents an actual method of gene delivery, which will be discussed in Section 13.2.2, Physical Delivery Methods. One can think of the physical delivery methods being brute force ways of getting genes into cells. Examples include the gene gun, microinjection, and electroporation, which will all be discussed in Section 13.2.2.

The chemical methods can be further subdivided into polymers and lipids. Typical polymers that can be used for gene delivery include, but certainly are not limited to, poly(L-lysine) (PLL), poly(ethylenimine) (PEI), and chitosan. Polymers used for gene delivery can be linear, hyperbranched, or dendrimeric. While PLL is a linear polymer, there are both linear and branched forms of PEI available. The branched form of PEI is *hyperbranched*, meaning that the branches do not occur in a regular fashion. When the branching is very well controlled and regular, we refer to the branched polymer as a *dendrimer*. Dendrimers can also be used for gene delivery, as is the case for starburst dendrimers. Poly(amido amine) (PAMAM) dendrimers are made in such a way that their sizes are easily controlled and their end groups can be functionalized.

Lipids have also been used widely for gene delivery. There are many different lipids that have been used to create DNA-containing capsules known as liposomes. Some are of the same ilk as what are already found in the plasma membrane, and some have been used for their pH-sensitive membrane-destabilizing properties. We will go into lipids in greater detail in Section 13.2.3, which discusses some of the main chemical delivery methods.

13.2   GENE METHODS IN GREATER DETAIL

13.2.1   Viral Delivery Methods

Why would a biotechnologist want to use a virus for gene delivery? First of all, viruses are known to produce relatively high transduction efficiencies. (Note the use of the word *transduction* here, which denotes gene delivery via a virus. *Transfection* is used to denote gene delivery via one of the nonviral gene delivery methods.) Viruses, in theory, will yield up to 100% transduction efficiency *in vitro*. Practically, however, the range of 80% to 95% is more realistic. It is often said that viruses naturally introduce genetic material into cells. The use of the word "naturally" may be based upon 4.5 billion years of evolution that have produced the efficient gene delivery machines that we encounter today both in the laboratory and in our everyday lives. Some viral applications yield permanent expression of the delivered genes. Retroviruses, which deliver RNA into cells, eventually have DNA copies of the delivered RNA inserted into the host's genome, which yields such permanent transduction.

Some of the viruses used for gene delivery are adenovirus, herpesvirus, adeno-associated virus, and certain retroviruses, which include lentivirus and FIV (feline immunodeficiency virus). We can use FIV to transduce human cells; humans won't acquire immunodeficiency syndrome from FIV, which is a very robust virus. Each of these viruses offers slightly different properties, so for a given application, one type of virus might be more suitable than another.

13.2.1.1 Retrovirus

The distinguishing feature of a retrovirus is that it delivers RNA into cells (Figure 13.2). Retroviruses are enveloped viruses that can be engineered to

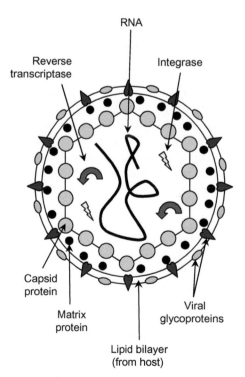

FIGURE 13.2 Basic structure of a retrovirus.

deliver up to 8000 bases of single-stranded RNA and can yield permanent transduction (meaning the delivered genes will be present inside the cell for the remainder of its life). The reason this is possible is because some viral genes are delivered as a part of retroviral transduction. With the retrovirus, the central dogma can be modified to include a pathway from RNA back to DNA, a process termed *reverse transcription*. In performing reverse transcription upon an RNA sequence, the result will be a DNA copy, referred to as *cDNA*. Do not confuse cDNA with genomic DNA: even though the two sequences may code for the exact same polypeptide, cDNA and genomic DNA are different. To understand this point, consider the following discussion of molecular biology.

Introns are eukaryotic DNA sequences that will be transcribed, but they do not code for a polypeptide (or even part of one). (Interestingly, the prokaryotic genome does not contain introns.) These noncoding sequences were at one time called "junk DNA." This term is a misnomer, however, because the removal of intron sequences can have deleterious effects upon the cell. Introns appear within a gene as spacers between exon sequences. To be clear, an exon is a stretch of DNA that codes for all or part of the final protein product. A single gene may have several exons that are separated by introns. Recall from Chapter 4 that the RNA that is first transcribed from such a stretch of DNA is termed the primary transcript, or pre-mRNA. The pre-mRNA will undergo the removal introns via splicing (Figure 4.13).

It is the presence of introns that explains why eukaryotic genomic DNA can have a different sequence from cDNA. This distinction should be kept in mind when looking up gene sequences in repository databases. The sequences that are reported will typically be cDNA sequences, which can present problems when trying to isolate genes from genomic DNA preparations using PCR and primers.

When delivering RNA into cells via retroviruses, reverse transcription of the RNA to produce cDNA takes place via the enzyme *reverse transcriptase*. The cDNA can then be integrated into the host genome via the enzyme *integrase*. Once the cDNA has been integrated into the genome, it will be replicated when the cell replicates, ensuring that both of the daughter cells will have a copy of the virally delivered gene. This is what is meant by "permanent transduction." In theory, the gene will be expressed forever (although in practice expression levels will decrease over time as the inserted sequence becomes part of the heterochromatin).

In theory, when we use an RNA virus for transduction, we can come back a year from now and, as long as the cell population is still alive, we should see expression of the delivered gene. This may appear to be the ultimate way to go for gene delivery, but it is not the case for a couple of reasons. First, recall our discussion of housekeeping genes and that the proteins they code for are needed at all times by the cell to stay alive. Integration via integrase is random, which means the cDNA can be inserted in the middle of a housekeeping gene. Such an insertion would render the housekeeping gene useless and would lower the amount of an essential protein, possibly leading to cell death. Another problem with permanent transduction again stems from random integration. If the cDNA were inserted into the middle of a tumor suppressor gene, the result could be the transformation of the cell into a tumor cell. Also, be aware that if the cDNA were inserted upstream of a proto-oncogene, it could serve to activate or mutate the gene into an oncogene, which will transform the cell. The phenomenon of random integration is a significant reason why retroviral gene delivery has not been used in the clinical setting. However, retrovirus-mediated gene delivery is a powerful tool in the laboratory that can be used to produce or enhance cell lines for basic scientific research.

Aside: When Fields Collide

One of the inherent problems that crop up when established fields merge is that of overlapping or redundant terminology. We've already discussed the insertion of exogenous DNA into a prokaryotic cell, a process termed "transformation" by molecular biologists and microbiologists. To the oncologist or cell biologist, however, "transformation" refers to the conversion of a mortal cell into an immortal cell. Another such word is "vector." In molecular biology, a vector is the main plasmid into which one might insert an exon. To a gene therapist, the vector is not the DNA, but rather the DNA carrier. An epidemiologist might call a person who spreads a disease a vector (e.g., Typhoid Mary). Confusing, yes, but reused words can also be celebrated as proof of scientific progress.

The main advantages of retroviral transduction are high delivery efficiency and the "permanent" transduction just described. A key disadvantage of retroviral use is the vector's inability to infect nondividing cells. However, in tissue engineering applications that utilize an *ex vivo* strategy, this disadvantage could be considered an advantage as somatic cells are largely protected from accidental transduction following the implantation of a construct containing transduced cells. There are also serious safety concerns with the retrovirus, such as the generation of replication-competent retroviruses and the production of the insertional mutations just described. Because of the substantial level of concern associated with retroviral use, alternative methods of gene delivery have been developed to reduce or eliminate these specific safety risks.

13.2.1.2 Adenovirus

The adenoviruses are a group of nonenveloped viruses that carry linear, double-stranded DNA. They occur naturally, infecting mucosal linings in humans and other mammals. There is no treatment for adenoviral infections, but a normal immune system is quite capable of combating them. This point is key to the gene therapist, because adenoviral vectors used for gene therapy will be recognized by the immune systems of patients who have combated infections by that specific adenoviral serotype before. (*Serotype* refers to serologically distinguishable cells or agents.) Even if they have not, a repeated dose of adenovirus-mediated gene delivery will be recognized and cleared.

Adenoviruses have icosahedral capsids (shells). In addition to their capsids and DNA cargo, these viruses are also made up of cement and core proteins. Cement proteins stabilize the capsid, and core proteins associate with the double-stranded DNA genome. The DNA is covalently attached to a terminal protein at each 5′ terminus. The structure of adenovirus is shown in Figure 13.3.

The adenoviral genome is about 36,000 bp. Up to 30,000 bp can be replaced by foreign DNA. When parts of the viral genome are removed, the virus is rendered replication-incompetent. It takes several rounds of processing to create such viruses. Propagation requires the use of a "helper" cell line that has been designed to express necessary proteins that are no longer represented in the viral genome. The final generation of the viruses has had most of the viral genes removed, leaving a vector that is termed "gutless." These gutless vectors contain only inverted terminal repeats and a packaging sequence surrounding the engineered gene that is to be delivered.

Adenoviral vectors can infect a broad range of human cells, including both dividing and nondividing cells. Adenoviral vectors are relatively stable and can be concentrated to high concentrations (titers). Adenoviruses do not cause integration of genetic material into host cell genomes, so insertional mutagenesis is not of concern, but at the same time, the lack of genomic integration means the resulting gene expression is transient.

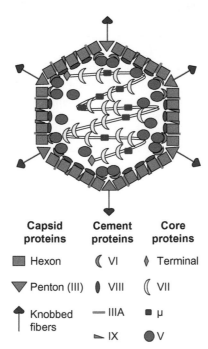

Capsid proteins	Cement proteins	Core proteins
■ Hexon	(VI	◆ Terminal
▼ Penton (III)	(VIII	((VII
↑ Knobbed fibers	— IIIA	■ μ
	⌐ IX	● V

FIGURE 13.3   The basic structure of an adenovirus.

13.2.1.3   Adeno-Associated Virus

The human adeno-associated virus (AAV) is a small, nonenveloped DNA virus with an icosahedral capsid. It delivers linear, single-stranded DNA that is approximately 4.7 kb long. AAV is naturally replication-deficient, with replication typically achieved via a helper virus such as adenovirus or herpesvirus. AAV can mediate long-term transgene expression and transduce a large range of cells. It can transduce nondividing cells, a feat that retrovirus cannot accomplish. They are replication-defective, nonpathogenic, and nonimmunogenic. The limitations of this method lie in the relatively small transgene size that can be delivered and the requirement of helper viruses for activation.

The infection cycle of AAV begins with the virus binding to host cell surface receptors and coreceptors. It enters the cell via clathrin-mediated endocytosis. Following endocytosis, acidification of the endosome is followed by the release of the AAV particle into the cytoplasm. AAV is able to integrate nonrandomly into the host genome. In humans, the site is on the q arm of human chromosome 19. The mechanism of subsequent nuclear entry is not fully understood, but the integration of the viral DNA into the host genome is dependent upon the presence of a helper virus.

The serotype AAV2 is the dominant prototype in AAV; the AAV vectors for gene therapy were once commonly developed from this serotype. In the laboratory, AAV vectors are constructed by deleting the genes *rep* and *cap* from the wild-type viral genome and using the newly created space for the engineered gene of interest. The recombinant DNA is then cotransfected with a separate plasmid that contains the *rep* and *cap* genes into host cells. Finally, the host cells are infected by an adenovirus to cause the cells to produce recombinant AAV vectors. After 2-3 days of incubation to allow for the production of virus particles, the cells are lysed and both recombinant AAV and adenoviruses are collected. AAV is then purified from the adenovirus-containing lysate by heat inactivation or gradient centrifugation. (For completeness, this is not the only way to produce AAV. Some use a method that circumvents the use of a helper virus.)

Keep in mind that AAV, like any virus, comes with all of the concerns regarding immunogenicity that we have already mentioned, so multiple serotypes of the virus can allow the gene therapist to use AAV more than once to deliver genes. AAV8 is a very promising serotype that has been used to address immunogenicity. Research has verified that most (tested) people are devoid of AAV8-neutralizing antibodies in their serum. Even in the presence of such antibodies, there is a lack of AAV8 inactivation on the first administration.

13.2.1.4   Herpes Virus

The herpesviruses are enveloped viruses with icosahedral capsids. They are able to deliver the largest payloads of any of the viruses used for gene delivery, which allows them to deliver multiple transgenes along with genes encoding specific transcription factors, which allows for tight control of transgene expression. Among all of the herpesviruses, the human herpes simplex virus 1 (HSV-1) has been used the most for gene delivery applications. It can be used to deliver engineered DNA payloads of up to 130 kbp. HSV-1 displays a natural tropism toward neurons, which suggests that it is especially suited for tissue engineering applications directed at neuropathy or diseases associated with the CNS.

Two types of HSV-1-based vectors have been developed: replication-defective vectors and amplicon vectors. As with retroviral and adenoviral vectors, replication-defective HSV-1 vectors are relatively easy to establish. Specific genes of the viral genome are deleted to make room for insertion of the engineered gene of interest, followed by transduction of a helper cell line that already produces the proteins encoded by the deleted viral genes. The result is a stock of replication-defective vectors. Amplicon vectors, on the other hand, are derived from plasmids that contain the HSV-1 origin of replication, the HSV-1 packaging signal, and the engineered gene of interest. The plasmid and an HSV-1 helper virus are then codelivered into a cell line that supports the growth of the helper virus. The helper virus typically has a feature that renders it replication-incompetent, such as a temperature-sensitive mutation that prevents replication at 37 °C.

The replication cycle of a herpesvirus begins with the initial binding of the viral envelope to cellular receptors, followed by the entrance of the capsid into the cytoplasm. The capsid is then able to bind to the nuclear membrane and release its DNA cargo into the host nucleus.

13.2.1.5   Baculovirus

Baculoviruses naturally infect insect host cells and have been used for the production of insect cell-based recombinant proteins. For example, the baculovirus-insect cell expression system has been used to produce EGF-collagen, a biomolecule that can induce cellular proliferation and therefore has been considered for tissue engineering applications. It can accommodate large inserts, which makes it possible for the biotechnologist to deliver multiple genes with the baculovirus. Due to its incapability of replication and absence of toxicity in mammalian cells, the baculovirus has emerged as a gene delivery system with ample transduction efficiency (75-85% in many cases). Baculovirus-mediated transduction efficiencies have been shown to be on par with those of adenovirus in human cardiomyocyte and fibroblast applications and better than those of adenovirus in human smooth muscle cells.

Each viral vector has its own strengths and shortcomings. It is incumbent upon each individual investigator to determine the vector that is most suitable for a particular application. Retroviral and adenoviral vectors have been traditionally the most widely studied and applied gene delivery vehicles in tissue engineering, with far fewer applications utilizing AAV, herpesviruses, and baculoviruses. Perhaps this is because the level of existing data involving retrovirus and adenovirus has made them more familiar entities, which has bred a sense of comfort among investigators. Because of the special advantages carried by other viral classes, viral researchers should continue to study and develop additional gene carriers to match suited carriers with specific tissue engineering applications. Capitalizing upon the benefits of specific viral gene delivery systems while reducing or eliminating side effects is a long-term goal for both gene therapists and tissue engineers.

Adenovirus, HSV-1, baculovirus, and others are used to achieve transient transfection. This means that the delivered gene will only be expressed for a limited period of time, usually on the order of 1-3 weeks. An advantage of these viral delivery methods is high transduction efficiency. Because the transduction is transient, we do not have to worry about random integration into the genome like we did with retrovirus. The main disadvantage of viral transduction is that the immune system will adapt to its presence and mount a response if the virus is injected into the body a second time.

Suppose for a moment that we use the influenza virus to deliver genes into a human. The patient will get sick if we do not first remove some of the viral genes. We remove part of the viral genome for two reasons. First, it would make little sense to inject a pathogenic virus into a patient. Ethically, injecting an agent that will make a person sick is fraught with problems. Legally, the Food and Drug Administration would be very unlikely to approve such an agent. Second, the viral heads of the different types of viruses have evolved to sizes that would just contain a given viral genome. Because the viral head is of finite size, there's only so much DNA that will fit inside. In removing some of the viral genes, we create space that can be filled with the researcher's genes of interest.

What could we deliver? We could deliver a gene coding for bone morphogenetic protein 4, a growth factor. This gene is 7101 bp in length. What could we not deliver? We could not deliver the dystrophin gene, which is over 2.2 million bp in length. This much mass simply would not fit within a viral head. We have already discussed the capacities of several viruses.

Even with attenuated viruses, we can only deliver them once into a patient. A second dose of the same virus would make the patient sick because the immune system will recognize the viral coat proteins, which are the same as the first time the virus was administered (refer to Figure 13.3). These proteins are capsid proteins (or envelope proteins for enveloped viruses). The first time that the body is exposed to these viral proteins, the immune system becomes educated and mounts what is known as a *primary immune response* to them. This response involves developing antibodies against the specific proteins, a process that takes time. The second time that the body encounters these foreign proteins, genes encoding antibodies against them will already exist as a result of the initial exposure. Because a blueprint for these antibodies will already exist, the immune response will be of greater severity and can be mounted in a shorter time. This is a *secondary immune response*. The secondary immune response should be avoided by the gene therapist because (1) it is very unpleasant for the patient and (2) the body will quickly be able to detect and destroy the viruses, thus reducing the effectiveness of the gene delivery treatment.

An advantage of viral delivery methods is high transduction efficiency and disadvantages include immunogenicity, and limitation to a single drug administration due to the secondary immune response. Most of the viruses yield transient transduction. Retroviruses yield permanent transduction, but permanent transduction is achieved through random genome integration, for which we have already discussed the significant disadvantages.

Aside: The Flu Vaccine

The same principle is used when we get inoculated against the flu every year. Viruses, such as the influenza virus, are constantly mutating and taking on forms that could possibly go undetected by our immune systems. The flu shot is a way of presenting these new viral proteins to the body so that, if we become exposed to the virulent (active) forms of these viruses, our immune systems have already been educated and can mount a quick response to keep flu symptoms to a minimum. *You do not get the flu from the flu shot!* It's possible that one might experience some mild symptoms while the body mounts its primary immune response, but the delivered vaccine is not virulent and cannot replicate, so a full-blown case of the flu is not possible from the vaccination.

An initial transduction with a recombinant virus is similar to getting a flu shot. The body will mount a primary immune response to the foreign particles. A second viral transduction (with the same recombinant virus) would be similar to a vaccinated person being exposed to live influenza virus. A secondary immune response, which is quicker and more intense than the primary immune response, will be mounted to clear the body of the viral particles as quickly as possible.

We are encouraged to get a flu shot every year because viruses are constantly changing. They change their capsid proteins so they can fly under the radar. From 1 year to the next, they may have a completely different set of proteins on display. The flu shot is given every year as opposed every week or every month because a complete cycle of viral epidemic takes about a year. Once it starts to get cold in the autumn, people's immune systems are slightly compromised. In addition, people spend more time indoors and in closer contact with other people. It's an opportune time for pathogens to invade. This is one reason why people tend to get sick when it gets cold—in the autumn or in the winter—although the potential to become ill due to a pathogen exists all year. Perhaps you'll make it through the autumn and winter and become exposed to the virus during the spring or summer. However, by this time, odds are that you will have been either vaccinated or exposed to the virus already, so a secondary immune response will keep you from getting quite so ill. If these viruses do not change their displayed proteins, you will run even less of a chance of getting very ill the following fall or winter because your body has been educated. To combat this, the viruses are always changing. Even though in the spring or summer there might be a new virus that the body has not seen, the weather will be warmer so people will be spending more time outside in conditions where viral concentrations are much lower.

At the beginning of the next flu season, it will be recommended that you get another flu shot to take care of possible exposure to these newly mutated viruses. One might ask how we know what the protein of the year is going to be. The answer is "We don't." However, agencies such as the Centers for Disease Control and Prevention in the United States constantly monitor what illnesses are making the population in general sick. Researchers will isolate and characterize the viruses, taking note of the serotypes that appear repeatedly. "This person is sick with strain 231, and this one has strain 238, and here is a 319, here are two more with 231 … We are seeing a lot of strain 231 lately. We think that by this fall this strain will be rampant, so it would be prudent to vaccinate the population against strain 231." That is a broad overview of how scientists pick out which viruses will be selected for vaccination programs each year. More than one protein is generally selected for the yearly inoculations, but there is an upper limit to the number of proteins to be used due to time and resource constraints.

13.2.2   Physical Delivery Methods

13.2.2.1   Gene Gun

The actual gene gun setup is very close to the earlier seemingly absurd description, except that the shot that it uses is much smaller than buckshot. It's not even bird shot (Figure 13.4). Microparticles on the order of ~0.5-1 µm are used, and they are typically made of gold. Gold can ionize to have a positive charge, and the positive charge will interact well with DNA or RNA thanks to the negative charge on every phosphate group in the polynucleotide.

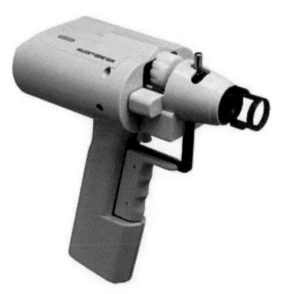

FIGURE 13.4   A gene gun.

The biotechnologist should be very aware that if there is going to be a charge on a gene delivery vehicle, then it almost certainly must be a positive charge to allow for interaction with the delivered polynucleotides. In this case, we would use gold microparticles because they ionize positively and they are biologically inert. We will coat the gold particles with DNA (for this example) and then load the coated microparticles onto a membrane. After loading the membrane into the gene gun, a sudden application of force is used to move the filter down the barrel of the gun. The force comes from a compressed, inert gas such as helium at a pressure on the order of 200-300 psi. The membrane, loaded with particles, will travel down the barrel of the gun until it reaches a position that prevents further movement of the membrane. However, when the membrane strikes the barrier within the barrel, the force of momentum will permit the gold nanoparticles to dislodge from the membrane and continue traveling through and beyond the barrel to interact with the cells and/or tissue. Some of the particles will travel completely through the outer layers of cells. As they pass through, some of the DNA can be stripped off of the gold by simple shear forces. Sometimes an entire particle will remain embedded within a cell. The result is that DNA is delivered past the plasma membrane into the cytoplasm or nucleus.

The gene gun is limited to exposed tissues, such as the epidermis, or cells in a plant leaf. Only limited penetration is attainable with the gene gun (100-500 µm), so it would be extremely difficult to transfect cells of an entire organ with this method. Yes, one could set the pressure of the gas to a higher level to impart greater momentum to the gold particles to get them deeper into a tissue, but the cells on the surface will undergo greater damage and the shockwave produced can dislodge cells from the extracellular matrix. There is a trade-off between tissue damage and depth of penetration using a gene gun.

13.2.2.2 Microinjection

Microinjection is simple to explain: itty bitty needles. The needle gets past the barrier of the cell (the plasma membrane) by poking a hole in it. If the procedure is done correctly, the membrane will close back up when the needle is removed, leaving behind minimal cellular damage. In the hands of a skilled operator, microinjection can yield close to 100% transfection efficiency. This technique can be used to transfer oligonucleotides to cell cytoplasms or nuclei or to transfer entire nuclei into enucleated eggs (as is the case with cloning). However, the use of microinjection to transfect hundreds or thousands of somatic cells *in vivo* would be an infeasible venture, in terms of both getting to and visualizing the cells of interest and the amount of time and effort required for the procedure.

Oftentimes, this technique is used with oocytes in suspension. These cells will be held in place by using a very small tube with an applied vacuum (Figure 13.5). Cells that are already adhering to a support can also be microinjected. Either way, it is important that the cell to be injected is in a position where it can be visualized with a microscope. The microscope has a CCD camera that is linked to a monitor to allow for real-time visualization.

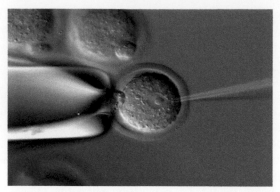

FIGURE 13.5   Microinjection of a frog oocyte.

The microinjection needle will be positioned just over the cell in the xy plane (at an angle) and then lowered slowly in the z direction until a slight depression can be seen on the cell surface. It looks much like lightly poking a water balloon with one's finger. After the correct needle position has been visualized, the needle is then moved slightly in the xz plane to prepare for a stab-and-inject movement that will allow the microinjector to introduce DNA into the cell using predetermined, computer-controlled settings. The injection will be made via positive pressure. The injection volume, injection pressure, and postinjection pressure are all tightly controlled. The volume of injection should be limited to only the amount of DNA solution that is to be introduced; otherwise, the cell could burst. The postinjection pressure is also important. This pressure is lower than the injection pressure but higher than the baseline pressure used at the beginning of the experiment. Without it, contents from the cytoplasm or nucleus would be forced back into the needle by a positive pressure created by the introduction of the injected material. Such a loss of cytoplasmic or nuclear material could spell death for the cell.

Aside: Dolly, the Sheep

Although not an example of microinjection, the case of Dolly, the sheep was the first published result of cloning using the nucleus from a somatic cell harvested from an adult donor. More recent examples of animal cloning via nuclear transfer were carried out by microinjection. Dolly is presented here because of historical significance.

The entire nucleus (including genomic DNA) from a somatic cell, specifically a mammary cell from an udder of a 6-year-old pregnant ewe, was put into an enucleated egg from a different breed of sheep via cell fusion. With a complete genome, the egg cell was then allowed to develop into a morula/blastocyst, which was then implanted into recipients. The result was a lamb, named Dolly (6LL3),

being born with a genome that was unrelated to the sheep that donated the enucleated egg cell. The mitochondria and mitochondrial DNA of Dolly presumably
came from both the donor and oocyte, but Dolly's genome (including both sets of
chromosomes) came solely from the donor of the somatic cell.

There were 277 cell couplings made using adult mammary cell nuclei. Of
these, 29 of the coupled cells progressed to the morula/blastocyst stage, being
implanted into 13 sheep. Of these 29 implantations, only 1 resulted in pregnancy.
This pregnancy went to term to produce a live lamb. Other lambs were produced
in the study by cloning using fetal- and embryo-derived nuclei. The impact of the
study was made by the discovery that cell differentiation does not involve irrevers-
ible DNA modifications that would prevent the cell from being transformed to a
totipotent state. ("Totipotent" is explained in Chapter 17.)

Dolly died on February 14, 2003, euthanized because of complications due
to lung cancer that was caused by a virus. She was 6 years old. Sheep of Dolly's
breed (Finnish Dorset) typically live 11-12 years, and the sheep who donated the
mammary nucleus was 6 years old at the time of cell harvest. It is interesting to
note that Dolly's life span was about right for a sheep of her breed if one considers that she was 6 years old at the time of nuclear transfer. Throughout her life,
Dolly had been plagued by health problems, including a weight problem, arthri-
tis, and cancer. However, it was also noted that her telomeres were prematurely
shortened, which may have been the result of shorter telomeres—a natural part of
aging—in the adult DNA donor.

13.2.2.3 Electroporation

A third physical gene delivery method is electroporation. This technique involves the delivery of a voltage (current) across the surfaces of cells, which will
result in the spontaneous creation of pores in the plasma membranes. When
the voltage is removed, the pores will close. When the voltage is applied and
the pores open up, current will travel through the cell. Recalling that DNA has
a net negative charge, it should be clear that DNA will migrate in the same
direction as electron flow in the applied current. When the current is removed,
the pores will spontaneously close, trapping some of the DNA within the cells.
Membrane-associated DNA aggregates also form in the presence of the voltage,
and these aggregates have been seen to remain at the membrane level for up to
10 minutes after the administration of the current. It is possible that these aggregates enter cells through endocytosis.

For mammalian cells, the process uses at least 200 V/cm for plated cells and
to 400 V/cm for cells in suspension, with pulses lasting a minimum of 0.5 ms.
Multiple pulses are usually delivered. It is possible that a small amount of cellular contents escapes during the procedure, especially ions and small proteins, if
the voltage is sufficient and the time of delivery is long enough. Electroporation
is very stressful to the cell. Although it can be used on most cell types, it is typically reserved for bacteria because of their high numbers and short cell cycles
(meaning they are expendable and have short recovery times).

13.2.3   Chemical Delivery Methods

The chemical gene delivery methods that entail polymers or cationic lipids offer several advantages for delivering genes into cells. First, the size of the carried DNA is not limited to something that will fit into a finite viral head. Artificial chromosomes with millions of base pairs have been delivered with both polymers and lipids. Second, there is potentially no threat of immune response, even with repeated injections of the complexes. Third, polymers and lipids are relatively inexpensive to produce and store. Finally, although non-viral gene delivery usually results in transient gene expression, a way to view this as an advantage is to consider that there is very low risk of random integration into the genomes of host cells. Of course, there are disadvantages as well. The efficiency of chemically mediated gene delivery (~25-40%) will be much lower than what is achieved with viruses. Some of the delivery vehicles are also somewhat *cytotoxic*, meaning they can cause cell death above certain concentrations.

13.2.3.1   Polymers

In forming *polyplexes* (polymer/DNA complexes) for gene delivery, one should be aware of different polymer architectures that can be used. There are linear polymers that consist of basic repeating units strung together (e.g., poly (L-lysine) (PLL)), and there are branched polymers. Of the branched, there are hyperbranched polymers such as poly(ethylenimine) (PEI) and PAMAMs. We will now take a look at these vectors and more in greater detail.

13.2.3.1.1   PEI

PEI is a polymer that has been used for years in common processes such as paper production, shampoo manufacturing, and water purification. The polymer can be produced in either linear or hyperbranched forms. In either form, each of the nitrogens has the potential to be protonated and therefore carry a positive charge. It is the pH of the environment that will determine the overall number of positive charges per PEI molecule. Some of the nitrogens will be more easily protonated than others, which is why PEI can serve as a buffer over a wide range of pH values.

Branched PEI comes from the polymerization of aziridine (Figure 13.6a). When aziridine is protonated, it is susceptible to ring opening at the amine via nucleophilic attack. A convenient nucleophile in this case would be a second aziridine molecule. As the ring from the first aziridine is opened, the positive charge will be transferred to the amine of the second aziridine ring. This cycle of ring opening can continue during propagation to yield different architectures— linear or branched—of (C–C–N) additions (Figure 13.6b). As shown in the figure, it's not guaranteed that any given nitrogen will yield a branch point during polymerization. Since this form of branching is neither controlled nor regular, it

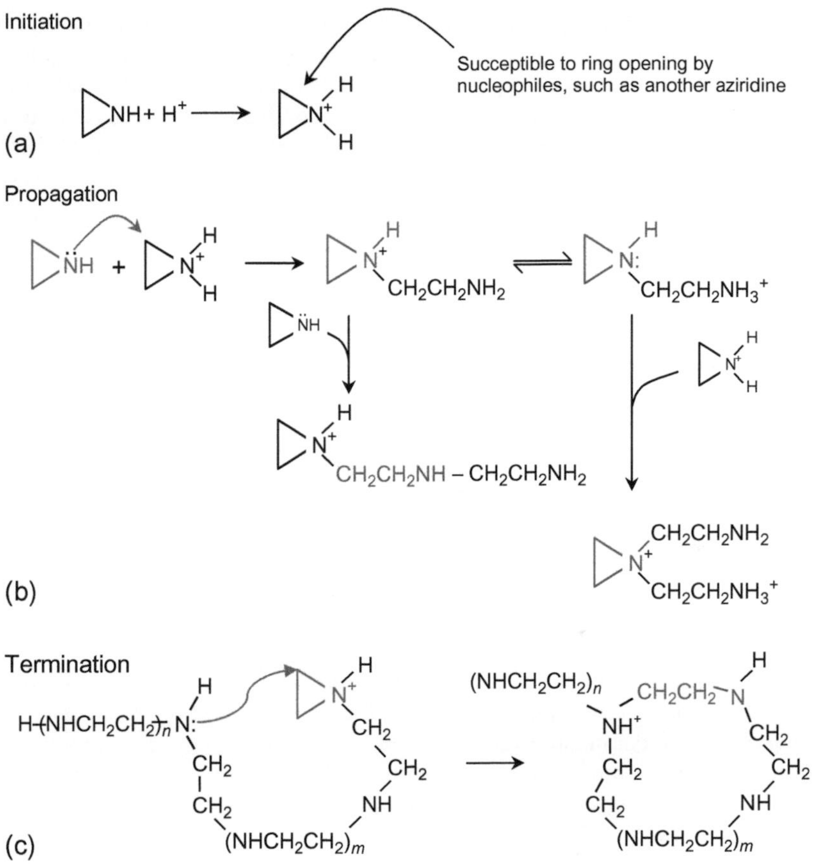

Initiation

(a)

Propagation

(b)

Termination

(c)

FIGURE 13.6   Polymerization of branched PEI. (a) Initiation occurs after the protonation of aziridine. (b) The ring from the initial aziridine (shown in black) is opened by nucleophilic attack by another aziridine (shown in red). Chain extension can occur in two different ways as shown. (c) Termination occurs when a ring is opened by a secondary amine in the same polymer.

is termed *hyperbranched*. At some point, a secondary amino group from within the polymer itself will act as a nucleophile for a terminal aziridine, forming a cyclic structure that can no longer be polymerized. This form of termination is termed back-biting termination (Figure 13.6c).

Again, note that each of the amines within a PEI molecule has the potential to be protonated. At physiological pH (7.2-7.4), PEI will be quite cationic. As such, it will readily interact with DNA. In fact, at physiological pH (and assuming there are enough amines present), there will be an excess of positive charges even after the PEI has complexed with DNA. This is important because positive charges are required for gene delivery complexes to gain entry into cells. Eukaryotic cells typically have a net negative charge on their exteriors. (Don't confuse this with the flipping of phosphatidylserines to the cell exterior during apoptosis. Although these negatively charged phospholipids are primarily found on the cytoplasmic side of the plasma membrane in cells that are not undergoing apoptosis, the exteriors of the cells still carry negative charge because of transmembrane proteins and other membrane constituents.) Having negative charges on the cell exterior means charge-charge interactions can occur between PEI/DNA complexes and the cell. The complexes will adhere to a portion of the plasma membrane and be endocytosed as was described earlier in this book. (As a quick check of your learning thus far, would the endocytosis be an example of phagocytosis or pinocytosis? The size of the gene delivery complexes will be on the order of 100 nm.)

One popular conjecture about what happens to PEI/DNA complexes following endocytosis is known as the proton sponge hypothesis, which entails the following steps:

1. PEI/DNA complexes are endocytosed.
2. Complex-containing endosomes fuse with lysosomes to create structures known as endolysosomes. (Using our knowledge of endosomal maturation, we will see that this fusion is not a requirement for the hypothesis to still be valid.)
3. V-ATPases will pump protons into the vesicular interior.
4. PEI will absorb many of these protons, preventing a change in pH. (Note that, in the case of an endolysosome, the lack of pH drop will prevent the activation of degradative enzymes such as nucleases.) However, protons will have still been introduced, so a charge gradient will have been established.
5. Cl^- will flow into the vesicle to neutralize the charge gradient. This, however, will set up an osmotic gradient.
6. Water will enter the vesicle to balance the osmotic gradient.
7. The vesicle will swell due to the entry of ions and water, leading to rupture.

This process will result in the release of PEI/DNA complexes into the cytoplasm. Note that, however, it is not necessarily the case that all endocytosed PEI/DNA complexes will follow this pathway. There is still some debate in the field regarding the cellular processing of such complexes.

13.2.3.1.2 Dendrimers

Dendrimers are branched polymers that are synthesized in a stepwise fashion to control both monodispersity (homogeneity of molecule sizes) and the exact number of branching layers, or "generations." Dendrimers can be synthesized by either divergent or convergent methods. For the divergent method, the dendrimer grows in a stepwise fashion outward from a multifunctional core molecule (Figure 13.7). Slight structural defects can occur in larger molecules, especially at higher generation numbers. For the convergent method, the dendrimer is constructed beginning with the end groups and progressing

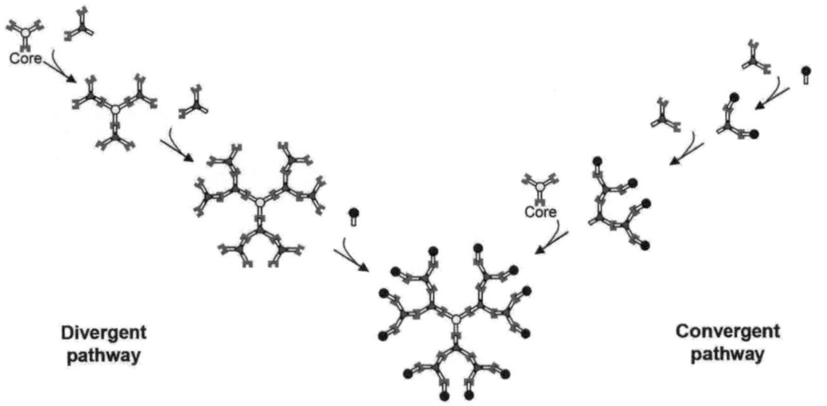

Divergent pathway

Convergent pathway

FIGURE 13.7   Dendrimers can be constructed via divergent or convergent pathways. In the divergent pathway (left), the dendrimer is extended outward from a multifunctional core molecule, often ending with a functionalized terminal group. The convergent method (right) begins at the outer ends and polymerization extends toward what will be the interior of the dendrimer, ending with the addition of the core molecule.

inward (Figure 13.7). Because defects will involve large sections missing from the molecule, defective structures can be more readily separated with this method. Unlike hyperbranched polymers, dendrimers are polymerized in a tightly controlled, stepwise fashion to produce relatively monodisperse sets of macromolecules. (For a discussion of monodispersity and polydispersity, see "Aside: Describing Polymer Distributions.") Dendrimer/DNA complexes are often called dendriplexes to preserve terminology that is analogous with lipoplexes and polyplexes. The low pK_as of the amines (3.9 and 6.9) afford the dendrimer the potential to buffer pH changes during acidification of the endosome.

There is an array of dendrimeric molecules that can be used for gene delivery. A commonly used dendrimer type is the PAMAM dendrimer (Figure 13.8a). Various generations have been used for gene delivery, with optimal generations differing by cell type, but generally between generations 5 and 10 (termed G5-G10). Spherical G5 PAMAM dendrimers have also been used *in vivo* with success. Other dendrimers, such as poly(propylenimine) dendrimers (Figure 13.8b), have also been used for gene delivery with transfection success also showing a dependence upon generation number. Presumably because of the increased density of positive charges at the dendrimer periphery, an increase in DNA binding was observed as dendrimer generation number went up. However, cytotoxicity was also seen to increase with generation number. This dichotomy is a common problem with all chemical delivery methods.

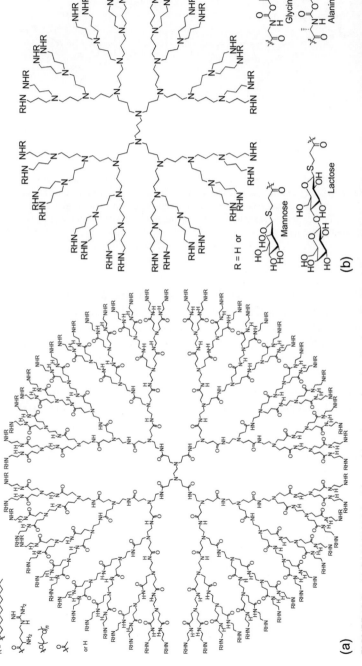

FIGURE 13.8 (a) G4 PAMAM dendrimer with end-group functionalizations. (b) G3 poly(propylenimine) dendrimer with carbohydrate or amino acid chain end-group functionalizations.

Aside: Describing Polymer Distributions

Polymerization is not an exact process. Not every polymer in a batch will have the same number of repeating units. Because of this, we need a way to describe just how pure a polymer solution is. This is accomplished via the *polydispersity index* (PI), which takes into account two different molecular weights:

$$\overline{M}_n = \frac{\sum_i N_i M_i}{\sum_i N_i}$$	The first molecular weight to consider is the *number average* molecular weight $\left(\overline{M}_n\right)$. This average is like the averages you are probably already used to. The molecular weight (M) for each polymer in a sample is added, and the sum is divided by the total number of molecules counted
$$\overline{M}_w = \frac{\sum_i N_i M_i^2}{\sum_i N_i M_i}$$	The second molecular weight to consider is the *weight average* molecular weight $\left(\overline{M}_w\right)$. This average is like a weighted average, where larger polymers are given more influence in the equation for molecular weight determination. The molecular weight for each polymer molecule is squared and added, and the sum is divided by the number of molecules and the sum of the molecular weights

Because squaring the values of the molecular weights will produce a stronger input from the larger molecules, the resulting weight average molecular weight will always be greater than the number average molecular weight, unless there is only one molecular weight in the polymer sample. As a result, $\overline{M}_w \geq \overline{M}_n$.

The polydispersity index is the ratio of the two molecular weights just described: $PI = \overline{M}_w / \overline{M}_n$. Note that $PI > 1$, except in the case where all of the molecules have the same molecular weight. In that case, $PI = 1$, and the polymer sample is said to be *monodisperse*.

Example 1: What are $\overline{M}_n$, $\overline{M}_w$, and PI for the following distribution of polymer sizes?

Answer:

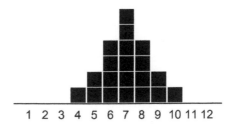

$$\overline{M}_n : \left(4+5+5+6+6+6+6+7+7+7+7+7+7+8+8+8+8+9+9+10\right)/20$$
$$= \left[1(4)+2(5)+4(6)+6(7)+4(8)+2(9)+1(10)\right]/\left(1+2+4+6+4+2+1\right)$$
$$= 140/20 = 7.00$$

$$\overline{M}_w : \left[1(4)^2 + 2(5)^2 + 4(6)^2 + 6(7)^2 + 4(8)^2 + 2(9)^2 + 1(10^2) \right] /$$
$$\left[1(4) + 2(5) + 4(6) + 6(7) + 4(8) + 2(9) + 1(10) \right]$$
$$= 1022/140$$
$$= 7.30$$

$$PI = 7.30/7.00 = 1.04$$

Example 2: Here, the previous distribution is still centered at 7, but it has been stretched out. What are the effects on $\overline{M}_n$, $\overline{M}_w$, and PI?

Answer:

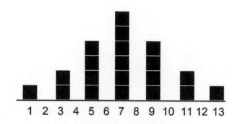

$$\overline{M}_n : \left[1(1) + 2(3) + 4(5) + 6(7) + 4(9) + 2(11) + 1(13) \right] / 20$$
$$= 140/20 = 7.00 \,(\text{The same as before.})$$

$$\overline{M}_w : \left[1(1)^2 + 2(3)^2 + 4(5)^2 + 6(7)^2 + 4(9)^2 + 2(11)^2 + 1(13)^2 \right] /$$
$$\left[1(1) + 2(3) + 4(5) + 6(7) + 4(7) + 2(11) + 1(13) \right]$$
$$= 1148/140$$
$$= 8.20 \,(\text{An increase over the previous value.})$$

$$PI = 8.20/7.00 = 1.17$$

(The distribution is more spread out. Polydispersity has increased.)

13.2.3.1.3 Chitosan

Chitosan is another polymer that has been well characterized for use in transfection. This natural, nontoxic polysaccharide lends DNase resistance to its cargo while condensing the DNA to form stronger complexes. The efficiency of chitosan is thought to rely upon its ability to swell and burst endolysosomes, which allows the delivered DNA to continue its path to the nucleus.

Chitosan is obtained by the alkaline deacetylation of chitin, which is the second-most abundant polysaccharide in nature. We can obtain it in large quantities from crab, shrimp, and lobster shells, considered by the seafood industry to be waste products. Chitosan is a polysaccharide composed of randomly distributed β-(1-4)-linked D-glucosamines and *N*-acetyl-D-glucosamines (Figure 13.9). It is *biodegradable* (is broken down inside the cell or tissue, usually by hydrolysis) and *biocompatible* (meaning the polymer and its degradation products are nontoxic) at "low" molecular weight (10-50 kDa).

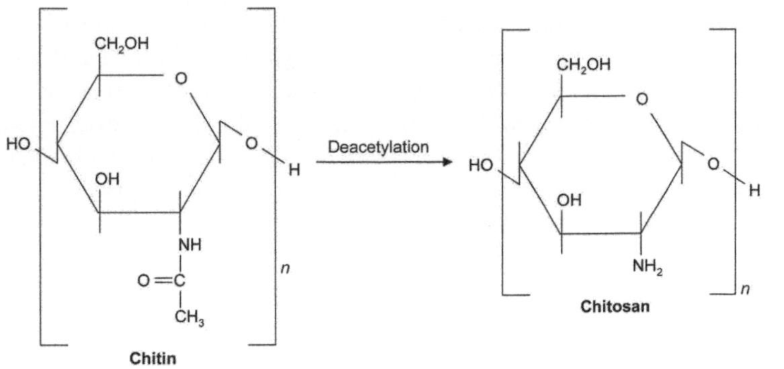

FIGURE 13.9   Chitosan is derived from chitin, which is found in shellfish shells.

Because of the adhesive and transport properties of chitosan in the GI tract, this polymer has been used for oral gene therapy applications. Chitosan/pCMVArah2 (Arah2 is the dominant anaphylaxis-inducing antigen in mice sensitized to peanuts) has been administered into AKR/J mice as an oral immunization method for peanut allergy. Chitosan/pDer p 1 (Der p 1 is a major triggering factor for mite allergy) has been investigated as a possible oral vaccine against allergies to dust mites. Chitosan is not limited to oral administration, however, and has been used for many gene delivery applications. Most notable are recent advances in the field of regenerative medicine.

13.2.3.1.4   PLL

PLL is a well-known polycation that has been widely studied as a nonviral gene delivery vector since the first reported formation of PLL/DNA complexes in 1975. Poly(L-lysine) is a polypeptide of the essential amino acid L-lysine. At physiological pH, each repeating unit of PLL carries a positive charge on the ε-amine of its side chain, a property that has been exploited to allow PLL to condense plasmid DNA to varying degrees depending upon salt concentration. Although the structure of PLL appears to be suitable for gene delivery, unmodified versions of this polymer are associated with low transfection efficiency and cytotoxicity.

PLL was one of the first cationic peptides used to mediate gene delivery. However, as the length of PLL increases, so does the cytotoxicity. Moreover, the polydispersity of PLL complicates modifications with ligands, making the chemical synthesis of PLL conjugates hard to control.

13.2.3.2   *Lipids*

Lipids, particularly cationic lipids, have been used for gene delivery in a process termed *lipofection*. Because of the aqueous environment inside cells and tissues, the hydrophobic tails of the lipids will coalesce to form hollow liposomes, the interiors of which can contain oligonucleotides for cellular delivery. The combination of more than one hydrophobic entity can often yield higher transfection efficiencies than using a single type of lipid.

In the lipids used for gene delivery, there is typically a glycerol backbone, one or two fatty acid tails, and a cationic head group. (It might help to think of the phospholipid structure presented in Chapter 1, although most lipids used for gene delivery will not have the phosphate group.) It is the head group that is varied according to the type of target cell and whether the experiment will be performed *in vitro* or *in vivo*. Sometimes, a molecule like cholesterol is used as part of the hydrophobic portion of the molecule. It is the hydrophobic part of the lipid that allows for self-assembly into liposomes. The morphologies of these structures can vary greatly, as we shall soon see.

13.2.3.2.1   Liposome Geometry

There are a number of structures that can result during the interaction of cationic lipids with polynucleotides to form *lipoplexes* (polynucleotide-containing liposomes). The shape of the lipoplex will be determined by the most thermodynamically favorable conformation, which is described by the *packing parameter* (*P*), which takes into account the ratio of certain size variables:

$$P = v / al_c$$

where v = the volume of the hydrocarbon, a = the effective area of the head group, and l_c = the length of the lipid tail.

We can use the packing parameter to predict the shape of the resulting lipoplex (Figure 13.10):

$$P < 1/3 \rightarrow \text{spherical micelle}$$

$$1/3 \leq P < 1/2 \rightarrow \text{cylindrical micelle}$$

$$1/2 \leq P < 1 \rightarrow \text{flexible bilayers and vesicles}$$

$$P = 1 \rightarrow \text{planar bilayers}$$

$$P > 1 \rightarrow \text{inverted micelles}\left(\text{hexagonal}\left(H_{\text{II}}\right)\text{phase}\right)$$

Lipids used for gene delivery typically have hydrocarbon tails of between 8 and 18 carbons. The tails are typically saturated, but sometimes, one double bond is present. The combination of hydrocarbon chains that is attached to glycerol can be asymmetrical, as in the phospholipids in the plasma membrane, or symmetrical. It has been shown that asymmetrical lipids with a shorter saturated tail and a longer unsaturated tail produce relatively high transfection efficiencies (vs. lipids with symmetrical tails).

The ionizable head group also plays a role in transfection efficiency. The head groups used in gene delivery can carry a single charge or multiple charges, and the net charge will typically be cationic. The following discussion presents some of the more common lipids used in gene delivery, separating them by the number of charges in the head group.

13.2.3.2.2   Monovalent Cationic Lipids

13.2.3.2.2.1   DOTMA N-[1-(2,3-dioleyloxy)propyl]-N,N,N-trimethylammonium chloride (DOTMA), was one of the first synthesized and

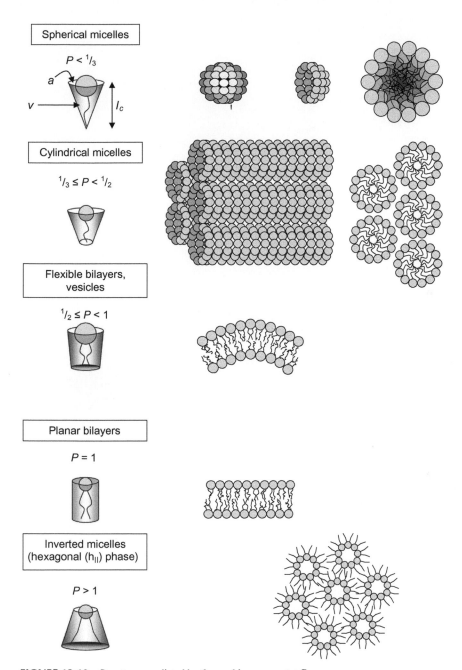

FIGURE 13.10   Structures predicted by the packing parameter *P*.

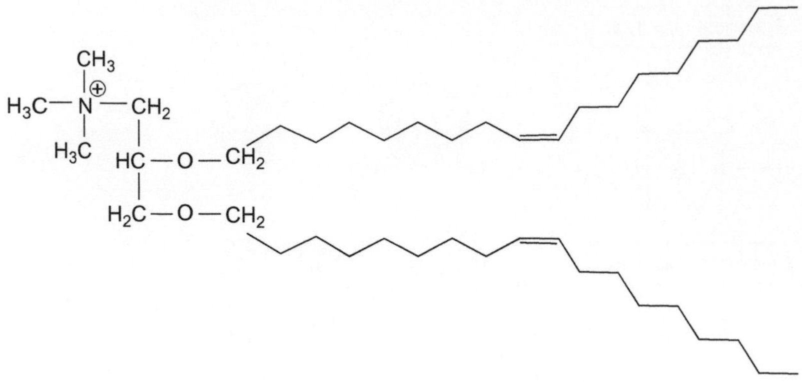

FIGURE 13.11   The structure of DOTMA.

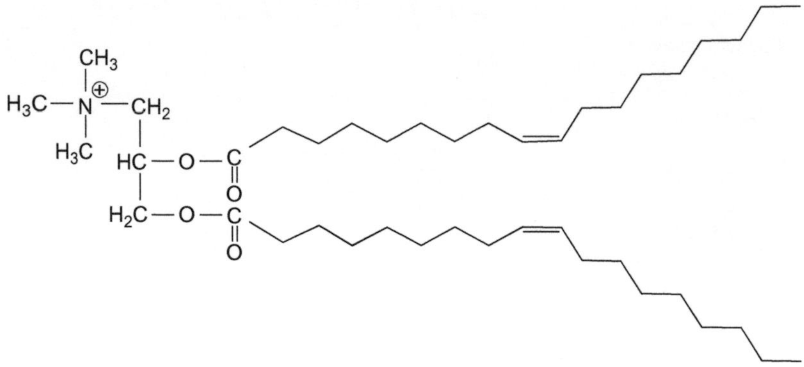

FIGURE 13.12   The structure of DOTAP.

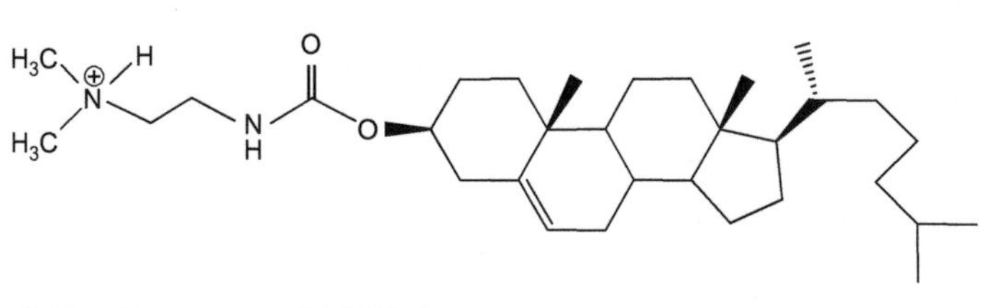

FIGURE 13.13   The structure of DC-Chol.

commercially available cationic lipids used for gene delivery. It contains two unsaturated fatty acid tails that are 18 carbons long with a double bond at carbon #9. The head group gets its charge from quaternary amine. The cationic head group provides the ability to entrap DNA or RNA in the liposomal structure. This feature, first described in the 1980s, significantly influenced and improved the potential of all nonviral gene delivery vehicles. The initial success of *in vitro* transfection of multiple cell lines with DOTMA sparked a movement to find

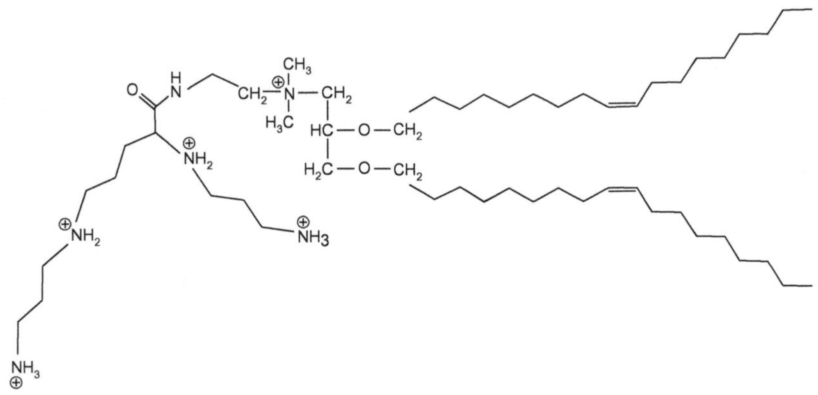

FIGURE 13.14 The structure of DOSPA.

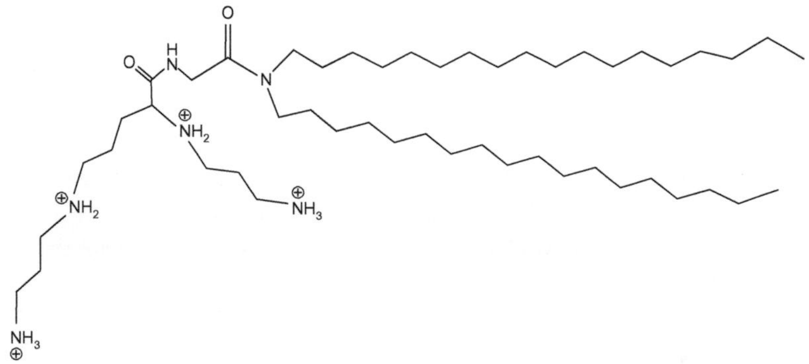

FIGURE 13.15 The structure of DOGS.

even more effective lipids for gene delivery. The most successful of these will be discussed later in this section.

DOTMA was paired with the helper lipid DOPE (discussed later) and sold under the trade name Lipofectin®. Following the commercial success of Lipofectin®, even more improvements in lipoplex formulation were sought with the intent of increasing transfection efficiency and lowering cytotoxicity. The structural changes included different combinations of side chains and attachments to the head group or head group modification via the replacement of one of the methyl groups ($-CH_3$) on the head group nitrogen with a hydroxyl ($-OH$). This latter modification was found to stabilize the resulting vesicles and increase transfection efficiency. Not surprisingly, the length of the hydrophobic tails also has an effect on transfection efficiency. Shorter tails appear to be more effective.

13.2.3.2.2.2 ***DOTAP*** [1,2-Bis(oleoyloxy)-3-(trimethylammonio)propane] (DOTAP), was first synthesized in 1990. The molecule consists of a quaternary

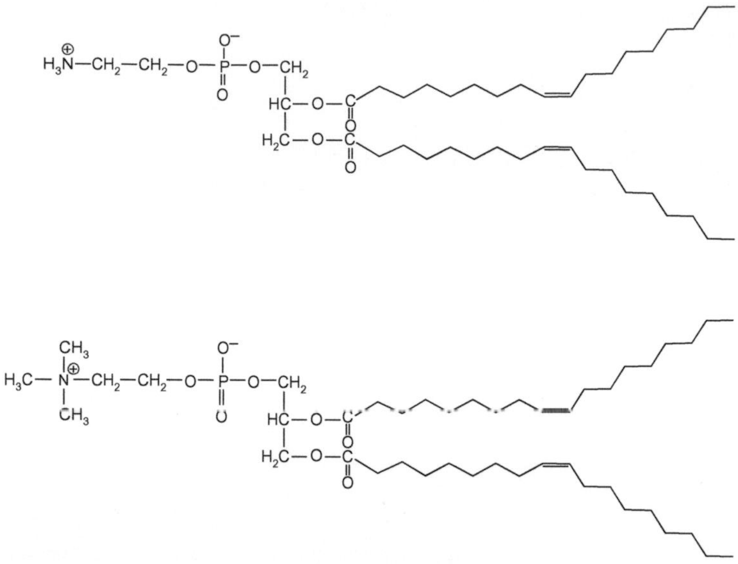

FIGURE 13.16   The structure of DOPE (top) and DOPC (bottom).

amine head group coupled to a glycerol backbone with two oleoyl chains. The only difference between DOTAP and DOTMA is that ester bonds link the chains to the backbone rather than ether bonds. It was originally hypothesized that ester bonds, which are hydrolyzable, could render the lipid biodegradable and reduce cytotoxicity.

The use of 100% DOTAP for gene delivery is inefficient, perhaps due to a high surface charge density after lipoplexes have been formed. DOTAP is completely protonated at pH 7.4 (which is not the case for other cationic lipids), so it is possible that more energy is required to separate DNA from the lipoplex after delivery into the cell. Thus, for DOTAP to be more effective in gene delivery, it should be combined with a helper lipid such as DOPE, as seems to be the case for most cationic lipid formulations.

13.2.3.2.2.3   *DC-Chol*   3β[N-(N', N'-Dimethylaminoethane)-carbamoyl]cholesterol (DC-Chol), was first described in 1991. DC-Chol contains a cholesterol moiety attached by an ester bond to a hydrolyzable dimethylethylenediamine. Cholesterol was reportedly chosen for its biocompatibility and the stability it imparts to lipid membranes (just like in our own cellular membranes), an idea that is supported by desirable transfection efficiencies and fairly low cytotoxicities.

In contrast to DOTMA and DOTAP, which are cationic because of quaternary amines in their head groups, DC-Chol gets its charge from a tertiary amine. Only half of these nitrogen atoms are protonated at physiological pH, which may result in better DNA dissociation following delivery into the cell. It may also provide lower cytotoxicity and lends the potential to act as a buffer when pH drops (such as in the late endosome/endolysosome).

13.2.3.2.3   Multivalent Cationic Lipids

13.2.3.2.3.1   DOSPA {2,3-Dioleyloxy-*N*-[2(sperminecarboxamido)ethyl]-*N*,*N*-dimethyl-l-propanaminium trifluoroacetate} (DOSPA), is another cationic lipid synthesized as a derivative of DOTMA. The structure is like that of DOTMA except for the addition of a spermine molecule ($C_{10}H_{26}N_4$) to the head group via some interesting chemistry (notice the carboxamido ethyl attachment). This cationic lipid, used with the neutral helper lipid DOPE (we will get to DOPE, I promise!) at a 3:1 ratio, is commercially available as the transfection reagent Lipofectamine®.

In general, the addition of the spermine functional group allows for a more efficient packing of DNA in terms of liposome size. The efficient condensation yields a smaller complex, possibly because of the four protonable nitrogens in the spermine. It has been shown that spermine can interact with one strand of dsDNA and wind around the major groove to interact with complementary bases of the opposite strand.

13.2.3.2.3.2   DOGS Di-octadecyl-amido-glycyl-spermine (DOGS), has a structure similar to DOSPA; both molecules have a multivalent spermine head group and two 18-carbon alkyl chains. However, the chains in DOGS are saturated, are linked to the head group through a peptide bond, lack a quaternary amine, and do not connect to a glycerol backbone. DOGS is commercially available under the name Transfectam®.

Much like the multivalent cationic lipid DOSPA, DOGS is very efficient at binding and packing DNA because of the spermine head group. DOGS is a multifaceted molecule in terms of buffering capacity. At pH values lower than 4.6, all of the amino groups in the spermine are protonated, while at pH = 8, only two are purportedly ionized, which promotes arrangement into a lamellar structure.

13.2.3.2.4   Neutral Helper Lipids

13.2.3.2.4.1   DOPE and DOPC When making lipoplexes for gene delivery, lipids with a cationic charge are desirable for interacting with DNA. However, mixing in a class of neutral lipids, known as *helper lipids*, will often improve transfection efficiency to a significant degree. Two very common helper lipids are dioleoylphosphatidylethanolamine (DOPE), the most widely used helper lipid, and dioleoylphosphatidylcholine (DOPC). (Look closely. These molecules are just specific forms of phosphatidylethanolamine and phosphatidylcholine, two plasmid membrane phospholipids we studied in Chapter 1.) Improved transfection efficiencies are thought to be due to a conformational change in the liposomes at low pH, to an inverted hexagonal packing structure (refer back to Figure 13.10). It has been shown that a hexagonal conformation allows for efficient escape of complexed DNA from endosomal vesicles via destabilization of the vesicle membrane. Fusion and destabilization of the lipoplexes during transfection are thought to occur due to the exposure of the endosomal membrane to invasive hydrocarbon chains.

DOPE also allows for closer contact and packing of DNA helices in liposomal formulations. It is thought that salt bridges form between the positively charged head groups of cationic lipids and the phosphate groups of DOPE molecules. This association would force the primary amine of DOPE to stabilize itself in the plane of the liposome surface, permitting closer interactions with the negatively charged phosphate groups of DNA. Conversely, DOPE could potentially cause the release of a negatively charged counterion from the positively charged head group of a cationic lipid, which would allow easier binding to DNA.

13.3   PREPARATION OF NONVIRAL GENE DELIVERY COMPLEXES

At this point, we have DNA at a known concentration, and we have selected a nonviral chemical delivery method. For the following example, we will use PEI, although the principles presented will apply to any chemically mediated gene delivery method.

In determining the amount of gene delivery vehicle to use for a given amount of DNA, it is important to consider the *charge ratio* because the negative charges of DNA will be interacting with the positive charges of the carrier. For this reason, it is less important to discuss the number of carrier molecules versus plasmid molecules. By taking into account only charges, we can alleviate the necessity of accounting for the number of bases in each plasmid. While it is true that in virtually all cases, the positive charges on carrier molecules are due to protonated amines, it should not be assumed that every amine will be protonated at physiological pH. Because a precise determination of protonation cannot be determined with a great degree of resolution, we will therefore instead pay attention to the number of amines used to interact with DNA molecules. Specifically, we will be concerned with the ratio of the number of carrier amines, which can carry positive charges, to the number of DNA phosphates, which carry all of the negative charges on the polynucleotide. This ratio is abbreviated as the *N:P ratio*.

Nonviral gene delivery complexes are usually delivered to cells in a solution form. The solution will contain two prime constituents (other than salts): the DNA or RNA to be delivered and the gene delivery vehicle. Let us begin by supposing that there will be no phosphates in the gene delivery vehicle. There will, of course, be plenty of phosphates in the polynucleotides. Although the number of phosphates will vary with the size of the polynucleotide to be delivered, there will only be one phosphate per (deoxy)nucleotide constituent. On the gene delivery vehicle side of things, there will typically be one nitrogen atom per basic repeating unit. ("Basic repeating unit" refers to the recurring portion of a polymer. For nonpolymers such as lipids, this discussion is still valid if we consider the entire molecule to be the repeating unit and the number of repeats is equal to one.) In the case of PEI, whether we are talking about the linear or branched form, the entire molecule can be subdivided into carbon-carbon-nitrogen repeats (Figure 13.17). Whether a particular nitrogen atom is a primary,

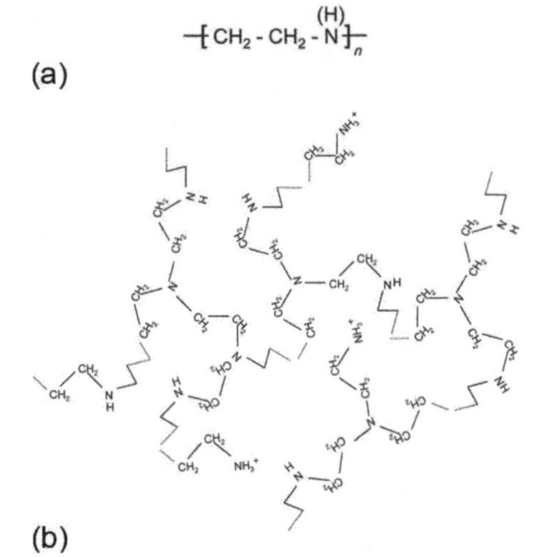

(a)

(b)

FIGURE 13.17 (a) Basic repeating unit of PEI. (b) Branched PEI, with 1°, 2°, and 3° amines.

secondary, or tertiary amine, it is a part of a C–C–N unit and has the potential to be protonated. The same is true for poly(L-lysine) (PLL) and chitosan, which are polymers that contain repeating units that contain one protonable amine each (Figures 13.9 and 13.18). (Note that in PLL, the α amino group in all but the terminal lysine will not be protonated because it will be part of a peptide bond upon polymerization.) Looking at basic repeating units, we can see that controlling the ratio of delivery vehicle nitrogens to polynucleotide phosphates—the N:P ratio—is a relatively straightforward process that can be handled by controlling the total mass used for each constituent, alleviating the need to factor in polymer sizes and polydispersities.

If we want to deliver DNA to a known number of cells, such as 100,000, we must determine the appropriate amount of carrier to use. This amount

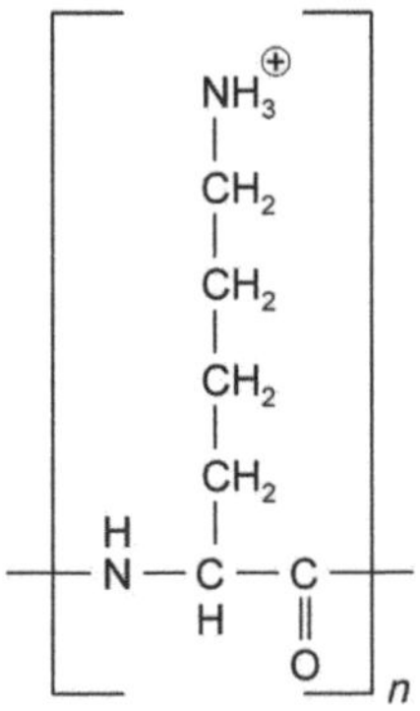

FIGURE 13.18 Basic repeating unit of PLL.

depends upon several parameters. An acceptable N:P ratio must be decided upon before the actual transfection procedure. This can often be determined by reading the scientific literature, but we can determine it on our own experimentally by trying out several values and selecting the optimum. Suppose that we have read several papers that state that an N:P ratio of 5.00:1.00 is optimal for a certain gene delivery vehicle in a given cell type. Keeping in mind that the N:P ratio is based upon either a known amount of gene delivery vehicle or a known amount of polynucleotide for a given experiment, using a standard amount of one of these constituents will ensure a set amount of the other. It is common to use a standard amount of DNA per transfection; for this example, let us deliver 2.00 μg. Since the N:P ratio involves a ratio of gene delivery vehicle repeating units to polynucleotide repeating units, we must factor in the molecular weights of these repeating units. The molecular weight for a repeating unit of the polymer to be used can be determined directly from the chemical formula. For PEI, consider the repeating unit to be $-[CH_2-CH_2-NH]-$, which has a molecular weight of 43.07 g/mol. The average molecular weight of the DNA base is approximately 308.00 g/mol (a more precise determination of this value is addressed by one of the problems at the end of this chapter.)

The molecular weight of a repeating unit of PEI is quite different from that of PLL. If we were separately to use these molecules to deliver a set amount of DNA to cell samples, assuming equal N:P, would we have to use more or less of the PLL (which has a larger repeating unit) relative to the PEI? We would have to use more PLL because the repeating unit is larger than that of PEI. Think of it in terms of delivering a mole of protonable amines: since the molecular weight of a mole of PLL repeating units is larger than a mole of PEI repeating units, a mole of protonable amines delivered via PLL will also be associated with greater mass. This formula we are creating is not restricted to polymers. If we use lipids, we will consider the size of each lipid molecule rather than the size of a repeating unit.

We must next consider the concentration of gene delivery vehicle that we have in solution. Pure polymer would be a solid, and it would be very impractical to weigh out microgram amounts for each transfection. Sometimes, the polymers are sold as very concentrated solutions, but the solutions are often too viscous to pipette with accuracy. It's much more common to have stock solutions from which one can draw during an experiment. The more concentrated the solution is, the less of it one will need to deliver a given number of protonable amines. This inverse relationship is why the dilution factor is (the concentration of the stock solution)$^{-1}$.

To make our formula more robust, we must also take into account the number of protonable amines contained in each repeating unit of the gene delivery vehicle. In practice, this number will typically be 1. However, if there were two primary amines per repeating unit, we would need to use fewer repeating units to get to the desired N:P ratio. This inverse relationship dictates that the number of positive charges per basic repeating unit should appear in the denominator of our equation.

We now have the following equation to help us determine how much gene delivery vehicle stock we will use to create one dose of transfection solution:

$$1 \text{Dose of Carrier Stock}$$

$$= (N : P)\left(\frac{DNA}{dose}\right)\left(\frac{MW_{CarrierRepeatingUnit}}{MW_{DNARepeatingUnit}}\right)\left(\frac{1}{Carrier Stock Concentration}\right)$$

$$\left(\frac{1}{Net \oplus Charge / Repeating Unit}\right)$$

The units for this formula are as follows:

$$\mu l / dose = [\ ]\left[\frac{\mu g}{dose}\right]\left[\left(\frac{g / mol}{g / mol}\right)\right]\left[\frac{\mu l}{\mu g}\right][\ ]$$

Using the values mentioned in the previous paragraphs, we can determine the amount of 4.307 mg/ml PEI stock solution to use per transfection:

$$1 \text{Dose of Carrier Stock} = 5.00 \times 2.00(43.07 / 308.00)(1.00 / 4.307)(1.00)$$
$$= 0.32 \mu l$$

While gene delivery vehicles are typically used to deliver plasmid DNA, they are not restricted to this single application. Recall that the definition of gene delivery given earlier in this text stated that it is the delivery of genetic material (polynucleotides) into cells to alter the function of the cells. Gene therapy is the use of gene delivery to achieve a therapeutic benefit for an organism. The material delivered could be plasmid DNA, linear DNA, single-stranded, or double-stranded RNA, no matter the size. There is some discussion in the community whether delivering small interfering RNA counts as gene delivery. Under the definition just given, there is little question. There is no dispute that using retroviruses to achieve permanent transduction via the delivery of RNA into cells is a form of gene delivery. Here, we assert that delivering siRNA to alter the expression of a protein also counts as gene delivery. Neither the RNA delivered by a retrovirus nor the double-stranded RNAs used as siRNA are genes, but they are both intended to alter the expression levels of a given protein. This is similar to delivering a gene in that the expression of a specific protein will be altered. One could argue that delivering an inhibitor of some protein could therefore also be considered gene delivery. However, our definition restricts us to the delivery of polynucleotides, so the delivery of proteins (even if they are transcription factors for a specific gene) does not count as gene delivery.

QUESTIONS

1. Chloroquine is a lysosomotropic agent, meaning it shuts off the proton/ATPases in lysosomes to prevent their acidification. Chloroquine has been used to enhance the transfection efficiency of certain nonviral delivery agents, presumably by preventing the activation of pH-sensitive degradative enzymes (acid hydrolases). However, chloroquine will reduce the transfection efficiency of lipoplexes that use the helper lipid DOPE. Explain this phenomenon.
2. Name two advantages of using DOPE in liposomal formulations.
3. What are M_n, M_w, and PI for the following distribution of polymer sizes?

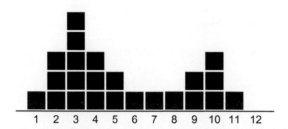

4. Refer to Chapter 4 for the structures of the nucleotide bases. Suppose that the percentages of A, T, C, and G bases in the genome of a certain cell type were all the same. Determine the average molecular weight of a DNA base in that genome.
5. Suppose we have two tubes. One contains DNA in a buffer solution. We put $2\,\mu L$ of this solution into a cuvette that contains $600\,\mu L$ of this buffer and find that it has an A_{260} of 0.343. In the second tube, we have a $10\,mg/ml$ solution of PLL.
   (a) What volume of the DNA solution will contain $4\,\mu g$ of DNA?
   (b) How much of the PLL solution will be needed to create one dose of gene delivery complexes containing $4\,\mu g$ of DNA at a 3:1 N:P ratio?
6. Suppose that we have an $8.614\,mg/ml$ PEI solution. What is the molarity of the solution?
7. Which of the listed options would be the best gene delivery vehicle or method for each application?

(a) Repeated deliveries to a patient with diabetes	Adenovirus	Retrovirus	Liposomes
(b) Repeated deliveries to skin cells in culture	Adenovirus	Retrovirus	Poly(L-lysine)
(c) Permanent delivery to cells in culture	Adenovirus	Lentivirus	Liposomes
(d) Delivery of RNA to a cell	Herpesvirus	Adenovirus	FIV
(e) Transfection of a frog oocyte	Microinjection	Electroporation	Gene gun
(f) Delivery to treat liver disease	Microinjection	PEI	Gene gun

8. Give three reasons (related to one principle) why permanent transduction via a retrovirus can harm a patient.

9. Suppose you want to use the polymer at the right to deliver 4 μg of plasmid DNA to 50,000 cells at 37° and pH 7.4. You will use N:P = 6.0 to deliver the 7000 bp plasmid. The polymer, for which $n = 1000$, is stored at 2.0 mg/ml in 90 ml of solution. You may use 300 for the molecular weight of a single DNA monomer. If we are to let the transfection proceed for 300 min, how much of your polymer solution will you use for the transfection procedure?

$$
\left[\begin{array}{c} \overset{\displaystyle NH_3^+}{\underset{\displaystyle |}{}} \\ CH_2 - CH - CH \\ \underset{\displaystyle |}{} \\ CH_2 \\ \underset{\displaystyle |}{} \\ NH_3^+ \end{array} \right]_n
$$

(Round atomic masses to the nearest integer.)

RELATED READING

GENERAL GENE DELIVERY

Advani, S.J., Weichselbaum, R.R., Kufe, D.W., 2003. Gene delivery systems. In: Kufe, D.W., Pollock, R.E., Weichselbaum, R.R., Kufe, D.W., Pollock, R.E., Weichselbaum, R.R., et al. (Eds.), Holland-Frei Cancer Medicine. sixth ed.. BC Decker, Hamilton, ON.

Munson, J.M., Godbey, W.T., 2006. Gene therapy. In: Bronzino, J. (Ed.), The Biomedical Engineering Handbook. third ed., Tissue Engineering and Artificial Organs, vol. 3. CRC Press/Taylor & Francis Group. Boca Raton, FL.

Verma, I.M., Somia, N., 1997. Gene therapy – promises, problems and prospects. Nature 389, 239–242.

Wolff, J., Lederberg, J., 1994. An early history of gene transfer and therapy. Hum. Gene Ther. 5, 469–480.

Zhang, X., Balazs, D.A., Godbey, W.T., 2011. Nanobiomaterials for nonviral gene delivery. In: Sitharaman, B. (Ed.), Nanobiomaterials Handbook. CRC Press, Taylor & Francis Group. Boca Raton, FL.

CLONING

Wilmut, I., Schnieke, A.E., McWhir, J., Kind, A.J., Campbell, K.H., 1997. Viable offspring derived from fetal and adult mammalian cells. Nature 385, 810–813.

Hosaka, K., Ohi, S., Ando, A., Kobayashi, M., Sato, K., 2000. Cloned mice derived from somatic cell nuclei. Hum. Cell 13, 237–242.

VIRAL DELIVERY METHODS

Buning, H., Braun-Falco, M., Hallek, M., 2004. Progress in the use of adeno-associated viral vectors for gene therapy. Cells Tissues Organs 177, 139–150.

Ding, W., Zhang, L., Yan, Z., Engelhardt, J.F., 2005. Intracellular trafficking of adeno-associated viral vectors. Gene Ther. 12, 873–880.

Gao, G.P., Alvira, M.R., Wang, L.L., Calcedo, R., Johnston, J., Wilson, J.M., 2002. Novel adeno-associated viruses from rhesus monkeys as vectors for human gene therapy. Proc. Natl. Acad. Sci. U. S. A. 99, 11854–11859.

Gonçalves, M.A., 2005. Adeno-associated virus: from defective virus to effective vector. Virol. J. 2, 43–59.

Grassi, G., Köhn, H., Dapas, B., Farra, R., Platz, J., Engel, S., Cjsareck, S., Kandolf, R., Teutsch, C., Klima, R., Triolo, G., Kuhn, A., 2006. Comparison between recombinant baculo- and adenoviral-vectors as transfer system in cardiovascular cells. Arch. Virol. 151, 255–271.

Hayashi, M., Tomita, M., Yoshizato, K., 2001. Production of EGF-collagen chimeric protein which shows the mitogenic activity. Biochim. Biophys. Acta 1528, 187–195.

Hu, Y.-C., 2006. Baculovirus vectors for gene therapy. Adv. Virus Res. 68, 287–320.

Jenkins, F.J., Turner, S.L., 1996. Herpes simplex virus: a tool for neuroscientists. Front. Biosci. 1, d241–d247.

Kay, M.A., Glorioso, J.C., Naldini, L., 2001. Viral vectors for gene therapy: the art of turning infectious agents into vehicles of therapeutics. Nat. Med. 7, 33–40.

Kost, T.A., Condreay, J.P., Jarvis, D.L., 2005. Baculovirus as versatile vectors for protein expression in insect and mammalian cells. Nat. Biotechnol. 23, 567–575.

Lachmann, R.H., 2004. Herpes simplex virus-based vectors. Int. J. Exp. Path. 85, 177–190.

Lilley, C.E., Branston, R.H., Coffin, R.S., 2001. Herpes simplex virus vectors for the nervous system. Curr. Gene Ther. 1, 339–358.

Matsushita, T., Elliger, S., Elliger, C., Podsakoff, G., Villarreal, L., Kurtzman, G.J., Iwaki, Y., Colosi, P., 1998. Adeno-associated virus vectors can be efficiently produced without helper virus. Gene Ther. 5, 938–945.

Russell, D.W., Kay, M.A., 1999. Adeno-associated virus vectors and hematology. Blood 94, 864–874.

Russell, W.C., 2000. Update on adenovirus and its vectors. J. Gen. Virol. 81, 2573–2604.

Smith, A.E., 1995. Viral vectors in gene therapy. Annu. Rev. Microbiol. 49, 807–838.

Zhang, X., Godbey, W.T., 2006. Viral vectors for gene delivery in tissue engineering. Adv. Drug Deliv. Rev. 58, 515–534.

PHYSICAL DELIVERY METHODS

Golzio, M., Teissié, J., Rols, M., 2002. Direct visualization at the single-cell level of electrically mediated gene delivery. Proc. Natl. Acad. Sci. U. S. A. 99, 1292–1297.

King, R., 2004. Gene delivery to mammalian cells by microinjection. In: Heiser, W.C. (Ed.), Gene Delivery to Mammalian Cells, Volume 1: Nonviral Gene Transfer Techniques. Part of Series: Methods in Molecular Biology, vol. 245. Springer, Secaucus, NJ, pp. 167–173.

O'Brien, J.A., Holt, M., Whiteside, G., Lummis, S.C., Hastings, M.H., 2001. Modifications to the hand-held Gene Gun: improvements for in vitro biolistic transfection of organotypic neuronal tissue. J. Neurosci. Methods 112, 57–64.

Teissié, J., Rols, M.P., 1993. An experimental evaluation of the critical potential difference inducing cell membrane electropermeabilization. Biophys. J. 65, 409–413.

CHEMICAL DELIVERY METHODS—PEI

Godbey, W.T., Wu, K.K., Mikos, A.G., 1999. Poly(ethylenimine) and its role in gene delivery. J. Control. Release 60, 149–160.

CHEMICAL DELIVERY METHODS—DENDRIMERS

Grayson, S.M., Godbey, W.T., 2008. The role of macromolecular architecture in passively targeted polymeric carriers for drug and gene delivery. J. Drug Target. 16, 329–356.

Zhong, Hui, He, Z.G., Li, Zheng, Li, G.Y., Shen, S.R., Li, X.L., 2008. Studies on polyamidoamine dendrimers as efficient gene delivery vector. J. Biomater. Appl. 22, 527–544.

Kukowska-Latallo, J.F., Bielinska, A.U., Johnson, J., Spindler, R., Tomalia, D.A., Baker Jr., J.R., 1996. Efficient transfer of genetic material into mammalian cells using Starburst polyamido-amine dendrimers. Proc. Natl. Acad. Sci. U. S. A. 93, 4897–4902.

Zinselmeyer, B.H., Mackay, S.P., Schatzlein, A.G., Uchegbu, I.F., 2002. The lower-generation poly-propylenimine dendrimers are effective gene-transfer agents. Pharm. Res. 19, 960–967.

CHEMICAL DELIVERY METHODS—CHITOSAN

Chew, J.L., Wolfowicz, C.B., Mao, H.Q., Leong, K.W., Chua, K.Y., 2003. Chitosan nanoparticles containing plasmid DNA encoding house dust mite allergen, Der p 1 for oral vaccination in mice. Vaccine 21, 2720–2729.

Ishii, T., Okahata, Y., Sato, T., 2001. Mechanism of cell transfection with plasmid/chitosan com-plexes. Biochim. Biophys. Acta 1514, 51–64.

Mansouri, S., Lavigne, P., Corsi, K., Benderdour, M., Beaumont, E., Fernandes, J.C., 2004. Chitosan-DNA nanoparticles as non-viral vectors in gene therapy: strategies to improve transfection efficacy. Eur. J. Pharm. Biopharm. 57, 1–8.

Mao, H.Q., Roy, K., Troung-Le, V.L., Janes, K.A., Lin, K.Y., Wang, Y., August, J.T., Leong, K.W., 2001. Chitosan-DNA nanoparticles as gene carriers: synthesis, characterization and transfec-tion efficiency. J. Control. Release 70, 399–421.

Roy, K., Mao, H.Q., Huang, S.K., Leong, K.W., 1999. Oral gene delivery with chitosan – DNA nanoparticles generates immunologic protection in a murine model of peanut allergy. Nat. Med. 5, 387–391.

Synowiecki, J., Al-Khateeb, N.A., 2003. Production, properties, and some new applications of chi-tin and its derivatives. Crit. Rev. Food Sci. Nutr. 43, 145–171.

CHEMICAL DELIVERY METHODS—PLL

Gonsho, A., Irie, K., Susaki, H., Iwasawa, H., Okuno, S., Sugawara, T., 1994. Tissue-targeting abil-ity of saccharide-poly(L-lysine) conjugates. Biol. Pharm. Bull. 17, 275–282.

Laemmli, U.K., 1975. Characterization of DNA condensates induced by poly(ethylene oxide) and polylysine. Proc. Natl. Acad. Sci. U. S. A. 72, 4288–4292.

Martin, M.E., Rice, K.G., 2007. Peptide-guided gene delivery. AAPS J. 9, E18–E29.

CHEMICAL DELIVERY METHODS—LIPOSOMES

Balazs, D.A., Godbey, W., 2011. Liposomes for use in gene delivery. J. Drug Deliv. 2011, 326497.

Boukhnikachvili, T., Aguerre-Chariol, O., Airiau, M., Lesieur, S., Ollivon, M., Vacus, J., 1997. Structure of in-serum transfecting DNA-cationic lipid complexes. FEBS Lett. 409, 188–194.

Chesnoy, S., Huang, L., 2000. Structure and function of lipid-DNA complexes for gene delivery. Annu. Rev. Biophys. Biomol. Struct. 29, 27–47.

Farhood, H., Gao, X., Son, K., Yang, Y.Y., Lazo, J.S., Huang, L., Barsoum, J., Bottega, R., Epand, R.M., 1994. Cationic liposomes for direct gene transfer in therapy of cancer and other dis-eases. Ann. N. Y. Acad. Sci. 716, 23–34.

Farhood, H., Serbina, N., Huang, L., 1995. The role of dioleoyl phosphatidylethanolamine in cat-ionic liposome mediated gene transfer. Biochim. Biophys. Acta 1235, 289–295.

Felgner, P.L., Gadek, T.R., Holm, M., Roman, R., Chan, H.W., Wenz, M., Northrop, J.P., Ringold, G.M., Danielsen, M., 1987. Lipofection: a highly efficient, lipid-mediated DNA-transfection procedure. Proc. Natl. Acad. Sci. U. S. A. 84, 7413–7417.

Felgner, J.H., Kumar, R., Sridhar, C.N., Wheeler, C.J., Tsai, Y.J., Border, R., Ramsey, P., Martin, M., Felgner, P.L., 1994. Enhanced gene delivery and mechanism studies with a novel series of cationic lipid formulations. J. Biol. Chem. 269, 2550–2561.

Ferrari, D., Peracchi, A., 2002. A continuous kinetic assay for RNA-cleaving deoxyribozymes, exploiting ethidium bromide as an extrinsic fluorescent probe. Nucleic Acids Res. 30, e112.

Gao, X., Huang, L., 1991. A novel cationic liposome reagent for efficient transfection of mammalian cells. Biochem. Biophys. Res. Commun. 179, 280–285.

Hsu, W.L., Chen, H.L., Liou, W., Lin, H.K., Liu, W.L., 2005. Mesomorphic complexes of DNA with the mixtures of a cationic surfactant and a neutral lipid. Langmuir 21, 9426–9431.

Hui, S.W., Langner, M., Zhao, Y.L., Ross, P., Hurley, E., Chan, K., 1996. The role of helper lipids in cationic liposome-mediated gene transfer. Biophys. J. 71, 590–599.

Israelachvili, J.N., 1991. Intermolecular and Surface Forces, second ed. Academic Press, London.

Jain, S., Zon, G., Sundaralingam, M., 1989. Base only binding of spermine in the deep groove of the A-DNA octamer d(GTGTACAC). Biochemistry 28, 2360–2364.

Leventis, R., Silvius, J.R., 1990. Interactions of mammalian cells with lipid dispersions containing novel metabolizable cationic amphiphiles. Biochim. Biophys. Acta 1023, 124–132.

Malone, R.W., Felgner, P.L., Verma, I.M., 1989. Cationic liposome-mediated RNA transfection. Proc. Natl. Acad. Sci. U. S. A. 86, 6077–6081.

Wasungu, L., Hoekstra, D., 2006. Cationic lipids, lipoplexes and intracellular delivery of genes. J. Control. Release 116, 255–264.

Zabner, J., Fasbender, A.J., Moninger, T., Poellinger, K.A., Welsh, M.J., 1995. Cellular and molecular barriers to gene transfer by a cationic lipid. J. Biol. Chem. 270, 18997–19007.

Zuhorn, I.S., Bakowsky, U., Polushkin, E., Visser, W.H., Stuart, M.C., Engberts, J.B., Hoekstra, D., 2005. Nonbilayer phase of lipoplex-membrane mixture determines endosomal escape of genetic cargo and transfection efficiency. Mol. Ther. 11, 801–810.

Zuidam, N.J., Barenholz, Y., 1997. Electrostatic parameters of cationic liposomes commonly used for gene delivery as determined by 4-heptadecyl-7-hydroxycoumarin. Biochim. Biophys. Acta 1329, 211–222.

Zuidam, N.J., Barenholz, Y., 1998. Electrostatic and structural properties of complexes involving plasmid DNA and cationic lipids commonly used for gene delivery. Biochim. Biophys. Acta 1368, 115–128.

Chapter 14

RNAi

14.1 COSUPPRESSION

Think back all the way back to the late 1980s (1990 is when the famous paper came out). There was a fellow named Richard Jorgensen who was interested in agricultural biotechnology, and he wanted to show that genetic engineering could be used in plants. He wanted to attract investors to his company, DNA Plant Technology Corporation, because, as with any company, money was needed for the company to thrive. In biotechnology, an important way to get money is through investors. As biotechnologists, this could mean one might go to a venture capital group to pitch an idea, and maybe the venture capital group will invest cash into the company in exchange for *equity* (percent ownership of the company).

As an entrepreneur, Dr. Jorgensen wanted to demonstrate genetic engineering in a way that investors (who quite often are not scientists) would understand *and be excited about*. (Remember, if you cannot excite somebody about your project, most likely nobody will want to give you any money for it.) By altering the appearance of some flowering plants, *"proof of concept"* would be established to demonstrate that protein expression patterns could be altered in plants via the delivery of RNA in this example. Proof-of-concept experiments do not necessarily produce the exact end result that a company or scientist is interested in, but they do show that the scientific idea is valid and can be built upon.

With the help of his research group, which included Carolyn Napoli and Christine Lemieux, Dr. Jorgensen set in on showing that the expression of a gene could be reduced or silenced through the delivery of RNA. He selected the chalcone synthase (CHS) protein, which plays a part in the production of a purple pigment in some plants, and he chose pink petunias as his model organism because changes in expression of the pigment would have been easily detectable. So the group delivered an antisense version of the *Chs* gene in the form of a plasmid, which contained a strong promoter (from the cauliflower mosaic virus) and a gene for kanamycin resistance. (A promoter, an exon, and an antibiotic resistance gene … note that we've already learned about these necessary components of an engineered plasmid.) As a control, they delivered a sense version of the *Chs* gene via a similar plasmid. And what did they get? In the flowers where the antisense plasmids were delivered, the group was successfully able

An Introduction to Biotechnology. http://dx.doi.org/10.1016/B978-1-907568-28-2.00014-9
Copyright © W T Godbey, 2014.

to knock down the expression of the purple pigment. However, in some of the plants receiving the sense version of *Chs*, what the group saw, instead of formerly pink flowers being induced to become more purple, was the production of beautiful *white* flowers! The plasmids somehow knocked out the expression of the purple pigment, and they did so more effectively than the plasmids coding for antisense mRNA. The experiment was repeated in dark purple petunias to investigate the strength of the gene suppression, and some white flowers were again produced. Flowers with very interesting combinations of purple and white patterns were also produced (Figure 14.1). Jorgensen termed this phenomenon "cosuppression," and it served to confound a significant portion of the scientific community for several years.

Dr. Jorgensen recounts the research in his own words:

To give you the full picture, we were doing two experiments - one that introduced an antisense Chs transgene, intended to silence endogenous Chs transcripts, and a second that introduced a sense Chs transgene that intentionally had been engineered to produce high levels of CHS protein expression, in order to over express CHS protein and possibly produce more pigment, assuming the endogenous Chs gene was rate-limiting to pigment synthesis (which we did not know for petunia, but was known for corn kernels and snapdragon flowers). The original purpose of these experiments was to try to produce plants with a visual alteration of a normal phenotype, merely as an illustration of plant genetic engineering in flower crops that we could use in fund raising efforts when talking to potential investors. We chose petunia as the target flower crop species because:

1) it was easily transformable with transgenes, at a time (1987) when very few plant species could be transformed,

2) a clone of the Chs gene was already available, and

3) it had flowers with anthocyanin-pigmented petals.

We chose a specific variety of petunia that had light pink flowers because we wanted to see if the sense Chs transgene could produce more deeply pigmented flowers (and whether the antisense transgene could reduce pigmentation). To our surprise, both sense and antisense transgenes reduced pigmentation, and the sense transgene was more effective at doing this than the antisense transgene! We did subsequent experiments with a deeply pigmented purple petunia line, and were still able to block all pigmentation in some of the transformed plants.

It was significant that our sense transgene was engineered for over expression of protein expression; otherwise we probably would not have discovered that a sense-oriented transgene could silence a homologous, endogenous gene. The frequency and degree of cosuppression by chalcone synthase transgenes are dependent on transgene promoter strength and are reduced by premature nonsense codons in the transgene coding sequence.

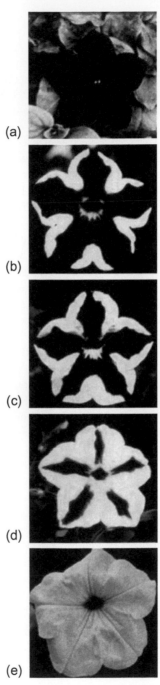

FIGURE 14.1 When Napoli, Lemieux, and Jorgensen transfected purple petunias such as the parent shown in (a) with a gene intended to amplify the production of a purple pigment, some very surprising results were obtained (b–e).

14.2   RNA INTERFERENCE

A few years later, two scientists by the names of Andrew Fire and Craig Mello performed a famous set of experiments to further investigate this type of suppression of gene expression, which by this point was referred to as "interference." What they found was the following: They injected single-stranded RNA into *C. elegans* and nothing happened. They injected antisense single-stranded RNA into these worms and not much happened. As a control, they injected double-stranded RNA and the worms no longer expressed the protein associated with that gene and the worms were curled and they twitched. The gene in question was involved with structural elements inside the worms, so that when the worms didn't have this structural element, they were no longer straight. The important result was that an effect was seen when double-stranded RNA was injected into these worms.

Let us take a moment to examine a more straightforward example regarding antisense technology. Single-stranded RNA can be injected into a cell such as a fertilized frog oocyte. The egg will divide, eventually forming a blastocyst. In the blastocyst, the gene in question will be expressed to the same degree as in the wild-type animal, unless the injected ssRNA was antisense to the mRNA for the corresponding gene. When antisense RNA is injected, there will be a reduction in the amount protein expression related to the gene. In terms of the central dogma, this should make intuitive sense: the cell takes a genomic message in the form of dsDNA and transcribes it into mRNA, which is then transported into the cytoplasm for translation. If one were to inject a sequence of single-stranded RNA that is complementary to this particular mRNA, would it not follow that the two complementary strands would bind together? The ribosome requires single-stranded RNA to perform translation, so it is predictable that this particular mRNA will not be translated and the cell will experience a reduction in the amount of the particular protein coded for by the gene.

If one were to deliver double-stranded RNA into the cytoplasm of the cell, as was done by Fire and Mello, there would likewise be reduced expression of the gene. This indicates that there is something more going on than simply preventing ribosome binding.

Jorgensen called this effect cosuppression. He delivered RNA coding for CHS, and not only was the delivered RNA suppressed but also the endogenous gene was suppressed. Fire and Mello, after determining the important role of double-stranded RNA, termed the suppression "*RNA interference*" (which we now abbreviate *RNAi*). For their discovery and characterization of gene silencing via double-stranded RNA, Fire and Mello were awarded the Nobel Prize in Physiology or Medicine in 2006.

Why would double-stranded RNA yield a reduction in the amount of protein produced from the associated gene? The answer has to do with three proteins, known as *Dicer*, *Slicer*, and *RISC*. If double-stranded RNA is delivered into a cell, the cell will destroy it. This may be the result of a primordial immune response: many naturally occurring viruses will infect cells via the delivery of double-stranded RNA. The cell can recognize the viral RNA because it is double-stranded—the only RNA produced by the cell itself will be single-stranded, so double-stranded RNA is an oddity that must have come from a foreign source. The cell will not only chop up the double-stranded RNA but also destroy any mRNA (even if it was produced by the cell itself) that is associated with the sequence of bases found in the double-stranded RNA, ostensibly because it may also be linked with a foreign source.

At the heart of RNA interference is RISC, which is an acronym for RNA-induced silencing complex. Cells may have many different types of RISCs, but there are at least three features that are common to all of them:

1. A subunit that serves as an RNA helicase
2. A subunit that binds to small single-stranded RNA
3. A subunit that acts as an endonuclease (the Slicer)

When double-stranded RNA is introduced into the cytoplasm, Dicer will bind to it and cleave it into smaller pieces, and RISC will associate with the smaller double-stranded RNA (Figure 14.2). The helicase in the RISC complex serves to separate this double-stranded RNA. The sense strand of this RNA will be discarded, leaving the RISC complex primed with a short, single-stranded, antisense RNA molecule. The primed RISC complex can now interact with other (single-stranded) RNAs, such as mRNA, and degrade the ones that pair with the sequence held by RISC.

If we designed the original double-stranded RNA based upon a specific gene that we wish to silence, then the antisense piece of RNA held by RISC will allow it to bind with the mRNA transcribed from the target gene. Next, a protein known as Slicer will degrade the bound mRNA strand. By reducing the number of mRNA sequences from the gene in question, the number of proteins expressed from that gene will be reduced.

It can be difficult to remember the functions of Dicer and Slicer. It may help to remember that *D*icer degrades *d*ouble-stranded RNA, and *S*licer severs *s*ingle-stranded RNA. (Confused? The television show NOVA produced an interesting 15 min feature in 1995 on RNAi that is a recommended supplementary material.) Although not as scientifically in depth as this discussion, the video does a good job of communicating basic concepts of how RNA interference works. It can be viewed at http://www.pbs.org/wgbh/nova/body/rnai.html.

Consider for another moment the RNAi process that began with double-stranded RNA and Dicer. Dicer degrades this RNA into much smaller fragments before RISC discards the sense strand. Depending on what is delivered into the cytoplasm, the step involving Dicer might not be necessary. If we were to deliver a small (19 to 24 nucleotide) single-stranded RNA sequence, the phenomenon of interference may still be observed. The RISC complex is able to

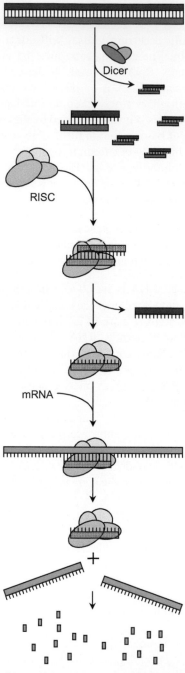

FIGURE 14.2 RNA interference hinges on the recognition of double-stranded RNA. If the dsRNA strand is sufficiently long, Dicer will cleave it into smaller fragments of small interfering RNA (siRNA). One of the smaller dsRNA fragments will be picked up by a RISC complex, and the sense strand (shown in blue) will be discarded. Any mRNA that contains a sequence that matches up with the retained antisense strand will be degraded by Slicer, a member of the RISC complex. Sliced mRNA fragments will be further degraded in the cytosol.

bind these small RNAs and continue with the process as already described. This RNA is called small interfering RNA, or *siRNA*.

14.3   mIRNA

We have now seen several examples of gene delivery. One can use it to deliver plasmids to express a protein of your choosing or to turn on a molecular cascade once the delivered plasmid is expressed, or one could deliver siRNA to knock down the expression of a gene.

Why don't we deliver a gene that will code for an RNA sequence that is antisense to a given mRNA to knock down gene expression and integrate it into the genome? Such a scheme has been tried, and it works to some degree to achieve permanent knockdown. However, a more effective route mimics something that already occurs in cell, a process that involves *microRNA* (*miRNA*). MicroRNAs have corresponding DNA coding regions in the genome. They are transcribed just like other genes, but contrary to the principles of the central dogma, they are not translated. The primary transcript of an miRNA, referred to as *pri-miRNA* (short for "primary-miRNA"), will contain one or several stem-loop structures (Figure 14.3). One (or more) of the stem loops will be cleaved by an enzyme complex called the microprocessor, which contains the protein Drasha. This occurs while the pri-miRNA is still in the nucleus. The result is a free-floating hairpin called a *pre-miRNA*. The pre-miRNA is then exported out of the nucleus and is acted upon by DICER, which cleaves the loop from the pre-miRNA to yield an miRNA duplex. Because of their similar structures at this point, the rest of the processing story is the same as for siRNA, involving the binding to RISC, separation of the double-stranded RNA, and retention of one of the strands to help RISC identify mRNA molecules to inactivate through cleavage via Slicer (compare to Figure 14.2). The cell uses this process to regulate protein expression, and biotechnologists have worked to achieve permanent knockdown of genes (*gene silencing*) by inserting DNA sequences into the genome of viral vectors to produce engineered miRNAs.

We have just examined several types of RNA interference: delivery of dsRNA, which can directly interact with Dicer (or RISC, if small enough, as is the case with siRNA); antisense RNA, which can pair with mRNA to create dsRNA, miRNA, a form of dsRNA that arises from inverted repeats in the gene that form stem-loop structures in the transcript that are further processed; and RNAi that arises from the overproduction of sense RNA. This last type of interference explains what was seen in the Jorgensen petunia experiments. It requires a host RNA-dependent RNA polymerase that recognizes and copies overexpressed transcripts to produce dsRNA molecules. This type of interference is beyond what we are focusing on in this chapter, although one could argue that it sparked the field of RNA interference.

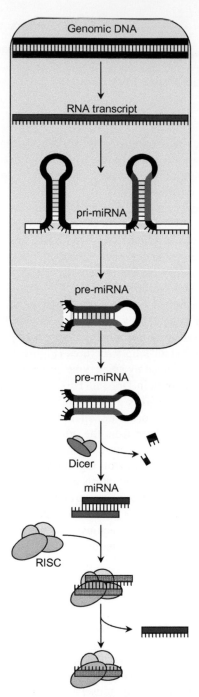

FIGURE 14.3 The origin of miRNA in the cell begins at the genome level. The primary RNA transcript contains inverted repeats that form stem-loop structures (pri-miRNA), which are cleaved into independent structures (pre-miRNA). Pre-miRNA is transported from the nucleus (shaded in green) to the cytoplasm, where it is acted upon by Dicer to produce miRNA. Notice that the structures of miRNA and siRNA cannot be distinguished at this point.

QUESTIONS

1. Explain the difference between siRNA and miRNA. Are Dicer and Slicer used for both siRNA and miRNA?
2. Considering that it only acts on double-stranded RNA, how is Dicer involved in the production of miRNA from single-stranded RNA?
3. What are the main functions of RISC that are common across species boundaries?
4. What are the differences and similarities between siRNA and miRNA?
5. How did RNA interference contribute to having white petunias?
6. Why might a mutated Dicer gene make a cell unhealthy, prone to viral infection, and potentially cancerous?
7. How does RISC determine which strand is sense and which is antisense?
8. Is it possible for miRNA to code for a protein?
9. A 19-24 bp strand of RNA was found in the cytoplasm, being held by an RNA-induced silencing complex. Is the strand siRNA or miRNA? Explain.
10. What is the difference between pri-miRNA and pre-miRNA?
11. Researchers want to fight a single-stranded RNA viral infection in a patient. The viral RNA has the sequence $(C)_{14}GGUGCA$. The cells that the viruses "attack" produce a vital protein that has an mRNA sequence that contains $(C)_{14}GGUGCA$. The researchers are considering using the siRNA $5'-UGCACC(G)_{14}-3'$. Should they use the siRNA or not? Explain.

RELATED READING

Béclin, C., Boutet, S., Waterhouse, P., Vaucheret, H., 2002. A branched pathway for transgene-induced RNA silencing in plants. Curr. Biol. 12, 684–688.

Cook, H.A., Koppetsch, B.S., Wu, J., Theurkauf, W.E., 2004. The Drosophila SDE3 homolog armitage is required for oskar mRNA silencing and embryonic axis specification. Cell 116, 817–829.

Denli, A.M., Tops, B.B., Plasterk, R.H., Ketting, R.F., Hannon, G.J., 2004. Processing of primary microRNAs by the microprocessor complex. Nature 432, 231–235.

Felekkis, K., Deltas, C., 2006. RNA intereference: a powerful laboratory tool and its therapeutic implications. Hippokratia 10, 112–115.

Fire, A., Xu, S., Montgomery, M.K., Kostas, S.A., Driver, S.E., Mello, C.C., 1998. Potent and specific genetic interference by double-stranded RNA in *Caenorhabditis elegans*. Nature 391, 806–811.

Gregory, R.I., Yan, K.P., Amuthan, G., Chendrimada, T., Doratotaj, B., Cooch, N., Shiekhattar, R., 2004. The microprocessor complex mediates the genesis of microRNAs. Nature 432, 235–240.

Hammond, S.M., Boettcher, S., Caudy, A.A., Kabayashi, R., Hannon, G.J., 2001. Argonaute 2, a link between genetic and chemical analyses of RNAi. Science 293, 1146–1150.

Heasman, J., Torpey, N., Wylie, C., 1992. The role of intermediate filaments in early Xenopus development studied by antisense depletion of maternal mRNA. Dev. Suppl. 1992, 119–125.

Jorgensen, R.A., 2003. Sense cosuppression in plants: past, present and future. In: Hannon, G.J. (Ed.), RNAi: A Guide to Gene Silencing. Cold Spring Harbor Laboratory Press, Cold Spring Harbor, NY, pp. 5–21.

Jorgensen, R.A., Doetsch, N., Müller, A., Que, Q., Gendler, K., Napoli, C.A., 2006. A paragenetic perspective on integration of rna silencing into the epigenome and its role in the biology of higher plants. Cold Spring Harb. Symp. Quant. Biol. 71, 481–485.

Kennedy, S., Wang, D., Ruvkun, G., 2004. A conserved siRNA-degrading RNase negatively regulates RNA interference in *C. elegans*. Nature 427, 645–649.

Liu, Y.P., Berkhout, B., 2011. miRNA cassettes in viral vectors: problems and solutions. Biochim. Biophys. Acta 1809, 732–745.

Napoli, C., Lemieux, C., Jorgensen, R., 1990. Introduction of a chimeric chalcone synthase gene into petunia results in reversible co-suppression of homologous genes in trans. Plant Cell 2, 279–289.

Peters, L., Meister, G., 2007. Argonaute proteins: mediators of RNA silencing. Mol. Cell 26, 611–623.

Que, Q., Wang, H.Y., English, J.J., Jorgensen, R.A., 1997. The frequency and degree of cosuppression by sense chalcone synthase transgenes are dependent on transgene promoter strength and are reduced by premature nonsense codons in the transgene coding sequence. Plant Cell 9, 1357–1368.

Que, Q., Wang, H.Y., Jorgensen, R.A., 1998. Distinct patterns of pigment suppression are produced by allelic sense and antisense chalcone synthase transgenes in petunia flowers. Plant J. 13, 401–409.

Torpey, N., Wylie, C.C., Heasman, J., 1992. Function of maternal cytokeratin in Xenopus development. Nature 357, 413–415.

Winter, J., Jung, S., Keller, S., Gregory, R.I., Diederichs, S., 2009. Many roads to maturity: microRNA biogenesis pathways and their regulation. Nat. Cell Biol. 11, 228–234.

Chapter 15

DNA Fingerprinting

DNA fingerprinting is a technology that is used to confirm identity. It is used by police crime labs to match evidence with suspects to help establish guilt or innocence, and it can also be used to determine paternity. Two different methods by DNA fingerprinting will be discussed here to show the evolution of the technology.

15.1 OLDER DNA FINGERPRINTING USES RFLPs

The first method involves restriction fragment length polymorphisms, or *RFLPs*. Imagine that you are a police detective who finds a large blood splatter at a crime scene. You collect a bit of the residue because you know that there is DNA in blood. Be careful here, though, because the DNA is not in mature red blood cells, which have no nuclei and therefore lack DNA. DNA is present in white blood cells, as well as any tissues that may be in the stain. (Other sources of crime scene DNA include semen, bone, skin, cells sloughed in the saliva, urine, feces, and cells at the root of a hair follicle.) Back in the laboratory (in this example, not only are you a detective but also a laboratory technician), you are able to isolate cells from the sample, after which you lyse them and collect the once-contained DNA. You have now isolated the genome of the person from whom the blood originated. You then expose the DNA to a set of restriction enzymes.

Consider a typical human somatic cell that has 23 pairs of chromosomes. Upon exposure to the restriction enzymes, these chromosomes will be chopped into very small fragments, small enough to run on a gel. The fragments will represent a very large array of sizes, so that, instead of bands, they will appear as a smear on the gel. In terms of restriction cuts and resulting fragment sizes (described earlier in this text), a smear by itself is of little use. However, probes can be used to further characterize the DNA in the smear. Similar to qPCR, the probes used in this technique are polynucleotides and will pair up with specific DNA sequences. The binding of one of these probes to its complementary sequence is known as *hybridization*.

Consider suspect 1, who has specific cut sites within a hypothetical stretch of DNA. In suspect 2, on the other hand, this same stretch of DNA contains a mutation in one of the recognition sequences for that restriction enzyme, so it

An Introduction to Biotechnology. http://dx.doi.org/10.1016/B978-1-907568-28-2.00015-0
Copyright © W T Godbey, 2014.

will no longer be cut at that point (Figure 15.1a). Looking at the figure for this particular stretch of DNA for the two suspects (Figure 15.1b), the lane corresponding to suspect 1 would contain four fragments after exposing the sequence to the specific enzyme, while the lane corresponding to suspect 2 would contain only three fragments. Keep in mind, however, that there is not a single, well-defined, and predicted pattern of bands for the human genome, but there is a characteristic pattern of bands for a given individual.

Keep in mind that we will be exposing the entire genome to the restriction enzyme, so the resulting gel would contain a smear and not distinct bands.

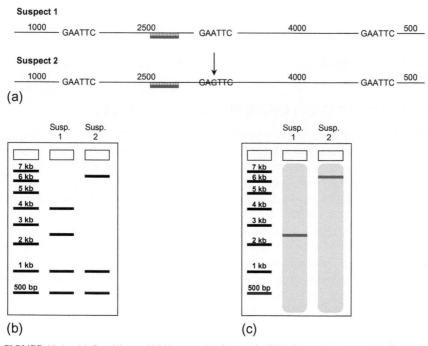

FIGURE 15.1   (a) Consider an 8000 bp stretch of genomic DNA from two suspects. In the DNA from suspect 1, there are three recognition sites for a certain restriction enzyme. In suspect 2, there has been a mutation in one of the recognition sites (denoted by an arrow) so that there are only two cut sites for the enzyme. (Fragment lengths, in bp, are shown by numbers over each line segment.) (b) When the stretches of DNA are cut with the enzyme and the resulting fragments are run on a gel, the difference between the banding patterns for suspects (a) and (b) can be seen as the 2500 and 4000 bp fragments are apparently combined in suspect (b) due to the mutated cut site. (c) However, we must consider that the entire genome is present in each DNA sample, not just an isolated 8000 bp fragment. After cutting the 3.08×109 bp genome with the restriction enzyme, there will be so many resulting fragment lengths that the gel would appear as a smear (represented by the gray areas on the gel). Exposure of the fragments to a probe (shown in red) allows for the hybridization of the probe to a specific area of the genome. Refer back to panel (a) to convince yourself that, although the probe will bind to the same place in each genome, the banding pattern when the probe is used will yield a different result for each suspect. When enough different probes are used, a distinct DNA fingerprint will be revealed for each individual.

After performing gel electrophoresis to obtain the smear, the genomic DNA is then blotted onto a membrane that is placed on top of the gel. Once bound to the membrane, the DNA is more accessible for further processing by means such as hybridization. The probe that is used in this technique is complementary to a specific DNA sequence, and it will bind at the same genomic location in the samples from both suspects 1 and 2. Although bound to the same genomic location, the probe will appear in different locations on the gel because of the selective degradation of the genomic DNA that has already taken place (Figure 15.1c). For a single band, it's a matter of whether the band appears high or low in the gel. The probe that is used in practice, however, is complementary to a sequence of DNA that is repeated many times throughout the genome. The result is not a single band, but rather a pattern of bands that is unique to each individual.

The probes that are used are radiolabeled, meaning they contain radioactive isotopes. To visualize them after the hybridization, the membrane is washed to remove unbound probe and then placed in a cassette with unexposed X-ray film. The film will be later developed to reveal the series of bands. Consider a murder scene. The way that the process takes place in a police crime lab is that the blood is rehydrated if necessary and the cells are isolated. The cells are lysed, and then the DNA is isolated, digested, and run out on a gel. The DNA is then blotted from the gel onto a membrane, which will be exposed to radiolabeled probes. (Blotting is performed by placing the membrane, which is a paper-thin piece of (typically) nylon, on top of the gel, followed by stacking paper towels on top of the membrane and weighting them down. The paper towels will wick the fluid from the gel like a sponge. The DNA in the gel will also be wicked, but it will be trapped in the membrane and not reach the paper towels.) The DNA-containing membrane is then washed and exposed to X-ray film. The samples that undergo this process include not only evidence taken from the crime scene but also samples taken from individuals who may be suspects. Of course, we will take blood or hair samples from suspects 1 and 2 and perhaps from the crime scene investigator since he collected the original sample, and certainly a sample will be taken from the victim. One cannot assume that all of the blood in a crime scene came from the victim; perhaps, there was a struggle or an accident whereby the murderer shed (blood) cells at the crime scene. The DNA from each of these samples will yield RFLPs that produce a characteristic pattern upon hybridization (Figure 15.2). If the set of bands from the crime scene sample matches up with those originating from one of the tested individuals, an identification will have been made.

15.2 NEWER DNA FINGERPRINTING USES STRs

The use of RFLPs is the older of the two techniques described here for DNA fingerprinting. The RFLP protocol requires a relatively large amount of DNA (>25 ng), which must be relatively undisturbed. The reason that the DNA must

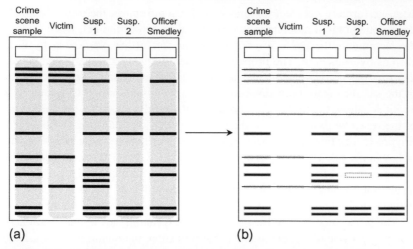

(a)   (b)

FIGURE 15.2   (a) RFLP pattern from a hypothetical DNA footprinting analysis of crime scene blood samples and DNA samples obtained from two suspects and the crime scene investigator, Officer Smedley. (b) To solve, eliminate all bands that are in common with both the crime scene sample and the victim. The perpetrator must have all remaining bands and only those bands. Note that suspect 1 has an extra band not reflected in the crime scene sample and suspect 2 is missing a band (shown in red). From the data, we can conclude that Officer Smedley satisfies the criteria. Either he is the killer (which, one would hope, is very doubtful), or he has shoddy data collection techniques. The suspects have been exonerated.

be fairly intact is that fairly large amounts of the probe must bind for visualization to be possible. If the DNA sample has been in a tomb for hundreds (or thousands) of years, which is ample time for DNA damage from ultraviolet light, oxidation, or spontaneous deamination, the quantity of undamaged fragments (or the ratio of undamaged/damaged DNA) may be too low to obtain a reliable band pattern upon hybridization. What used to be a fragment of 1000 bp may have degraded into many fragments of under 100 bp each. There would be no way of determining whether these fragment lengths resulted from restriction enzyme cuts or DNA damage.

The requirement of having relatively undisturbed DNA is a significant restriction. One could not identify, say, the last Russian czar from a collection of burned bones using RFLP technology. However, the identification of the remains of Czar Nicholas II was indeed made using a more modern technology, which involves *short tandem repeats* (*STRs*). STRs are 2-5 bp DNA sequences that are repeated several times in succession. For example, "GATAGATAGATAGATAGATAGATAGATAGATA" is an example of repeated GATA sequences, which is one of the main STR markers used for DNA fingerprinting. STRs occur throughout the genome. While specific repeated sequences are common to all humans, the lengths of the repeated sequences are quite different. Consider a specific genomic locus that contains an STR: D7S820, which resides on chromosome 7. This STR contains repeats of the tetramer GATA. Keeping in mind that you got half of your genome from your mother and half of your genome from your father and that each of them was donated, at this locus, an STR, you might find that, at the D7S820 locus, your genome contains alleles of perhaps 6 and 11 GATA repeats, as opposed to your friend who might have 9 and 14 repeats at the same locus. So, even though you and your friend both possess the D7S820 STR, the lengths of your D7S820 alleles differ from those of your friend. In general, STR sequences can range from 3 to 21 repeats (some might argue a higher range), depending on the specific STR locus. (For D7S820, the range is 6 to 14 repeats.)

Since we have two alleles for every genomic feature, including STR locations, how could differing repeat lengths yield any type of valuable information? The striking feature of STRs is that they are flanked by known sequences. These sequences are the same for you as they are for your friend, or even the author of this book. Even though the lengths of our STRs are different, the flanking sequences are the same. That means we can use PCR to amplify the regions between these known sequences—meaning we can amplify the STRs—to levels that are sufficient for analysis, allowing one to determine the two lengths of a given STR that are in a given genome. Odds are that the two lengths will be different for you and your friend, as well as your classmates, or the author of this book. Since it is true that the number of repeats in an STR is a finite number, so there are a finite number of combinations for their lengths (a range of 3-21 yields 190 combinations); there is a small probability that you will share the same pair of STR lengths as another given individual for a given location in the genome. However, there are thousands of STR locations in the genome, and most of these sites have their own identifying flanking sequences.

CODIS, which stands for Combined DNA Index System, is a database maintained by the US Federal Bureau of Investigation that has records of core STR markers for over 1×10^7 individuals, and the number of records is steadily rising. In the United States, for STR evidence to be admissible at trial, it must match up with 13 agreed-upon loci (Figure 15.3). In the United Kingdom, ten STR loci are typically used for identification purposes (using the National DNA Database, or NDNAD). Taking D7S820 as a standard, for which there are $(14-6+1)[(14-6+1)+1]/2=45$ combinations of allele lengths (recall that there are 6 to 14 repeats per allele for this STR, and know that the number of combinations of n objects taken 2 at a time$=n(n+1)/2$ when order is not important (*e.g.,* 6 and 11 repeats would show up as equivalent to 11 and 6 repeats on a gel)), then a specific DNA fingerprint using 13 similar STRs would be unique for one out of $(45)^{13}=3.10 \times 10^{21}$ individuals, a specificity greater than one encompassing the entire population of the Earth. The actual specificity is much lower, however, because allele lengths are not distributed randomly over a range, meaning some numbers of repeats are more likely than others. For the D7S820 locus, 10 and 11 repeats are the most common lengths (at least for a population sample taken in northeastern Brazil).

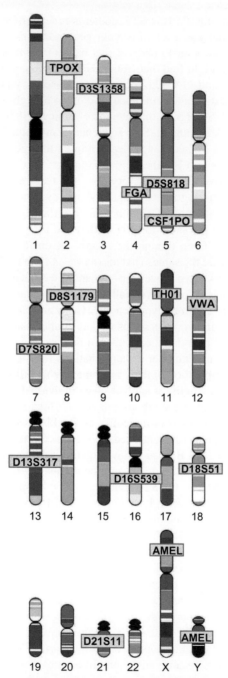

FIGURE 15.3 Names and relative positions of the 13 core CODIS STR loci on human chromosomes. Note that although it is not one of the core loci, AMEL is included here because it is often included to determine gender. *Relative positions taken from http://www.cstl.nist.gov/strbase/fbicore.htm.*

In comparison with the DNA fingerprinting technique that uses RFLPs, the STR system is far more sensitive. Only ~1 ng of DNA is needed, and the integrity of the DNA is not as much of an issue. When using PCR to amplify the STR regions, as long as the primers can bind to one of the sides of an unbroken DNA strand, that strand will be amplified exponentially and allow for accurate analysis. In theory, only one unbroken stretch of DNA in the amplicon region is needed. With RFLPs, no amplification takes place so only the DNA that is collected can be visualized, and if it is damaged, it will yield poor or confounded results.

QUESTIONS

1. Name two advantages of using STRs rather than RFLPs for DNA fingerprinting.
2. If a certain STR can range from 8 to 13 repeats per allele, how many unique combinations of this STR are possible?
3. Why wouldn't a microinjected plasmid appear in DNA fingerprinting? What if the injection had been performed on an oocyte that then developed into a complete organism?
4. Describe some differences between DNA fingerprinting and DNA footprinting.

RELATED READING

Butler, J.M., 2007. Short tandem repeat typing technologies used in human identity testing. Biotechniques 43, ii–v.

http://www.cstl.nist.gov/strbase/intro.htm.

Ferreira, da Silva L.A., Pimentel, B.J., Almeida, de Azevedo D., Pereira, da Silva E.N., Silva dos Santos, S., 2002. Allele frequencies of nine STR loci—D16S539, D7S820, D13S317, CSF1PO, TPOX, TH01, F13A01, FESFPS and vWA—in the population from Alagoas, northeastern Brazil. Forensic Sci. Int. 130, 187–188.

Seringhaus, M., 2009. The evolution of DNA databases: expansion, familial search, and the need for reform. In: DePaul University College of Law 9th Annual CIPLIT Symposium & 12th Annual Niro Distinguished Lecture, October, 3, 46 pages.

Chapter 16

Fermentation, Beer, and Biofuels

The harnessing of cells for the production of beer and biofuels involves processes that are perhaps surprisingly similar. However, before we can fully appreciate these two applications that take advantage of fermentation performed by cells, normal pathways of cellular metabolism must first be discussed.

Glycolysis is the cellular breakdown of the sugar glucose for energy. In-depth discussion of the details of the glycolytic pathway can be found in any good biochemistry textbook and will not be included here. However, there are certain details that are important for the understanding of biotechnological applications that utilize cellular respiration that we should appreciate.

16.1 GLYCOLYSIS

16.1.1 The Embden-Meyerhof Pathway

In your body, glucose is the preferred energy source. This is true for most living organisms. While the metabolism of many different sugars can be used for energy, the involved pathways all seem to converge upon the glycolytic pathway at some point. The pathway shown in Figure 16.1, known as the Embden-Meyerhof pathway, is the glycolytic pathway used by humans. It takes place in two phases: the investment phase and the payoff phase.

The investment phase refers to the steps of glycolysis that require energy input. The energy is obtained through the hydrolysis of a high-energy phosphate in an ATP molecule. The investment phase includes the initial steps. For instance, the first step is the conversion of glucose into glucose 6-phosphate, which can be written as

$$\text{glucose} + \text{ATP} \rightarrow \text{glucose 6-phosphate} + \text{ADP}$$

Notice that the ATP serves two purposes. It supplies the phosphate that is attached to the initial glucose molecule, and it supplies the energy needed to drive

An Introduction to Biotechnology. http://dx.doi.org/10.1016/B978-1-907568-28-2.00016-2

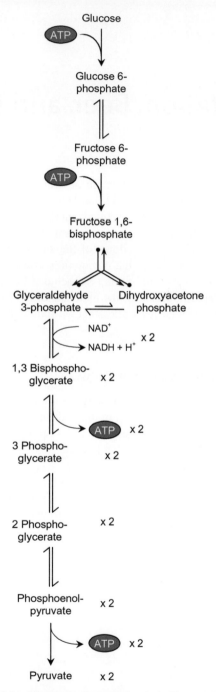

FIGURE 16.1   The Embden-Meyerhof pathway of glycolysis. The overall ATP yield is 2 mol ATP per mole of glucose. Specific steps involving ATP are shown.

the reaction forward. After this initial step, the glucose of glucose 6-phosphate undergoes isomerization from a 6-carbon sugar into the 6-carbon sugar fructose as fructose 6-phosphate.

$$glucose 6 - phosphate \rightarrow fructose 6 - phosphate$$

In the next step, another ATP is invested to yield a fructose to which two phosphates have been attached, one on carbon 1 and one on carbon 6:

$$fructose 6 - phosphate + ATP \rightarrow fructose 1, 6 - bisphosphate + ADP$$

One might ask, "If the purpose of glycolysis is to yield energy in the form of ATP for the cell and the initial steps require the hydrolysis of two ATP molecules, why would the process ever occur?" The answer lies in the fact that reactions can be coupled. Despite the investment of two ATP molecules in the investment phase, the yield of the payoff phase will be four ATP molecules, leaving the cell with a net gain of two ATP molecules for the conversion of one glucose molecule into two pyruvates.

Moving ahead in the cascade, the payoff phase of glycolysis has two ATP-yielding reactions. The first is the conversion of 1,3 bisphosphoglycerate, which can be viewed as a deprotonated glyceric acid with two high-energy phosphates attached, into 3-phosphoglycerate, which only holds one high-energy phosphate. The energy from this cleavage of a high-energy phosphate is used to form an ATP molecule:

$$1, 3 bisphosphoglycerate + ADP \rightarrow 3 - phosphoglycerate + ATP$$

Further molecular conversions occur to set up the removal of the final high-energy phosphate to produce another molecule of ATP. The reaction is as follows:

Phosphoenolpyruvate + ADP → Pyruvate + ATP

From this cursory introduction to glycolysis, it may appear that two ATP molecules were invested to yield a payoff of two ATP molecules. However, notice from Figure 16.1 that the 6-carbon molecule fructose 1,6-bisphosphate is split into two 3-carbon molecules, dihydroxyacetone phosphate and glyceraldehyde 3-phosphate. Each of these 3-carbon products will progress separately down the remainder of the glycolytic pathway to yield two ATP molecules and one pyruvate molecule each, for a total of four ATP and two pyruvate molecules. The net *yield* is 2 ATP molecules and 2 pyruvate molecules (and 2 NADH molecules).

The above discussion is by no means all of glycolysis. Enzyme names, kinetics, and individual reaction mechanisms are all important. The fate of pyruvate is also a key concern. For instance, in the presence of oxygen, not only can additional energy be extracted from the pyruvate molecules, but also the two molecules of NADH produced in forming 1,3-bisphosphoglycerate can themselves be used for energy. Metabolism in the presence of oxygen is known as aerobic respiration and in the absence of oxygen is known as an aerobic respiration. In anaerobic conditions, the pyruvate molecules will undergo fermentation to form products such as lactate in humans or ethanol in yeast.

Aerobic respiration yields far more energy than does anaerobic respiration. The pyruvate is broken down into acetyl Co-A, which can then enter the citric acid cycle (also known as the Krebs cycle or the TCA cycle) to produce carbon dioxide plus more ATP (or the energy-equivalent GTP), more NADH, and $FADH_2$. These latter two molecules can then enter the electron transport chain to eventually yield more ATP molecules while converting oxygen into water. Note that the difference in energy production from one mole of glucose is a net of 32 *versus* 2 mol of ATP for aerobic *versus* anaerobic respiration. That is why aerobic respiration is the preferred form of glucose metabolism in our bodies: the energy needs of the body can be met with a smaller amount of starting material. However, when the biotechnologist harnesses microbes for the production of fermentation products, oxygen supply will need to be eliminated to force the organisms into anaerobic respiration and subsequent fermentation.

16.1.2   The Entner-Douderoff Pathway

The glycolytic pathway used in humans (and yeast), the Embden-Meyerhof pathway, is not the only glycolytic pathway that is used in nature. The

Entner-Doudoroff pathway is used by certain prokaryotes, such as *Zymomonas* and *Pseudomonas*, to break a molecule of glucose into two molecules of pyruvate. While this pathway begins with the same conversion of glucose into glucose 6-phosphate, it is distinctly different from the previous pathway in the next three steps, which produce 2-keto-3-deoxygluconate (KDPG) (Figure 16.2). From here, KDPG, a 6-carbon molecule, is broken down to the 3-carbon molecules glyceraldehyde 3-phosphate (G3P) and pyruvate, two molecules we have already seen:

$$2\text{-keto-}3\text{-deoxygluconate} \rightarrow \text{glyceraldehyde3-phosphate} + \text{pyruvate}$$

The G3P is converted to pyruvate using the same steps as in the Embden-Meyerhof pathway. The net result of the Entner-Doudoroff pathway is one glucose molecule broken into two pyruvate molecules with a net yield of 1 ATP (plus 1 NADH and 1 NADPH). The energy investment is cut in half because only one step uses ATP for phosphorylation. The lower ATP yield is due to the direct production of one pyruvate molecule, without ATP production, from the 6-carbon intermediate KDPG. This means that only one molecule of G3P is produced per glucose, so the pathway from G3P $\rightarrow \rightarrow \rightarrow$ pyruvate will only be utilized once per glucose molecule, thereby cutting the number of ATP produced in the payoff phase in half.

Notice that the difference in ATP production between the two pathways has implications in terms of glucose utilization. Suppose that a certain number of cells need a certain amount of energy from ATP to stay alive. To get that energy, the cells will burn glucose. Cells that utilize the traditional (Embden-Meyerhof) pathway will produce two moles of ATP per mole of glucose, while cells that utilize the alternative (Entner-Doudoroff) pathway will produce one mole of ATP per mole of glucose. Since the energy needs of the two cultures are the same, it would take twice as much glucose to yield the same required amount of energy using the alternative pathway. The result will be that cells such as *Zymomonas* could be expected to break down glucose at a faster rate, thereby producing pyruvate at a faster rate. (The rate of production of pyruvate will become more important after fermentation is discussed.)

From an engineering standpoint, it might appear that using *Zymomonas* for the production of ethanol might be superior to using yeast. However, there are more factors to consider, as we shall soon see.

16.2 FERMENTATION

Let us now consider what can be done with the pyruvate molecules produced by glycolysis. Human cells, under anaerobic conditions, will convert pyruvate into lactate. Certain microbes might convert pyruvate into acetyl CoA, which will then serve as a gateway to many other potential fermentation products. For instance, acetyl CoA can become phosphorylated and then used to yield acetate. Acetate is the conjugate base of acetic acid, which is found in the vinegar you

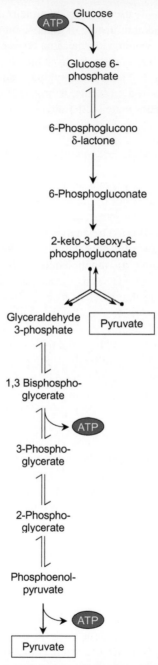

FIGURE 16.2 The Entner-Douderoff pathway of glycolysis. The overall ATP yield is 1 mol per mole of glucose. Two moles of pyruvate will also be generated. Reactions involving NADH (and NADPH) have been omitted for clarity.

might put on your salad. Acetyl CoA can also be converted into acetaldehyde, which can be further reacted to produce ethanol, another fermentation product of commercial value. Fermentation can also be used by certain microbes to produce acetone, a solvent commonly used in fingernail polish remover. From acetone, there is a chemical pathway to produce isopropanol, also known as rubbing alcohol.

Fermentation processes are performed by cells to extract energy from a starting material such as glucose. Fermentation processes do not consume oxygen, hence the extrapolation by some people that fermentation must occur in oxygen-free environments. (A complete absence of oxygen is not a strict requirement for fermentation to occur, but oxygen is not involved in the chemical reactions.) The hallmarks of fermentation processes are as follows:

1. Energy is produced.
2. Oxygen is not consumed.
3. The NADH/NAD$^+$ ratio is unchanged by the process.
4. The hydrogen to carbon ratio is unchanged between reactant and product.

Consider the fermentation of glucose into lactic acid in humans (Figure 16.3). Recall that two molecules of pyruvate will be formed by the breakdown of one molecule of glucose via glycolysis. As a result of anaerobic respiration, pyruvate will be converted to lactate. This step serves to remove the NADH that was produced earlier in the glycolytic pathway. In the presence of oxygen, the NADH would be able to enter the electron transport chain, which would be indirectly responsible for the hydrogenation of oxygen to form water. In the absence of oxygen, the electron transport chain will be halted so NADH will not be used to produce energy. If left unchecked, the concentration of NADH would increase and turn off certain glycolytic enzymes, meaning the breakdown of glucose to obtain energy will be halted. To prevent this potentially deadly effect, NADH will be siphoned off by fermentation to produce lactate. Notice that the conversion of pyruvate to lactate will alter the NADH/NAD$^+$ ratio

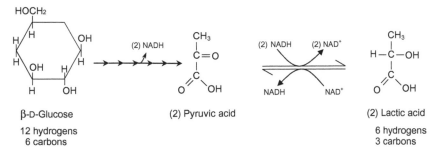

β-D-Glucose
12 hydrogens
6 carbons

(2) Pyruvic acid

(2) Lactic acid
6 hydrogens
3 carbons

FIGURE 16.3 Fermentation of glucose into two lactic acid molecules. Notice that the net number of NADH=0, and the ratio of H/C=2 for both glucose and lactic acid.

(recall the hallmarks of fermentation), so this individual step alone should not be considered fermentation. Rather, the conversion of glucose into two molecules of lactate is a fermentation process. The NADH/NAD$^+$ ratio is unchanged, as is the ratio of hydrogen to carbon atoms in proceeding from a glucose to two lactate molecules.

NADH, in aerobic respiration, is an energy source. It passes an electron pair, through a hydride ion, into a chain whose members serve as a series of electron carriers. As the electrons are passed from one set of carriers to another, changes in free energy are harnessed to pump protons out of the mitochondrion to create a charge gradient across the inner mitochondrial membrane. The gradient eventually becomes large enough to drive protons back into the mitochondrion via a proton ATPase, but the ATPase is driven in reverse so that it forms ATP rather than hydrolyzing it.

Since NADH serves as an energy source, it should make sense that the cell uses the concentration of NADH (monitored by the ratio [NADH]/[NAD$^+$]) to help determine its own energy state. If there is an excess of NADH, the cell behaves in a way consistent with being in an energy-rich state; all of its energy needs are met. Energy-producing processes such as glycolysis and the citric acid cycle will be inhibited, and the cell will resort to other pathways designed for energy storage.

Consider that you are a cell trying to stay alive in hard times (in terms of energy production), perhaps in an environment with insufficient oxygen. Without fermentation, you would experience a buildup of NADH, which would halt glycolysis, the very thing that you do not need to happen during hard times. Cells have a solution to this problem, however, which uses NADH as part of a side reaction: the reaction of pyruvate going to lactate (pyruvate acid going to lactic acid). For every pyruvate that is converted, one NADH will be converted back to NAD$^+$. This means the net NADH/NAD$^+$ ratio will be unchanged because one NADH is produced for every pyruvate that is produced. This fulfills the third hallmark of fermentation, that is, that the net NADH/NAD$^+$ ratio remains unchanged. Recall that the first hallmark of fermentation was that oxygen is not consumed. It was the lack of oxygen that forced the cell to go through this process in the first place! The third hallmark of fermentation is that the hydrogen to carbon ratios of the initial reactant and final product are unchanged. As shown in Figure 16.3, the ratio of hydrogen to carbon atoms in glucose is the same as it is for lactic acid. Note that this third hallmark focuses on the process as a whole, not an individual reaction.

For humans, fermentation is the production of lactic acid from glucose. As already mentioned, other organisms produce far different products via fermentation: ethanol, acetone, isopropanol, and butanol are all fermentation products. Several of these pathways are shown in Figure 16.4. The biotechnologist can harness microorganisms to produce each of these products, sometimes on an industrial scale. For instance, yeast can be used to produce ethanol as part of the beer-making procedure.

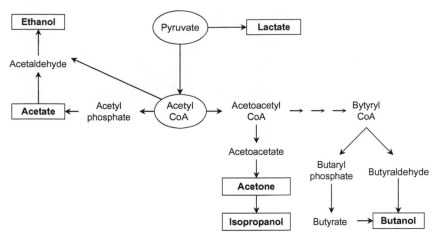

FIGURE 16.4 Rudimentary metabolic map of some common fermentation products, shown in boxes. This particular map shows products attainable from gas fermentation with *Clostridia*.

16.3   THE PRODUCTION OF BEER

(Special thanks to Graham Satterwhite, beer home-brewing expert)

Beer has been around for 5000-7000 years. Ancient evidence of beer's existence can be found on a Sumerian tablet from 4000 BC and again in a ~4000-year-old Sumerian poem written to honor their patron goddess of brewing. Beer is mentioned in poetry, it comes up in sonnets, and in scientific and religious texts, so evidently there is some kind of allure to it. Now, as budding biotechnologists, we will be able to appreciate beer for how it incorporates fermentation, cell respiration, and cell culture into a product with mass appeal. The production of beer is perhaps the oldest biotechnology known to man.

The process of beer production uses fermentation, carried out by yeast, to produce ethanol. The fermentation uses carbohydrates that are found in grains as the primary fuel source. The process is carried out by yeast glycolytic enzymes using the Embden-Meyerhof pathway (Figure 16.1).

Keep in mind that grains are just seeds. There's a reason why seeds have been chosen—they contain dense energy stores and nutrients for the baby organism that's going to grow from the seed, much like the yolk of an egg serving a developing chicken embryo. One of the most common types of seeds used for producing beer is barley, although other grains such as wheat can be used.

16.3.1   Malt

The first step in the brewing process utilizes enzymes that are found in the seeds themselves to break down the complex carbohydrates (starches) that are also stored in the seeds. The products will be simple sugars, and the process is known as *malting* (Figure 16.5). The seeds are allowed to germinate because, in

FIGURE 16.5   Malted barley.

normal development, the growing young plant is going to need to access these complex carbohydrates for its own energy. The energy needs of a developing organism are relatively large, so early on, the plant embryo will produce the enzymes needed to break down the complex carbohydrates that are stored in the endosperm into simple sugars. The reason that a brew master would want such production of simple sugars is that they will be used as fuel for the yeast.

The process of germination is not allowed to continue to completion because the developing plant would use the sugars for its own energy needs. To prevent this, heat is used to halt germination. The process is allowed to progress long enough for the embryo to produce the starch-degrading enzymes, but once the enzymes are present, germination can be halted because only the enzymes are needed to produce the simple sugars. (Recall that enzymes catalyze chemical reactions; a living organism is not needed for an enzyme to function.) The sugars that will be produced in the next step include glucose, fructose, and the disaccharide maltose (which is made up of two glucose molecules). Maltose is typically the sugar of interest for beer production.

16.3.2   Wort

The product of malting will be used to produce a nutrient broth, similar in principle to the LB broth used for *E. coli* growth. This nutrient broth, known as *wort*, will be used as a culture medium for yeast. It is produced by adding water to the malt, which is then mashed (Figure 16.6). (*Sour mash* is slightly different. Microbes are permitted to partially degrade the mash, which will acidify it, which will give the resulting product a sour taste.) The malt enzymes will then catalyze reactions to form maltose, glucose, and other sugars. After this maturation of the mash, it will exist as a slurry that contains seed debris, proteins, and both simple and complex carbohydrates (Figure 16.7). The solid portion is separated out by gravity, perhaps by using a mash tun that uses a porous support bed to hold the solid grain remnants as the liquid slowly flows through.

FIGURE 16.6   Mashing.

FIGURE 16.7   Mature mash.

The collected liquid is then boiled with hops (Figure 16.8). Hops give a characteristic bitter taste to the final beer product, but historically, they were once used because of their antiseptic qualities. Beer used to be transported around the world via ships, and the presence of a hop extract in the beer would prevent microbes from metabolizing it during transit. (As a side note, Indian pale ales—IPAs—have a fairly high alcohol content for a similar reason. Made in India, they once had to be shipped around the Horn of Africa to get to their final destinations, and the trip could take months. The higher alcohol content served as a preservative.) Hops are also used to lend a distinctive aroma to the final product. After the wort has been boiled with the hops, it is cooled and aerated. Aeration is necessary to enrich this nutrient broth with oxygen for the next step.

16.3.3   Yeast Cultures

The third step is to prepare the raw beer. A yeast culture is established in the aerated wort. Two-stage growth will be employed. The first stage will be aerobic

FIGURE 16.8   Dried leaf hops.

growth of the yeast, and the second stage will switch to anaerobic conditions. The purpose of the aerobic growth is to allow the yeast population to increase rapidly. With oxygen present, the yeast will be able to extract more energy per sugar molecule, which means the energy requirements of the yeast will be more easily met, thus allowing them to proceed through the cell cycle unhindered. The wort contains many simple sugars from the breakdown of starches. When the yeast are first added to the aerated wort, there are much room for expansion (recall contact inhibition), plenty of nutrients, and plenty of oxygen. These conditions allow them to grow and divide very quickly. The final products of aerobic metabolism of the sugars will be water and carbon dioxide, which is the source of carbonation in beer. As the growth curve for the yeast culture enters the late log phase, where nutrients and room for expansion are becoming the limiting factors, the culture will be capped to prevent further access to oxygen. The culture will continue to grow and metabolize the sugars until all of the oxygen is consumed, at which point the microbes will switch to anaerobic metabolism and utilize a fermentative pathway. The remaining sugars will be broken down into carbon dioxide and ethanol.

In Chapter 5, we discussed the growth curve, which certainly applies to yeast cultures in beer brewing. Within the growth curve, the beer brewer must try to minimize the lag phase, which is critical to brewing. The longer the lag time is, the greater the risk of contamination via microbial infection. Brewers will often make a starter yeast culture to minimize the lag phase (Figure 16.9). A starter culture is like a minibeer that is used to increase the pitch rate (of yeast into the wort). The sugar selection for the starter culture is critical. If anything other than dry malt extract is used for the starter substrate, the lag time will increase despite an increase in the final yeast numbers because the yeast will have to produce enzymes to metabolize different sugar types. If your starter culture uses, for example, sucrose, then time will be needed when eventually you ask your yeast to metabolize the maltose that is in the wort. If the starter culture is made with dry malt extract, then there will be no need for the yeast to make new sugar-metabolizing enzymes when they are added to the wort.

FIGURE 16.9   A mature yeast starter culture.

The preceding is the rudimentary process for making beer. There are far more intricacies that can be employed to alter the color, clarity, foam, bitterness, carbon dioxide content, overall taste, aroma, and even the texture of the final product. For instance, the amount of foam (head) formed after pouring the beer is related to the amount of protein in the liquid. The amount of protein can be controlled during the malting process. The addition of a protease such as papain will shorten the polypeptides and make them less able to form a large, stable layer around bubbles as they form, which translates to less head on the beer. While a good amount of foam is considered a positive thing, too much protein in the beer will result in cloudiness, especially when the beer is cold. This is related to proteins partially coming out of solution to the point where they become visible. Letting the proteases work too long can result in flat or foamless beer, but not letting them work long enough will result in cloudy beer. The amount of foam in the final product can also be controlled by regulated venting of the carbon dioxide gas during the growth of the yeast culture.

Additional carbonation can also be added via an in-line carbonator. (Notice the return of our old friend, papain, in the previous paragraph. While papain is used here to produce a clear, amber beer, it is also routinely used in establishing primary cell cultures from explants (Chapter 7). This enzyme is present in the leaves, roots, and fruit of the papaya plant.)

16.3.4   Skunky Beer

Let us now briefly revisit the Entner-Doudoroff pathway (Figure 16.2), a gly-colytic pathway that is used by certain microbes such as *Zymomonas*. Recall that, in that pathway, for every glucose molecule and investment of one ATP molecule would yield 2 ATP and 2 pyruvate molecules. This is in contrast to the pathway used by humans, which requires an investment of 2 ATP molecules to yield 4 ATP and 2 pyruvate molecules per glucose. The net ATP yields are 1 ATP molecule for the Entner-Doudoroff pathway and two ATP molecules for the Embden-Meyerhof pathway (the pathway used by humans and yeast). In Section 16.1.2, an example was given for two imaginary cell cultures that had a set energy requirement for the maintenance of normal functions. Let us suppose that the amount is 10 (arbitrary) units of ATP equivalents per minute. To get that amount of energy, the cell would have to use $10/2 = 5$ units of glucose if it used the Embden-Meyerhof pathway for glycolysis, while it would require twice as much glucose ($10/1 = 10$ units) if it used the Entner-Doudoroff pathway, assuming anaerobic respiration. The point here is that the cell would have to burn twice as much glucose just to stay alive, which implies that glucose would be metabolized twice as quickly to produce the necessary energy for survival. One might wonder whether this pathway could be applied to make beer production more efficient. Instead of using yeast, could *Zymomonas* be used to ferment the sugars in wort into ethanol? *Zymomonas* would break down the sugars faster than yeast; therefore, the rate of ethanol production should be increased. From a business standpoint, this one set of facts might justify a switch from yeast to *Zymomonas*. The problem is that *Zymomonas* will produce other side products such as acetaldehyde. (Acetaldehyde is a 2-carbon aldehyde referred to in the brewing industry as ethanal.) A beer drinker might care about this because acetaldehyde is a molecule that contributes to the ill effects often described as a hangover. It is produced during ethanol metabolism in the liver. Having this molecule present in the beer would help ensure that the beer would provide a more significant hangover to the consumer, which is not a very good market-ing ploy. In addition to acetaldehyde, glucose metabolism in *Zymomonas* also yields hydrogen sulfide, also known as sewer gas. *Zymomonas* happens to be the number one contaminant responsible for beer spoilage, imparting an undesir-able taste and smell to the product. Returning to the business model, not only is *Zymomonas* bad because it produces a poor product, but also it will outcompete the yeast for glucose by metabolizing it more quickly, resulting in a cell culture made primarily of *Zymomonas* and not yeast. Not only should *Zymomonas* not be added to the wort, but also active steps should be taken to avoid its introduction via contamination.

Zymomonas contamination is often the result of scale-up. When an individual goes from a home brew to an industrial scale of production, adequate adjustments must be made for cleaning the equipment because the probability of a contamination is increased. During the brewing process, a layer of matter will collect on the bottom of the casks. *Zymomonas* easily lives in that layer because the material is energy-rich. Once the *Zymomonas* culture is established, it is very hard to kill. To prevent this from happening, much of the brewing equipment must be sterilized.

16.4 FERMENTATION TO PRODUCE BIOFUELS

16.4.1 Ethanol: A Biofuel with Problems

In addition to being used as a beverage ingredient, ethanol can be used for energy. Many larger cities in the United States require that gasoline contains up to 10% ethanol. The purpose of using ethanol in gasoline is not to reduce greenhouse gases such as carbon dioxide—current corn ethanol technologies release about the same amount of greenhouse gases as does gasoline. Ethanol is used as a means to stretch existing oil and gas supplies and for corn-producing countries such as the United States to reduce their dependence upon foreign oil.

The ethanol that is used in gasoline is a bioproduct—it is produced by yeast fermentation of sugars that are derived from corn starches. The process is strikingly similar to the production of the ethanol in beer: a mash is produced, corn starches are broken down into simple sugars via added enzymes (α-amylase), and yeast fermentation converts the sugars into ethanol. In fact, the fermented corn mash is called "beer." After the fermentation, the ethanol concentration is 8-12%. Concentration via distillation will bring the purity up to 92-95%, and most of the remaining water will be removed from the vapor phase by molecular adsorbents to yield >99% ethanol.

In addition to not reducing greenhouse gases during combustion, ethanolic fuels come up short as being environmentally friendly when one considers that the production of ethanol involves the use of fossil fuels. Corn must be harvested, typically using tractors, and transported, commonly by trucks burning diesel fuel, to the processing facility. Heat for cooking the corn mash and for distillation is generated by coal, oil, or natural gas. In addition, fossil fuels must be used as the ethanol product is trucked to facilities that mix it with gasoline. Other problems that distance ethanol from the status of a perfect biofuel include its heat of combustion, for which gasoline automobiles are not optimized, and the fact that the corn that is used could instead be used as a food supply for people. Riots stemming from food shortages from Haiti, to Egypt, to Indonesia in 2008 can be traced back to lowered food supplies that were the result of corn crops being earmarked for ethanol production in the United States, which drove the price of corn up (Figure 16.10).

Some well-meaning environmentalists have argued that ethanol should be used in a pure form as a fuel source for cars. One major problem with an ethanol-based system for automobiles lies in the transport of the ethanol. Compare ethanol to oil—oil can be transported via pipeline, but ethanol cannot. Ethanol readily mixes with water, including the water that is in the air we breathe. Humidity presents a problem because the ethanol company must then spend money and energy to transport the unwanted water that is mixed with the ethanol product. Even worse, when you put this ethanol from the pipeline into your car, upon combustion, a significant amount of energy will be used to heat the water that's in the fuel instead of being used to make the car move. Removal of the water at this point would be economically infeasible, requiring another distillation step that would need more fuel energy.

Another major problem with using pure ethanol as a fuel for the mass automobile market is that ethanol cannot be dispensed through traditional pumping systems. One reason for this is that ethanol vapor will erode metal components that are used in conventional pumps, making small cavities. This effect is known as *cavitation*. Because of the high volatility of ethanol, this is a significant potential problem. An alternative type of pump could be used, such as a centrifugal pump (Figure 16.11). This type of pump operates like a fan, spinning between 1750 and 3500 RPM. However, the spinning requires energy, which must be weighed against the energy benefits of the ethanol fuel. Another common type of pump to consider is the positive displacement pump. This pump works like a piston to force liquid through tubes via pressure. If there is vapor in the liquid (and there will be in the case of ethanol), then part of the energy used to move the liquid through the tubing is going to be spent in compressing the vapor. Energy will be wasted in this fruitless compression. Once again, the overall energy yield of the ethanol fuel will be lowered at the pumping site. One way to get around this problem is to cool the ethanol so that less ethanol will be in the gas phase, which would mean that there would be less vapor to compress with the positive displacement pump. The problem with this solution is that the cooling process will again use energy and cost money.

One undeniable advantage of using ethanol as a fuel is that it can originate from a renewable source. However, one must consider the ramifications of messing with the food supply. The planet can only support a finite number of people, on the order of 1×10^{10}. This number is based, in part, on the amount of food that can be produced. If some of this food is instead used to produce energy, then the total number of people that can be supported will be reduced.

16.4.2   Biobutanol

Some of the end products in the biochemical fermentative pathways are ethanol, acetone, isopropanol, acetate, butyric acid, and butanol. Of these, butanol might

updated 10:42 p.m. EDT, Mon April 14, 2008

Riots, instability spread as food prices skyrocket

BANK INFORMATION CENTER

Food riots hit West and Central Africa
13 March 2008

"A global rise in food prices has led to several outbreaks of civil unrest in West and Central Africa over the last few weeks. In Cameroon, Reuters reports that at least 24 people have been killed; over 1600 people have been arrested since February 25, and already 200 have been tried and given jail sentences of up to three years."

FP
Foreign Policy

Global food-riot watch
Fri, 01/18/2008 - 1:21pm

"On Monday, 10,000 Indonesians demonstrated outside the presidential palace in Jakarta after soyabean prices soared more than 50 per cent in the past month and 125 per cent over the past year, leaving huge shortages in markets.

The rapid soyflation, a result of surging Chinese demand, rising oil prices, and ethanol production in the United States, has also hit other parts of the region, such as Japan, South Korea, and Hong Kong ..."

FIGURE 16.10   Bad things happen when one messes with the food supply.

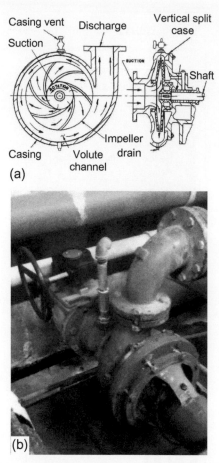

FIGURE 16.11 (a) Schematic of a centrifugal pump (from http://navalfacilities.tpub.com/mo230/mo2300161.htm). (b) Centrifugal pump in use.

serve as a biofuel with properties superior to ethanol. Butanol is a 4-carbon molecule, as opposed to the 2-carbon molecule ethanol. This makes butanol less volatile. Also, compared with ethanol, butanol is far less hygroscopic, meaning it will pull far less water out of the air. Butanol and water don't mix very well because of the longer carbon chain, which makes butanol less hydrophilic than ethanol. This has very strong implications, including the ability to transport butanol via pipelines. Also, because the longer carbon chain renders butanol less volatile, conventional pumps can be used to dispense the fuel. The alternative pumping systems just described can also be used because there will be less vapor associated with butanol so the problem of spending energy to compress butanol in the gas phase will be greatly reduced. Inside the engine, a reduction

TABLE 16.1 Energy Considerations for Butanol versus Ethanol

	Butanol	Ethanol
Energy content (kBTU/gal)	110	84
Vapor pressure (PSI at 100 °F)	0.33	2.0
Useful air/fuel ratio	11.1	9.0

In comparing butanol to ethanol, one can see that butanol carries more energy per unit volume, is less volatile, and can be mixed with a greater amount of air for combustion.

in the amount of water in the fuel means a greater amount of energy will be available to make the vehicle move because energy will not be wasted in heating that water. Finally, every mole of butanol contains more chemical potential energy than a mole of ethanol because of the two extra hydrocarbons that can be oxidized. This means that, because of the density of the two compounds, butanol carries more energy per unit volume than does ethanol. Specific parametric values are given in Table 16.1. In short, butanol would be a more attractive biofuel than ethanol because of its higher energy potential and the possibility of using traditional pumps and transport systems.

Notice that the pathways shown in Figure 16.4 provide for many products in addition to butanol. If all we were interested in was the production of butanol, then these alternative pathways would serve to reduce the total yield of our desired product. One approach aimed at increasing butanol yield is to prune this reaction tree. For instance, by knocking out the gene that codes for the enzyme acetate kinase, the pathway from glucose to acetate will be eliminated. This theoretically should increase the use of all of the other pathways in the tree. Because the cell is not spending time, energy, and resources on the production of acetate, the potential for greater butanol production should be realized.

Additional branches could also be trimmed to increase the output of butanol by the same concept. This has been achieved in the laboratory with some success. For example, a significant increase in the amount of butyrate produced by *Clostridium* has been achieved by knocking out the gene coding for phosphotransacetylase (plus two other mutations). While these results are very encouraging, the butanol yield from these microbes is still not sufficient for industrial-scale use. A yield of 12% from glucose would make this method of butanol production economically feasible on a large scale.

16.4.3   Cellulose

There is another potential energy source—cellulose—that is being investigated as a feedstock for biofuel production. Cellulose is found in grasses, paper, and plant

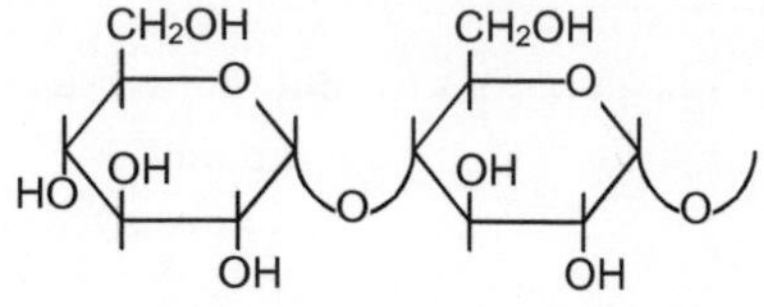

(a)

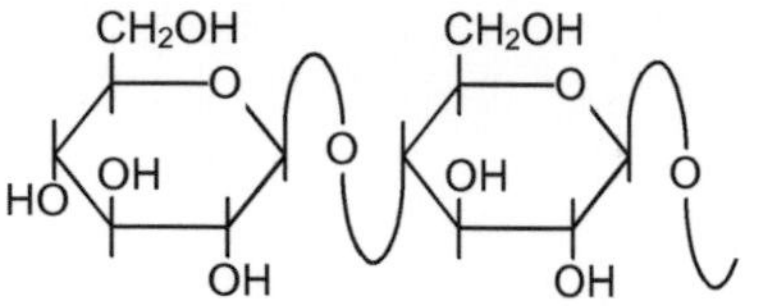

(b)

FIGURE 16.12   The structures of two polymers of glucose: starch and cellulose. (a) For starch, the glucose repeating units are connected by ($\alpha 1 \rightarrow 4$) linkages. (b) For cellulose, the glucose repeating units are connected by ($\beta 1 \rightarrow 4$) linkages. Humans do not produce an enzyme to help break these bonds.

husks, all of which are thrown away by humans as waste materials. Cellulose, like starch, is a polymer of glucose molecules. The major difference between these two polymers lies in how the glucose molecules are linked together. Starch uses α-acetal linkages, while cellulose uses β-acetal linkages (Figure 16.12). Although humans can digest starch, cellulose is useless to us as a food source because we lack the enzyme to break, specifically, the glc($\beta 1 \rightarrow 4$)glc (β-acetal) linkages. The enzyme that we would need is called cellulase. By incorporating a gene coding for cellulase into the genome of certain microbes, some biotechnologists are making great strides at converting what was once a waste material into a feedstock for bio-fuel production. By breaking cellulose down to glucose, greater potential would exist for producing other biofuels, including ethanol and butanol.

QUESTIONS

1. Suppose that you have two vats of identical glucose-containing solutions that you are storing in your shed. They each become contaminated on Monday: one with a microorganism that utilizes the Embden-Meyerhof pathway of glycolysis and the other with a microorganism that utilizes the Entner-Douderoff pathway. On Tuesday, unaware that the solutions are con-taminated, you sneak into your shed to enjoy a sweet beverage. Ugh! Neither solution tastes like you expected. Compare the tastes of the drinks in terms of sweetness.
2. What are the four defining characteristics of fermentation?

3.
   a. What would happen if we put yeast into a vat with wart and seal it so no oxygen is available? Draw the growth curve.
   b. What would happen if we put yeast into a vat with wart, allowed it to grow to the mid-log phase, and then sealed the vat it so no oxygen is available? Draw the growth curve.
4. Compare and contrast the growth curves of two different starter cultures of yeast after each is added to wort for beer brewing. The first starter culture was grown in a solution containing glucose as the energy source, and the second was grown in a solution containing malt extract.
5. What is the point of each environmental condition when growing alcohol-making bacteria first in an aerobic and then in an anaerobic environment?
6. Consider the process of beer making. If, in making the malt, we let the process proceed for three times longer than we should, predict the effect on the final beer product.
7. "Ethanol derived from corn and used in gasoline is good for the environment because it reduces CO_2 emissions." Give three reasons why this is not the case.
8. Name two advantages of using butanol over ethanol as a fuel source.
9. What would happen if, instead of a human producing lactic acid, he produced ethanol as a fermentation product?
10.
   a. Does the human body produce ethanol via fermentation?
   b. Can Clostridia produce lactic acid via fermentation?
   c. Can any living organism produce the powerful solvent acetone?
11.
   a. Why cannot yeast produce beer that is 50% alcohol?
   b. Yeast are used in producing liquors that are over 50% alcohol. How is this possible?

RELATED READING

Farrell, A.E., Plevin, R.J., Turner, B.T., Jones, A.D., O'Hare, M., Kammen, D.M., 2006. Ethanol can contribute to energy and environmental goals. Science 311, 506–508.

Lehmann, D., Hönicke, D., Ehrenreich, A., Schmidt, M., Weuster-Botz, D., Bahl, H., Lütke-Eversloh, T., 2012. Modifying the product pattern of *Clostridium acetobutylicum*: physiological effects of disrupting the acetate and acetone formation pathways. Appl. Microbiol. Biotechnol. 94, 743–754.

Liew, F.M., Köpke, M., Simpson, S.D., 2013. Gas fermentation for commercial biofuels production. In: Fang, Zhen (Ed.), From: Liquid, Gaseous and Solid Biofuels – Conversion Techniques, Chapter 5.

Mosier, N.S., Ileleji, K., 2006. How fuel ethanol is made from corn. Purdue Extension Bioenergy Series (ID-328). http://www.ces.purdue.edu/bioenergy.

Chapter 17

Stem Cells and Tissue Engineering

It makes sense to begin this chapter with the definition of a stem cell. There happens to be some controversy over this definition, which varies between laboratories. For the purposes of this text, we will define a stem cell as being a cell that has the potential to be differentiated down any of the three germ cell lineages.

Germ cell lineage is related to embryology, or the development of an embryo. Considering that the cells in your body have basically the same DNA (not counting mutations), it is reasonable to imagine that your bone cells have the same DNA as your liver cells, which have the same DNA as your nerve cells. These three cell types behave very differently, but they still have the same DNA. The reason for their common genomes is that they all came from the same originating cell, a fertilized egg. Since all of the cells in your body were derived from the same fertilized egg, then it is reasonably straightforward to see why all of the cells in your body should have the same DNA (for the most part). After an egg cell has been fertilized, it will divide into two cells, which will divide into four, which will divide into eight. The cells will continue to divide, although asynchronously, so the number of cells in the growing mass will not necessarily be a power of two. At ~96 h after conception, the cells will have formed a solid ball of about 32 cells, known as a *morula* (from the Latin word *morus*, meaning mulberry, because of its appearance) (Figure 17.1). From this point, the cells will continue to divide and develop into a *blastocyst*, which is a hollow sphere of about 150 cells (Figure 17.2). The outer layer of the blastocyst, known as the *trophoblast*, surrounds a fluid-filled cavity known as the *blastocoel* (ˋblas-tə-sēl) and an *inner cell mass*. It is the cells of the inner cell mass that will develop into the embryo and further on into the fetus.

In a process known as *gastrulation*, cells in the inner cell mass will continue to proliferate and will reorganize into the *gastrula*, which has three distinct layers: the endoderm, the mesoderm, and the ectoderm (Figure 17.3). The *endoderm* is the inner layer of the gastrula. Cells in this layer will go on to form the internal organs and the lining of the digestive tract. The middle layer of the gastrula is the *mesoderm*. Cells from this layer will form the muscles (including heart), bones, and cartilages. The outer layer of the gastrula is the *ectoderm*. These cells will form the skin and nerves.

An Introduction to Biotechnology. http://dx.doi.org/10.1016/B978-1-907568-28-2.00017-4
Copyright © W T Godbey, 2014.

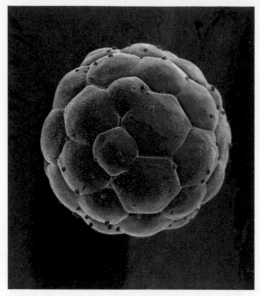

FIGURE 17.1   Is it a morula or a berry? (It's a morula.)

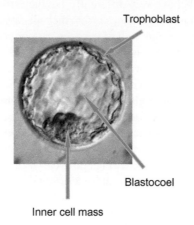

FIGURE 17.2   A blastocyst, which has the shape of a hollow spheroid. Layers are indicated.

Returning to the definition of the stem cell, it should now be clear that "the three germ cell lineages" refer to cells of endodermal, mesodermal, and ectodermal origins. On a smaller scale, a mesoderm progenitor cell can differentiate to produce muscle cells (skeletal, smooth, or cardiac muscle cells), cartilage cells (chondrocytes), and/or bone cells (osteoblasts, osteoclasts, and osteocytes) because the cell types all belong to the same germ cell lineage: the mesoderm. Once a stem cell has differentiated into a specific germ cell lineage,

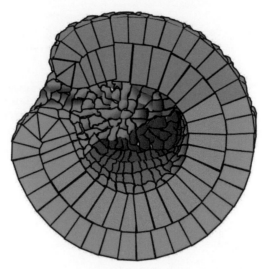

FIGURE 17.3 Hypothetical cross section through a gastrula. Locations of the three germ cell lineages are as follows: pink, endoderm; blue, mesoderm; green, ectoderm. Within the ectoderm, further delineation is indicated via shade: light green, epidermis; middle green, neural tissue; dark green, notochord.

it is relatively difficult to change its differentiation pathway to yield a cell of a different germ cell lineage.

17.1 POTENTIAL

A fertilized egg can develop into a complete organism. After it has divided into two cells, if the cells are separated, they can each develop into a separate complete organism. Even at the eight-cell stage, each of the cells has the potential to form a complete organism by itself. In other words, the potential of each of those cells is that they can make a complete organism. The term used in to describe this potential is *totipotent*. At or shortly after the eight-cell stage, the cells begin to lose the ability to create an entire organism, but they still retain the ability to become members of any of the three germ cell types. This plurality of potential is the basis for the term *pluripotent*. It is the cells of the inner cell mass in the blastocyst that give rise to embryonic stem cells. As the inner cell mass develops into the gastrula, the cells will commit to one of the three germ cell lineages. For instance, a mesodermal cell can be expected to produce skeletal muscle, smooth muscle, chondrocyte, osteoblast, osteoclast, etc. Such a mesodermal cell has the potential to become any of multiple types of cells under the same germ cell lineage; it is said to be *multipotent*. These levels of cellular potential are laid out in Figure 17.4.

The mesoderm is also called mesenchyme, and cells from the mesenchyme are also called mesenchymal cells. This brings us to the MSC, a type of cell that has been studied intensely for applications to tissue engineering and

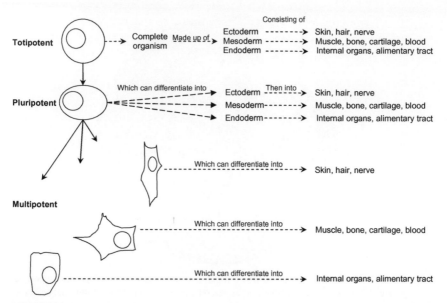

FIGURE 17.4 Levels of cellular potential and some of the cell types that can differentiate from the cellular progenitors.

regenerative medicine. Historically, "MSC" used to stand for "mesenchymal stem cell." However, as the nature of stem cells became better understood and definitions were adjusted, it was realized that MSCs did not fit the definition of a stem cell—they were already committed to a mesodermal lineage and (at the time) could not be driven down the other two germ cell lineages. Multipotent cells are not stem cells. As a result, MSCs could no longer be called stem cells. But the abbreviation "MSC" was already well accepted in the scientific community, and this term could not simply disappear. For a while, "MSC" was no longer an acronym, although the term still referred to this particular cell type. Eventually, "MSC" was accepted to stand for "mesenchymal stromal cell." This is an interesting example of one of the problems associated with a rapidly developing field: the language associated with the field must develop alongside of the field. New words must be invented, and definitions shifted. No matter what definition a specific laboratory will use for "stem cell," an MSC can be universally called a mesenchymal stromal cell.

Note—it is predicted that eventually, the technology will exist to drive MSCs to cell types belonging to the ectodermal and endodermal lineages. At that time, "MSC" will once again refer to a stem cell.

17.2   AN ALTERNATE VIEW OF STEM CELLS

Once again, consider that every somatic cell in your body has basically the same DNA, excluding things like mutations and recombinations. (*Somatic cell* refers to a body cell, as opposed to *germ cell*, which refers to a reproductive cell such as a sperm or egg cell, which has only half of the number of chromosomes as compared with somatic cells.) From your bone cells, to your liver cells, to your nerve cells, the genomes of each are basically the same despite these cell types looking and behaving very differently. What makes a bone cell a bone cell as opposed to a liver cell lies largely in the set of genes that the cell expresses. In theory, if the cell could turn off the bone-specific genes and begin to express the genes necessary to be a liver cell, then the bone cell could become a liver cell or any other cell type for that matter. This would make the original bone cell a stem cell, would it not? Your bone cells, however, do not have this ability. This brings us to an alternative view of what it means to be "stem": rather than a stem cell being a particular cell type that one can pick out of a body, hold, point to, or identify via a set of surface markers, many believe that "stem" refers to a property that a cell may or may not have. Certain cells would then be said to have a stem property. Our original definition of a stem cell would not change using this alternative view, but thinking of "stem" as a property makes the logic tree less convoluted as more and more cell types have been driven into alternate lineages via controlling the gene expression.

17.3   USING STEM CELLS

Now that we have a definition for stem cell, we can turn our attention to the differentiation of stem cells. One of the ideas for stem cell therapy entails the cells being grown *en masse* in an undifferentiated state, prior to the need for an engineered clinical application. Suppose that a patient comes into the clinic in need of a new liver. Currently, many patients will die waiting on a donor liver, a wait that can last for years. The number of available organs for transplant is low, and it is difficult to find a match from the limited supply. One goal of stem cell therapy is to have a large supply of stem cells that are available for differentiation into whatever tissue (or organ) a patient may need, be it liver, kidney, bone, etc. The popular press often misrepresents the true intent of these therapies, focusing upon totipotent cells. Totipotent cells are not used for stem cell therapy. Similarly, embryonic stem cells are the focus of the moral objections to stem cell therapy. Developing embryos are not the only source of stem cells; adults also have a supply of stem cells distributed throughout their bodies. While bone marrow and amniotic fluid have both been used to produce adult stem cell cultures, stem cells can be found throughout the body in very low numbers. A current challenge is the isolation of these cells, which make up a very low percentage of the population in even a stem cell-rich tissue. There is no universal stem cell marker, although there are certain markers (membrane proteins, glycoproteins, and the like) that appear on stem cells in combinations that render them distinct from other cell types.

Let us now return to the concept of a somatic cell being reprogrammed to become a different cell type. This has, in fact, been achieved with fibroblasts from a mouse. In their landmark paper from 2006, Takahashi and Yamanaka were able to generate pluripotent cells from embryonic and adult fibroblasts via the addition of four growth factors to the growth medium (Oct3/4, Sox2, c-Myc, and Klf4, if you are keeping score.) Following the injection of these cells into developing blastocysts, these modified cells took part in embryonic development. The cells were said to have been induced into pluripotency, hence the term *induced pluripotent stem cells* (iPSCs). For this breakthrough, Shinya Yamanaka, along with John Gurdon, who also worked with the reprogramming of cells to take on the stem character, won the Nobel Prize in Physiology or Medicine in 2012.

One problem with stem cell culture is the maintenance of the cells in an undifferentiated state. Many times, teratomas will form in the culture vessel. A *teratoma* is a collection of tissue that contains differentiated cells from all three lineages. Teratomas can have teeth, hair, muscle, bone, etc.—you name it. Some can even resemble complete organisms. They are not functioning organisms, though; they are merely a complex amalgamation of tissues. They are not a good source of tissues or even of cells because of the array of differentiation found within. In terms of stem cell culture, a teratoma can be considered a contaminant. When iPSCs were transplanted just under the skin in nude mice, tumors containing a variety of tissues from all three germ layers were formed. These tumors were teratomas.

One might ask whether an adult stem cell would have less potential to divide than an embryonic stem cell. Both types of stem cells display a characteristic that is common to all stem cells: they can renew themselves through cell division, even if they've been inactive for long periods of time. Stem cells do not have a Hayflick limit: they are immortal. Do not confuse this with being transformed. There is a fine line between a stem cell and a cancer cell, and there is concern in the field about cultured stem cells transforming into cancer cells after implantation. Similar to cancer cells, stem cells are able to renew their telomeres during cell division, but stem cells achieve this through a telomerase-independent pathway.

17.4   TISSUE ENGINEERING AND REGENERATIVE MEDICINE

After stem cells have been obtained, they will often be expanded through culturing, often for the purposes of tissue engineering or regenerative medicine. Most tissue engineering and regenerative medicine applications involve the implantation of cells, often housed in a porous scaffold, into a host (Figure 17.5). The difference between these two fields lies mainly in where cell proliferation within the scaffolds occurs. If the scaffolds are seeded with cells that are allowed to proliferate in the laboratory prior to implantation, the application is an example of tissue engineering. If the scaffolds are implanted and development of the tissue occurs within the body, the approach is an application of regenerative medicine. In both cases, the scaffolds mature inside a bioreactor.

FIGURE 17.5   In the early days of tissue engineering, Charles Vacanti seeded biodegradable, po-rous scaffolds in the shape of an ear with cartilage cells from a cow and implanted the constructs un-der the skin of athymic mice. While these particular constructs were not to be used for implantation onto a human, they did demonstrate the feasibility of growing cells on constructs to produce tissues.

If the bioreactor is an apparatus, then we are referring to tissue engineering; if the bioreactor is a body, then we are referring to regenerative medicine. (The cartilage cells seeded onto the scaffold in Figure 17.5 were allowed to incubate for one week before implantation; therefore, that experiment was an example of tissue engineering.)

The goal of tissue engineering is the restoration of tissue structure or func-tion by using biological components. Certain applications of tissue engineering have already been realized in the clinical setting, such as engineered skin. For burn victims, the application of engineered skin to wound sites helps to prevent dehydration and infection. During the early days of this technology, the skin was created from an initial biopsy of skin cells, often from the foreskins of male babies following circumcision. A more recent research uses cells from the patient's own body. From one biopsy, an *explant* culture can be established and expanded to an area that would cover a tennis court. (No, not in a single sheet!) An explant involves a small amount of tissue. The tissue is mechanically and/ or chemically treated to break down extracellular matrices to free individual cells. The resulting mixture is then plated into a bioreactor to allow the cells to adhere to a surface and multiply. After a culture has been established in this manner, *selective media* can be used to ensure that only the cell type of inter-est can grow. Culture expansion is when the cells are passed from one flask or bioreactor to multiple flasks (bioreactors), thus expanding the total surface area of that particular culture.

Engineered vaginal tissue has also made it to clinical trials. The methods employed during the underlying research were very similar to what has already been described: Biodegradable scaffolds were seeded with epithelial cells, and the constructs were cultured in a perfusion bioreactor prior to implantation. In the rabbit model, these constructs were implanted as a total vaginal replace-ment, with the constructs being accepted and integrated into the host tissues.

The histological characteristics of the constructs were similar to normal vaginal tissue after several months.

Arterial tissue has been engineered in multiple settings. Figure 17.6 shows a construct that was seeded with a mix of endothelial, smooth muscle, and fibroblast cells. Tissue-engineered blood arteries employ more than one cell type

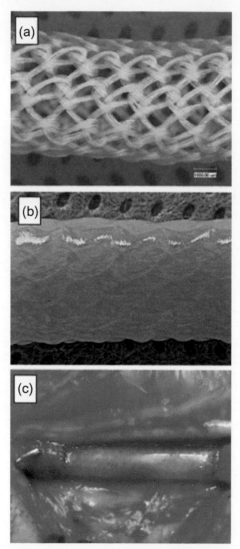

FIGURE 17.6 Example of engineered artery. Biodegradable, macroporous mesh (pore size ~1 mm) (a) before and (b) after embedding into a fibrin/cell matrix. (c) Fibrin-based vascular graft after implantation in the arterial circulation (ovine carotid model). *From Jockenhoevel and Flanagan (2011): http://www.intechopen.com/books/tissue-engineering-for-tissue-and-organ-regeneration/cardiovascular-tissue-engineering-based-on-fibrin-gel-scaffolds.*

by necessity. Niklason et al. produced engineered arteries using pulsatile media flow to induce migration and orientation of smooth muscle cells seeded into the scaffolds. After incubation, a second layer of cells, this time endothelial cells, was seeded onto the luminal side of the vessel construct. The problem of vessel rupture due to poor mechanical properties was addressed by this approach, purportedly due to the smooth muscle cells and the matrix proteins they produced.

Another tissue-engineered vascular model that has been investigated used three cell types during construction. The support for the construct was made, not from a degradable synthetic scaffold, but from dehydrated fibroblasts wrapped temporarily around an inert tubular support. A sheet of smooth muscle cells was layered around the support, and this was covered with a sheet of fibroblasts to provide an external structural layer. The constructs were allowed to incubate for at least eight weeks, after which a layer of endothelial cells was seeded onto the dehydrated fibroblasts on the luminal side of the constructs. The constructs had the mechanical strength needed to handle the loads generated by flowing blood, purportedly due to the adventitial layer that was made from the seeded fibroblasts and the matrix they produced.

This section is by no means an exhaustive list of tissue engineering and regenerative medicine applications. Virtually every tissue and most organs of the body are under investigation for eventual clinical applications. Further discussions can be found in textbooks that are readily available on the subject.

17.5   BIOREACTORS

Bioreactors can range from very simple to very complex. A cell culture dish or flask, made of treated polystyrene, placed in an incubator with controlled temperature and humidity can count as a bioreactor (Figure 17.7). The polystyrene can be coated with collagen or poly(lysine) or can be treated electronically via corona discharge to provide the culture surface with an electric charge to promote cell adhesion. The culture vessel for simple systems will not be airtight so that gas exchange can occur, especially for oxygen and carbon dioxide.

17.5.1   Incubators

The incubator is an important piece of the bioreactor system. The standard incubator participates in three functions: controlling the temperature, osmolarity, and pH of cell media.

Temperature: Temperature is controlled via a heater and simple thermostat. Just like you getting uncomfortable if the temperature gets too hot or too cold, cells too are affected by suboptimal temperatures. If it is too cold, metabolism and other cellular processes will slow in part because energy has been removed from the system and chemical reaction rates will be affected. If the temperature is too high, in addition to experiencing altered reaction kinetics and protein folding, the cell may begin to transcribe and translate a new set of genes that

FIGURE 17.7   Polystyrene tissue culture dishes and flasks in an incubator.

code for products known as heat shock proteins. The optimal temperature for cells differs depending upon the organism. For example, mammalian cells including human cells are kept at 37 °C in the incubator, the same as the normal human body temperature, while yeast cells are kept at 30 °C.

Osmolarity: If cell medium were to evaporate, it is water that would leave the liquid and enter the gas phase. Salts, sugars, and proteins will be left behind in the liquid medium resulting in an increased osmolarity in the medium. The way to prevent this is to prevent water loss through evaporation. While sealing the medium-containing culture vessel will prevent water loss, it will also affect gas exchange, which will result in altered pH, the induction of hypoxia, and an eventual switch to anaerobic/fermentative metabolism. A safer way to prevent water loss is by increasing the partial pressure of water over the medium to the saturation point. While the value for the amount of water needed to reach saturation in an incubator of a given size and temperature can be calculated, such calculations are not necessary. Saturation can be achieved by putting a pan of water inside the incubator. As long as water is kept in the pan, the water levels in the atmosphere of the incubator will remain at saturation. Evaporation from cell media will be slowed or stopped, preserving the osmolarity of the media.

(Of course, a pan of water in a nonsterile 37 °C environment will eventually allow for the growth of microorganisms such as algae, so the pan should be cleaned regularly and an algicide should be mixed into the water.)

pH: Recall the reaction

$$CO_2 + HOH \leftrightarrows H_2CO_3 \leftrightarrows HCO_3^- + H^+$$

This reaction is used to control the pH of most cell media in an incubator. First, CO_2 is pumped into the incubator and maintained at a concentration of 5%. This amount is much higher than the concentration of CO_2 in the air (~0.039%), so the above reaction will be driven to the right since there is plenty of water in the humidified incubator atmosphere. The pH of the cell media is buffered by including sodium bicarbonate as one of its ingredients (~0.37% (w/v)), and the medium is typically titrated to pH = 7.2 before use. The bicarbonate will serve as a buffer to hold the pH of the medium fairly constant. However, as the cells metabolize the sugars in the medium, they will produce their own CO_2 and eventually lower the pH of the medium over several days. The mammalian cell media should generally be changed twice per week.

17.5.2 Static and Dynamic Cultures

Tissue culture plates and flasks come in various sizes, with growing areas ranging from <1 to 225 cm² and larger (Figure 17.8). The size of a tissue culture flask is limited by the size of the incubator in which it will be placed. To help maximize the amount of growing area without requiring the purchase of enormous incubators, improvements to flask design have been made. For instance, some tissue culture flasks have multiple layers of growing surface within the same outer shell to increase the growing surface area per incubator unit volume (Figure 17.9). Three-dimensional growing surfaces, in the form of scaffolds, also exist and allow for the culture of a greater number of cells in a given volume. Three-dimensional scaffolds also allow for more complex culturing that eventually yields a tissue with more than one cell type.

Cells growing in tissue culture plates, in flasks, or on scaffolds suspended in a bottle of medium are all examples of *static culture*. There is no stirring or flow involved. A more complex bioreactor utilizes *dynamic culture*. In such cultures, the media may be stirred slowly as is illustrated in Figure 17.10. This is used for cultures of nonadherent cells and often for cells that are growing in three-dimensional scaffolds. The rate of mixing can be controlled by controlling the RPM of the stir bar. Mixing allows for better gas exchange, since pockets with high waste concentrations will not be allowed to accumulate around the cells. Mixing also allows nutrient and oxygen concentrations to be evenly distributed throughout the culture medium.

Another example of a dynamic culture system is the perfusion bioreactor. Such systems utilize a pump that allows for the medium to flow over culture surfaces. Incorporating flow into the culture conditions is used in both two-dimensional and

FIGURE 17.8   Tissue culture plates and flasks come in many sizes.

FIGURE 17.9   Multilayered tissue culture flask.

FIGURE 17.10   A spinner flask, an example of a dynamic culture bioreactor. *Photo courtesy of Prof. Taby Ahsan, Tulane University.*

three-dimensional cultures. For two-dimensional cultures, cells grow on an interior surface of a chamber, which has an inlet and outlet for medium to pass over the cells (Figure 17.11). The three-dimensional system is very similar, except that the medium either is pumped through the cell-containing scaffolds or is dripped on top of the scaffolds and allowed to flow through via gravity (Figure 17.12). These bioreactors are usually closed systems, allowing for the medium to be passed over the culture surfaces several times before it's replaced with fresh medium.

Advantages to having the medium flow over the cells are realized through the effects that are brought about by having the cells exposed to controlled and directional shear, which aid in the differentiation of pluripotent and multipotent cells. MSCs grown in static culture with no growth or differentiation factors will either not differentiate or (more likely) differentiate down multiple pathways to produce a heterogeneous culture. MSCs grown in the presence of shear stress will tend to differentiate down pathways of cells that provide structure, such as bone cells. Muscle cells will also tend to line up in a parallel fashion when exposed to shear (Figure 17.13). Recall that a tissue is more complex than simply having multiple cells of the same type growing on the same particular scaffold; it is a collection of cell types that interact as a unit. Including the parameter of shear is useful for creating tissues that are exposed to shear in their normal functioning.

17.6   POLYMERIC SCAFFOLDS

Supports that are used for three-dimensional cultures are known as scaffolds. Most scaffolds currently in use for tissue engineering and regenerative medicine are made of biodegradable and biocompatible materials. *Biodegradable* means that the material will be broken down under physiological conditions, such as when a polymer is hydrolyzed (broken apart by water). *Biocompatible* means that the degradation products of the original material are not toxic to the cells.

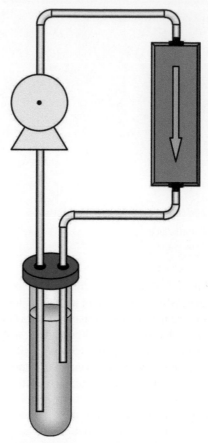

FIGURE 17.11   Scheme for a 2-D perfusion bioreactor, which uses a pump to bathe cells growing on a surface.

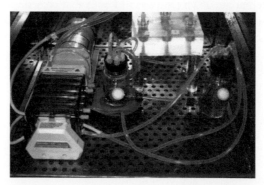

FIGURE 17.12   A 3-D perfusion bioreactor set up in an incubator. Cells are seeded into scaffolds, which are placed in perfusion cassettes (top of the figure). A pump is located on the left-hand side of the figure.

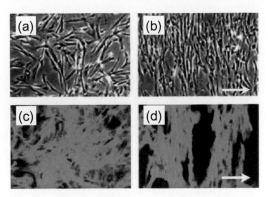

FIGURE 17.13 Fluid shear stress induces alignment in vascular smooth muscle cells. Cells were cultured on slides (a) under static conditions or (b) exposed to 20 dyn/cm² shear stress. (c) F-actin (a structural protein) in cells cultured under static or (d) flow conditions. Arrows indicate the direction of the fluid flow (Lee et al., 2002).

17.6.1 Homopolymers

Consider glycolic acid. Several of these molecules can be conjugated as shown in Figure 17.14 to create the homopolymer poly(glycolic acid). (A *homopolymer* is a polymer made of a single type of molecule or *repeating unit*.) Lactic acid can also be used to make a homopolymer—in this case, poly(lactic acid) (Figure 17.15). During the polymerization of either of these molecules, a H and OH will come off as water. This implies that the reaction could go backward so that the polymers could be broken apart by water, a process known as *hydrolysis*. Poly(glycolic acid) and poly(lactic acid) are typically abbreviated PGA and PLA, respectively.

The consistency of PGA is like a piece of felt (Figure 17.16). It's biodegradable via hydrolysis, and glycolic acid is biocompatible (as are oligomers of glycolic acid). When PGA is implanted into an animal, it will be placed into an aqueous environment, which means that it will degrade over time. This fits well into an overarching goal of tissue engineering: after the construct is seeded with cells and implanted, it will start to degrade. However, as the cells grow and

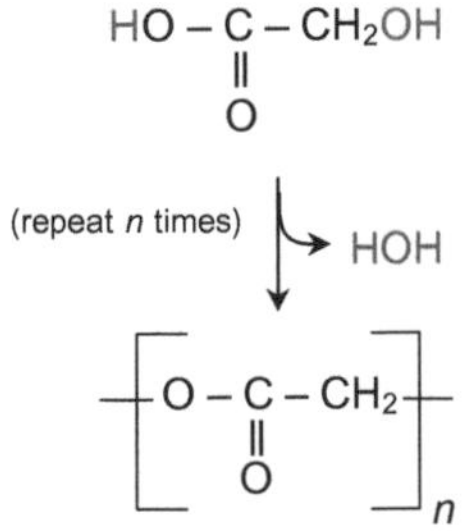

$$HO - \underset{\underset{O}{\|}}{C} - CH_2OH$$

(repeat n times) $\searrow$ HOH

$$\left[O - \underset{\underset{O}{\|}}{C} - CH_2 \right]_n$$

FIGURE 17.14 Glycolic acid is polymerized to form poly(glycolic acid).

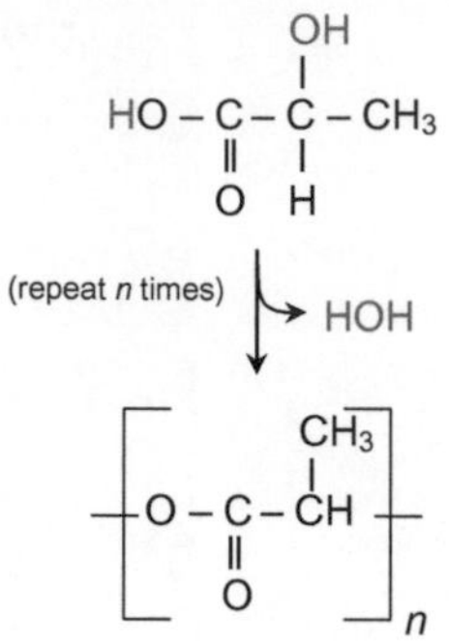

$$
\begin{array}{c}
\text{OH} \\
| \\
\text{HO}-\text{C}-\text{C}-\text{CH}_3 \\
\| \quad | \\
\text{O} \quad \text{H}
\end{array}
$$

(repeat *n* times) $\searrow$ HOH

$$
\left[\; \text{O}-\overset{\displaystyle \overset{\text{CH}_3}{|}}{\underset{\underset{\text{O}}{\|}}{\text{C}}}-\text{CH} \;\right]_n
$$

FIGURE 17.15   Lactic acid is polymerized to form poly(lactic acid).

FIGURE 17.16   PLA is a flexible polymer that can be spun into different geometries such as sheets or open cylinders. The polymers shown here could serve as scaffolds for engineering arteries.

proliferate, they will secrete their own extracellular matrix. The perfect system would have the rate of extracellular matrix construction occur at the same rate as PLA degradation, thus keeping the mechanical properties of the construct constant as the construct matures into a tissue. Over time, it would be desired that the implant consists only of cells and extracellular matrix, with no scaffold material remaining. The perfect final product would have cells in the same orientation, with the same distribution of cell types and the same amount of extracellular matrix as normal tissue (or organ).

With lactic acid, there are a carboxylic acid and an alcohol side group that can be used for polymerization. As before, polymerization of two units causes the release of a water molecule. This implies that PLA is hydrolyzable (and therefore biodegradable). PLA has different properties compared with PGA. While PGA is like a felt, being easily shaped or deformed, PLA is relatively stiff but more able to hold its shape under compressive forces, making it more suitable for applications that require load bearing. Sometimes, PGA scaffolds will be formed into the desired shape and then coated with PLA to help the construct maintain its shape.

17.6.2   Copolymers

Not all polymers are homopolymers. If two or more repeating units are used to make the polymer, the product is said to be a *copolymer*. There are many different possible arrangements of the repeating units—too many to describe here. However, some rudimentary architectures are shown in Figure 17.17.

The repeating units of a copolymer can come together in a random fashion, forming a *random copolymer*. However, important structural characteristics can emerge if the repeating units are joined in nonrandom ways. If each repeating unit is repeated in a regular, alternating fashion, the copolymer is an *alternating copolymer*. Quite often, small blocks of homopolymers can be covalently linked to produce *block copolymers*. These are given more descriptive names to indicate how many blocks were joined, such as diblock or triblock copolymers. Polymers can also be branched. One method used to achieve branching is to take one polymer and graft it into the chain of another polymer, forming a graft copolymer. Although Figure 17.17e shows a homopolymer being grafted onto a second homopolymer, one or both of these units can be a copolymer.

An example of a copolymer used in tissue engineering is poly(lactic-co-glycolic acid) (PLGA), which is a copolymer of the two homopolymers we have already mentioned, PLA and PGA. Different ratios of the PLA and PGA can be used to yield PLGA constructs with different properties, such as predetermined compressive strengths and degradation rates. Stiffer than PLA but more pliable than PGA, PLGA has been used in several applications, such as for bone and cartilage engineering.

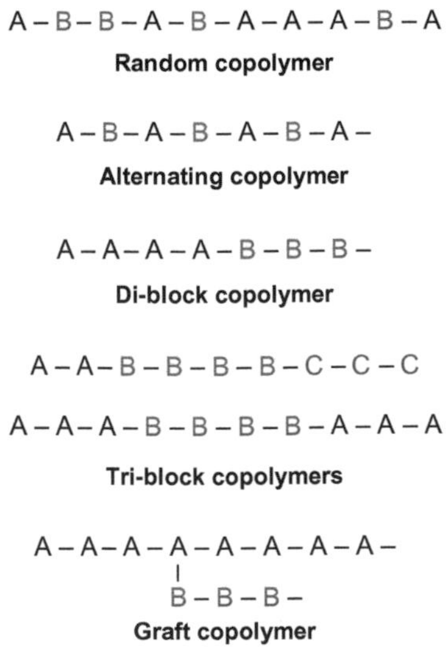

A – B – B – A – B – A – A – A – B – A

Random copolymer

A – B – A – B – A – B – A –

Alternating copolymer

A – A – A – A – B – B – B –

Di-block copolymer

A – A – B – B – B – B – C – C – C

A – A – A – B – B – B – B – A – A – A

Tri-block copolymers

A – A – A – A – A – A – A – A –
|
B – B – B –

Graft copolymer

FIGURE 17.17   Examples of copolymer architectures.

17.7   BRINGING IT ALL TOGETHER: A TISSUE ENGINEERING APPLICATION

Let us now turn to the example of engineered blood vessels. A naive view of blood vessels is that they are a series of tubes that run throughout the body for the purpose of circulating blood cells, which carry oxygen, to all of the cells in the body. However, replacing a vessel such as the radial artery with a piece of plastic tubing would not be sufficient to replace the function of this artery. While the plastic tubing would prevent the loss of blood through its walls and could be made pliable enough to allow for movement of the host's arm, it would not allow for the leakage of immune cells or macromolecules as may be needed by the body along the site of the implant. The foreign material of the plastic might also initiate a foreign body response, which would be deleterious to the host. New blood vessels sprouting from the implant would likewise not be possible, and gas exchange in the area of the implant would be impeded. A superior alternative would be the seeding of a porous scaffold with cells that could eventually proliferate to create arterial tissue.

The structure of an artery is relatively straightforward. As shown in Figure 17.18, there are three distinct layers to the vessel: the *intima*, the *media*, and the *adventitia*. The intima is the layer of cells that surround the lumen of the artery (the layer that is intimate with the blood). It is composed of endothelial cells. The medial layer is composed of smooth muscle cells, and the adventitia is composed mainly of connective tissue such as collagen. An engineered construct of an artery can be made starting with a porous scaffold. The pores allow cells to migrate into the interior of the scaffold. The *porosity*, or percentage of a given volume of scaffold that is made up of pores, will increase as the scaffold material is hydrolyzed. However, the deposition of extracellular matrix should make up for the loss of scaffold material. An ideal construct would have the supporting scaffold polymer degrade at the same rate that the cells establish extracellular matrix.

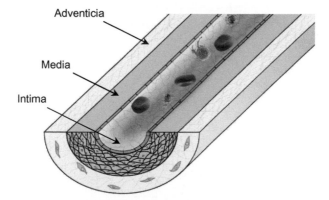

FIGURE 17.18   Structure of an artery, with three layers as indicated. The intimal layer is composed of endothelial cells, and the medial layer is composed of vascular smooth muscle cells. The adventitia is structural, composed mainly of connective tissue made of the cells within.

For the specific application presented here, tubular PGA scaffolds were seeded with smooth muscle cells and cultured for eight weeks with a specialized perfusion bioreactor. The smooth muscle cells were seen to migrate into the interior of the PGA matrix and provided a smooth surface for a second seeding of cells onto the construct. This second seeding utilized endothelial cells, applied only to the luminal surface of the construct. Further culturing produced intact vessels, some of which were successfully implanted into small pigs. While the production of tissue engineering constructs with more than one cell type was an imported procedural step for the time (1999), the striking feature of this piece of work was that the bioreactor utilized pulsatile flow of the cell medium to produce constructs that more closely mimicked a native artery. This approach has since undergone well over a decade of development, with improvements being made to both chemical and physical aspects of the culturing system. An important lesson from this application is that a good bioreactor will necessarily mimic the conditions of the eventual implantation site. One should be aware of as many parameters as possible, both chemical and physical, to allow constructs to develop with the desired complexity of form and function.

QUESTIONS

1. Why doesn't the water in the incubator need to be carefully measured?
2. If the water in a cell culture incubator pan evaporates, what will happen?
3. How is pH controlled in an incubator?
4.
    a. Cancer cells are immortal, which means that they are able to circumvent the Hayflick limit. Describe how they are able to do this.
    b. Stem cells are also immortal. Are they able to circumvent the Hayflick limit in the same or a different way than cancer cells?
5. Dr. Richtofen has a patient without telomeres. What kind of doctor do you expect him to be?
6. What is the difference between a primary cell culture and a cell line? Could you ever have a primary cell line? Explain your answer.
7. Which is the most remarkable?
    a. A hematoprogenitor cell becoming a muscle cell
    b. An embryonic stem cell becoming a kidney cell
    c. An MSC becoming a liver cell
    d. An adult stem cell becoming a nerve cell
    Explain your answer.
8. Which is the most remarkable?
    a. A totipotent cell becoming/producing a multipotent cell
    b. A multipotent cell becoming/producing a pluripotent cell
    c. A pluripotent cell becoming/producing a multipotent cell
    d. A totipotent cell becoming/producing a pluripotent cell
    Explain your answer.

9. True or False (write a reason for every answer of "false"):

T	F	Pluripotent cells are stem cells
T	F	MSCs are stem cells of the mesenchymal (mesodermal) lineage
T	F	A stem cell is a cell that can, for example, become a muscle cell and then become a nerve cell
T	F	A pluripotent cell can become an entire organism
T	F	Blood cells belong to a mesenchymal lineage (i.e., they originate from the mesoderm)
T	F	A multipotent cell from the ectoderm could be differentiated into the skin or nerve cells
T	F	Embryonic stem cells have been isolated from blastocysts
T	F	Scientists have isolated totipotent cells from the bone marrow
T	F	A cell that produces a teratoma is (was) a stem cell
T	F	A cell that produces a teratoma is (was) a totipotent cell

RELATED READING

Atala, A., Kasper, F.K., Mikos, A.G., 2012. Engineering complex tissues. Sci. Transl. Med. 4, 160rv12.

Cao, Y., Vacanti, J.P., Paige, K.T., Upton, J., Vacanti, C.A., 1997. Transplantation of chondrocytes utilizing a polymer-cell construct to produce tissue-engineered cartilage in the shape of a human ear. Plast Reconstr Surg. 100, 297–302.

Dahl, S.L., Kypson, A.P., Lawson, J.H., Blum, J.L., Strader, J.T., Li, Y., Manson, R.J., Tente WE, DiBernardo L., Hensley, M.T., Carter, R., Williams, T.P., Prichard, H.L., Dey, M.S., Begelman, K.G., Niklason, L.E., 2011. Readily available tissue-engineered vascular grafts. Sci Transl Med. 3, 68ra9.

De Filippo, R.E., Bishop, C.E., Filho, L.F., Yoo, J.J., Atala, A., 2008. Tissue engineering a complete vaginal replacement from a small biopsy of autologous tissue. Transplantation 86, 208–214.

Godbey, W.T., Atala, A., 2002. In vitro systems for tissue engineering. Ann. N. Y. Acad. Sci. 961, 10–26.

Jockenhoevel, S., Flanagan, T.C., 2011. Cardiovascular tissue engineering based on fibrin-gel-scaffolds, tissue engineering for tissue and organ regeneration. In: Eberli, D. (Ed.), InTech. http://www.intechopen.com/books/tissue-engineering-for-tissue-and-organ-regeneration/cardiovascular-tissue-engineering-based-on-fibrin-gel-scaffolds.

Lee, A.A., Graham, D.A., Dela Cruz, S., Ratcliffe, A., Karlon, W.J., 2002. Fluid shear stress-induced alignment of cultured vascular smooth muscle cells. J. Biomech. Eng. 124, 37–43.

Lee, E.J., Vunjak-Novakovic, G., Wang, Y., Niklason, L.E., 2009. A biocompatible endothelial cell delivery system for in vitro tissue engineering. Cell Transplant. 18, 731–743.

Moran, J.M., Pazzano, D., Bonassar, L.J., 2003. Characterization of polylactic acid-polyglycolic acid composites for cartilage tissue engineering. Tissue Eng. 9, 63–70.

Niklason, L.E., Gao, J., Abbott, W.M., Hirschi, K.K., Houser, S., Marini, R., Langer, R., 1999. Functional arteries grown in vitro. Science 284, 489–493.

Solan, A., Dahl, S.L., Niklason, L.E., 2009. Effects of mechanical stretch on collagen and cross-linking in engineered blood vessels. Cell Transplant. 18, 915–921.

http://stemcells.nih.gov/info/.

Takahashi, K., Yamanaka, S., 2006. Induction of pluripotent stem cells from mouse embryonic and adult fibroblast cultures by defined factors. Cell 126, 663–676.

The Nobel Prize in Physiology or Medicine, 2012. Press Release. Nobel Media AB. http://www.nobelprize.org/nobel_prizes/medicine/laureates/2012/press.html (08.10.2012).

Chapter 18

Transgenics

18.1 ICE-MINUS BACTERIA

The first experiments involving transgenics, as applied to plants, did not produce transgenic plants; they made transgenic bacteria that were then sprayed onto plants. The bacteria in question were *Pseudomonas syringae*, also known as ice-forming bacteria. These bacteria contain a protein in their outer membranes that serves as a nucleation site for water as it freezes. When the temperature drops down to around 32° and frost forms, it will form first on and around these proteins on the bacteria. The bacteria reside all over the plant, which means that the plant will be covered with frost and will die or suffer extensive damage (Figure 18.1).

Experiments were performed to knock out the gene that coded for the nucleation site protein. *Knockout* refers to the genetic technique of mutating a specific gene in the genome by targeted recombination. The mutation is intended to render the gene useless. By knocking out the gene that codes for this membrane protein, a strain of *P. syringae* was created that is known as *"ice-minus."* These bacteria were then sprayed onto crops to help retard the formation of frost crystals.

18.2 BT PLANTS

"Bt" in this case stands for *Bacillus thuringiensis*, a Gram-positive bacteria that can be found in soil. *B. thuringiensis* is known to produce a toxin to aid in its own survival. Biotechnologists have been successful in incorporating the gene for this toxin into many types of plants, including cotton, corn, soybeans, peanuts, and potatoes (Figure 18.2). These plants are known as *Bt plants*. Bt plants express this gene in great enough quantities that crystals of this toxin will form. Caterpillars, beetles, or other similar pests that consume these crystals while eating the plants will die. (The crystals are harmless to humans and birds.) In this way, farmers have been able to grow crops that produce their own pesticides without the need of spraying chemicals, thus allowing the crops to be classified as "organic." The advantages of Bt plants include better crop yield because they are not eaten by pests and cheaper crop production because the expense of spraying is not needed. By the year 2000, over 50% of the soybean crops in the United States were Bt plants.

An Introduction to Biotechnology. http://dx.doi.org/10.1016/B978-1-907568-28-2.00018-6
Copyright © W T Godbey, 2014.

FIGURE 18.1   Certain bacteria produce proteins that serve as nucleation sites for ice crystal formation. Spraying plants with ice-minus bacteria can inhibit the formation of ice crystals to a small degree. *Photo by George Hodan, from http://www.publicdomainpictures.net.*

FIGURE 18.2   (a) Lesser cornstalk borer larvae extensively damaged the leaves of this unprotected peanut plant. (b) After only a few bites of peanut leaves with built-in Bt protection, this lesser cornstalk borer larva crawled off the leaf and died. *Photos by Herb Pilcher, from Suszkiw (1999).*

Just as bacteria can develop resistance to certain drugs, insects can develop resistance to certain pesticides. As is also a common response in evolution, different strains of *B. thuringiensis* have developed different forms of toxin for protection. Biotechnologists have identified and isolated genes for these different toxins and have incorporated genes for multiple toxin forms into single plants in a process known as pyramiding, to produce *pyramided plants*.

In the United States, the first Bt plants were the NewLeaf potato, produced by Monsanto Co. and sold in supermarkets in 1996. Although the first patent for Bt technology was filed in 1988, a legal fight between Mycogen Seeds, Inc. (an affiliate of Dow AgroSciences, LLC) and Monsanto delayed the awarding of patent protection until 2005. The patent was ultimately awarded to Dow, and Monsanto suspended its Bt potato program, which had expanded to include several varieties of potatoes.

18.3   HERBICIDE RESISTANCE

Broad-spectrum herbicides are chemicals that kill virtually all plants. Consider the notion that if a plant were developed that were resistant to a broad-spectrum herbicide, then crops of this plant could be sprayed with the herbicide and all weeds and competing plants would die except for the plants of the desired crop. Glyphosate is the active ingredient in most broad-spectrum herbicides, including a product known as Roundup™. It has the structure shown in Figure 18.3. Glyphosate acts to inhibit the activity of 5-enolpyruvylshikimate 3-phosphate synthase (EPSP synthase), which is involved in the biosynthesis of the aromatic amino acids and tetrahydrofolate, ubiquinone, and vitamin K. These particular pathways are not used in mammals, fish, birds, or insects, so EPSP synthase presents an attractive target for killing plants.

With glyphosate serving as such an effective herbicide, engineering plants that are resistant to this agent is the goal of the above biotechnology for enhanced crop production. One approach to obtaining plants with glyphosate resistance is to have them make more EPSP synthase than an application of glyphosate could inhibit. This can be achieved by inserting additional copies of the EPSP synthase gene into the plant genome, along with multiple enhancers in front of the inserted genes. Another approach is to give the plants a gene coding for a slightly different EPSP synthase, one that is resistant to glyphosate. The result

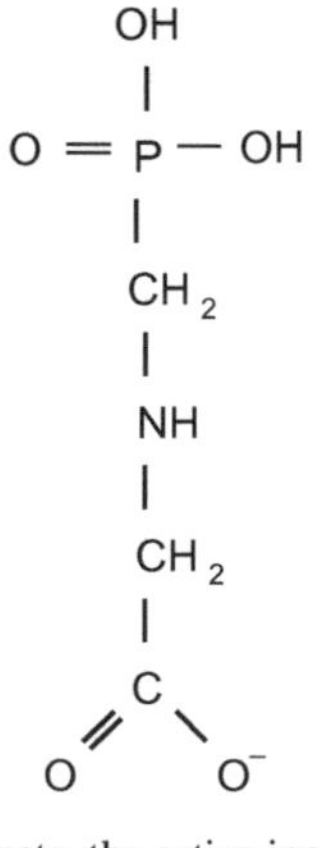

$$
\begin{array}{c}
\text{OH} \\
| \\
\text{O} = \text{P} - \text{OH} \\
| \\
\text{CH}_2 \\
| \\
\text{NH} \\
| \\
\text{CH}_2 \\
| \\
\text{C} \\
{}^{\displaystyle /\!/} \quad {}^{\displaystyle \backslash} \\
\text{O} \qquad \text{O}^-
\end{array}
$$

FIGURE 18.3   The structure of glyphosate, the active ingredient of Roundup™.

in this case was *Roundup Ready plants*, which were produced by Monsanto (who also produced Roundup). A specific example of a Roundup Ready plant is *soybean*, a plant that is resistant to glyphosate (Figure 18.4).

Questions have arisen as to whether the produce of Roundup Ready plants is affected by exposure of the plants to glyphosate. One line of reasoning is that herbicides and pest control agents that are sprayed onto crops must gain FDA approval prior to use with the food supply. These agents have been deemed safe, so a plant that has been exposed to these agents is supposedly safe for consumption if it has been properly washed. A counterexample to this reasoning is DDT, which is a very effective insecticide that is still used in parts of the world to control mosquitoes to combat malaria. One problem with DDT is that it is not metabolized very quickly, so prolonged consumption of food that has been sprayed with DDT can lead to a buildup of the chemical in fat cells, eventually leading to toxic effects. The presence of DDT is not limited to plants, though, as it will be present in birds that may consume some of the crops and in fish due to runoff of rainwater into rivers and lakes. These multiple points of entry into the food chain led to DDT levels that eventually adversely affected humans, leading to premature births and low birth weights. This counterexample is a vivid illustration that a governmental agency is not infallible, despite its best testing efforts. However, this is only a singular counterexample and is not intended to cause one to doubt

FIGURE 18.4   Roundup Ready™ soybeans. *Photo from http://www.flickr.com/photos/24004923@ N06/3810996291/in/set-72157622010267260/.*

every single pesticide in use today. The debate whether to eat only organic produce or not cannot be resolved, since arguments against existing pesticides and herbicides are based upon future results that have not been realized.

Keep in mind that the only difference between organic foods and regular crops is whether or not pesticides or herbicides have been used during cultivation. A problem with organic crops is that they will, by necessity, cost more to produce. Without pesticides, there will be increased crop loss due to pests. The monetary amount of crop loss is expected to exceed the amount of money that would have been spent on pesticides. As a result, the price of the produce will be inherently higher if the grower is to break even or realize a profit.

18.4 TOMATOES

18.4.1 The Flavr Savr Tomato

Most adults like tomatoes, and there's nothing quite like a homegrown tomato picked ripe. The problem is that finding a fresh, ripe tomato at the grocery store is very difficult. The reason for that is that tomatoes sold at most grocery stores are mass-produced on farms and then shipped to long distances. Ripe tomatoes are kind of squishy, and if they are loaded into crates and transported hundreds of miles in trucks, then these squishy, fragile tomatoes will be bruised, ruptured, or completely squashed during the trip. To combat this problem, mass-produced tomatoes are picked before they are ripe because green tomatoes are hard, so they can be packed into crates and shipped. Prior to being stocked on a grocery store shelf, they will be exposed to ethylene gas to finish the ripening process on site. Over time, consumers have become used to mediocre tomatoes and are generally satisfied with what is offered in the grocery store, that is to say, until they have the good fortune to eat a homegrown tomato, after which they wonder why they can't get such good tomatoes in the store. Researchers at Calgene decided that, if they used *antisense technology* (a gene therapy method we will cover very soon), they could target and suppress the gene coding for polygalacturonase (PG). PG is an enzyme that digests proteins in the cell wall. The idea was that by *knocking down* expression of this gene (i.e., causing a lower expression of the gene), there would be a reduction in protein digestion in the cell wall. It is this specific type of protein digestion that is responsible for the softening of tomatoes during ripening. The hypothesis was that all of the ripening processes would still occur to yield a great tasting tomato that was still firm enough for shipping.

Box GM.1: As an Aside, Let Us Briefly Look at Antisense Technology

We already know from the central dogma that the transcription of DNA produces messenger RNA that can then be translated into a protein. Antisense technology utilizes the insertion of a new "gene" into the genome, complete with a promoter and an "exon." The RNA product of transcription of this inserted sequence will be complementary (antisense) to the mRNA sequence of the targeted gene. The result will be a double-stranded RNA sequence that will not undergo translation, but rather will be degraded by the cell. (If you have read Chapter 14 "RNAi," in this book, then you probably have a better idea of how this technology works than the Calgene inventors did in 1987, when they were producing their first genetically modified tomato plants with suppressed PG.)

There was one problem with the scenario: the product didn't meet the expectations. The idea was good, but the product wasn't satisfactory because the company went about the process in the wrong way. First of all, the tomatoes were bland. Second, the tomatoes still bruised easily. So, now, we have a tomato that still can't be shipped when it's ripe, and even it could be shipped, it didn't taste good. Third, the tomatoes cost about twice as much to produce because the plants yielded about half as many tomatoes per vine versus normal tomato plants. Half the tomatoes mean double the cost per fruit.

One reason for the lower production numbers per vine was due to logistics. Calgene was a California company, and research and development on the Flavr Savr tomatoes took place in California. However, when the project progressed to mass production, the fields used for the scale-up were located in Florida. The environment at the Florida farms was more humid, and the soil was sandier (Figure 18.5). Sandy

FIGURE 18.5   Not all dirt is the same. Shown here are samples of soil from (top) California and (bottom) Florida.

soil caused the root systems to develop differently so that the plants were not as hardy. The higher humidity in Florida was responsible for a greater amount of fungus in the air. Fungal infections were a problem for the Flavr Savr crops, which again raised the cost of production.

More problems for the product stemmed from opinions. Around this time, American attitudes toward genetically modified foods were not favorable. Public perception was that if somebody ate one of these "Frankenfoods," terrible things could happen to the consumer.

At this point, we have a product that doesn't taste as good as what is already on the shelf, costs more to produce, and the public doesn't accept in the first place. Monsanto eventually bought out the company. There were many legal disputes between Calgene and Monsanto regarding the patent rights. Monsanto claimed that they had patent protection for the process of genetically modifying plants, while Calgene claimed that that protection only applied to what Monsanto had actually produced as opposed to all genetically modified plants. Calgene argued that they were suppressing PG, which was not covered by Monsanto's patent. Of course, the matter went to court, and nothing goes through the courts quickly. So while Calgene was paying massive legal bills to try to fight Monsanto, their product was not doing very well in the marketplace. The death blow came when Monsanto bought Calgene, Inc., and the Flavr Savr tomato was no longer pursued.

That, however, is not the end of the story for genetically modified tomatoes ….

18.4.2   Safeway Double-Concentrated Tomato Puree

There was another company called Zeneca, based in the United Kingdom, that worked with GM tomatoes. (Zeneca went on to become AstraZeneca.) The Zeneca product found greater success than did the Calgene tomatoes in part because of better business decisions, plus a public perception of GM foods that was more favorable in Great Britain than in America.

Zeneca aimed their tomato project at developing a bulkier product. The ripening process was not to be altered. The tomato had a lower water content, so by its very nature, the tomato was firmer and therefore easier to ship whether green or ripe. The lower water content also meant that the product had greater viscosity, which was more suited for purees and soups. The tomatoes in fact tasted good and did make it to the market.

Business acumen came into play through marketing. Zeneca was aware that the consumer who buys a tomato from the store expects a certain product, and these tomatoes weren't that product. However, the consumer who buys puree just expects something red and gloppy that tastes good. The GM tomatoes were therefore used for sauces and purees.

The tomatoes were grown and developed in California around 1994. By 1996, they were sold at certain locations of Safeway and Sainsbury's, two grocery store chains in the United Kingdom (Figure 18.6). At Safeway, they were

FIGURE 18.6   Cans of puree from genetically modified tomatoes were sold in the United Kingdom under the Safeway and Sainsbury's monikers. *Photo credit: Adrian Dubock, from http://blogs.kqed. org/pressroom/whats-in-your-next-meal/tomato-paste-made-with-ge-tomatoes-in-the-mid-1990s/.*

marketed as the Safeway Double-Concentrated Tomato Puree. It was a more viscous product so there were lower production costs. Normally, to make a tomato puree, the tomatoes are picked and chopped up, and then they must be stewed and reduced, which means that they are simmered to boil away some of the water content before packaging. Since the Zeneca product had a lower water content in the first place, the amount of boiling time was reduced, so the amount of energy needed for processing was also reduced. These lower costs presented an opportunity to the company: the puree could be marketed in the same-sized cans as the competitors' products but at a cheaper price, or it could be sold in larger cans at the same price. The company realized that offering the product as a store brand with a cheaper price might create the perception that the product was of lower quality. However, by making the can larger but charging the price already accepted by the public, the company could create the perception of great value.

By 1999, the Safeway GM tomato puree had 60% of the market share. While this sounds like a great success, consider the following question: if as recently as 1999, the Safeway product possessed 60% of its market share in the United Kingdom, then why are we all not eating GM tomato puree now? The answer is that something went wrong. First of all, there was a problem with public perception. The grocery stores did not try to hide the fact that they were selling a genetically modified product. In fact, customers visiting the store were presented with a flyer that touted the new product. "Hey, we've got genetically modified foods that taste great; this is science at its best. It tastes great *and* it is a better value. You should try some today!" People did try it and realized that it was a quality product with better value. However, around that time, news of the Flavr Savr came out. The Flavr Savr was an example of a GM tomato product that didn't taste good and failed in America. Because of the Flavr Savr, public opinion changed: "I've heard about these tomatoes. They're awful!" Public perception dictated that sales would go down. At that point, the market share for Safeway Double-Concentrated Tomato Puree went from 60% to just 25% better than normal tomato purees. It was still very good that the GM product sales were outpacing the traditional products, but keep in mind that the GM tomatoes were grown on small farms, while their competitors came from large conglomerates, so a large competitive advantage was required if the small-farm product was going to compete successfully. The smaller advantage that existed after the Flavr Savr debacle represented a turning point for both the product and the company.

Around the same time, not only had public perception of GM foods started to change in the United Kingdom, but also public perception of store brands turned to regard such products as having lower quality. Since these tomatoes were being sold under the Safeway moniker, people started to assume that they were inferior. This further contributed to decreasing market share. Because of these two unfortunate events, neither of which had anything to do with the reality of the fine tomato product in the cans, not enough profit was generated to sustain the project. Public perception changes all the time, and something that has nothing to do with a given product can bring down that product or the entire company associated with it.

A moral to this story is that if you're forming a new biotech company, make sure that you invest in a good business staff. You might perform the best science in the world, but if you can't get people to buy it or the product, the company will fail. Some would suggest that over 50% of your budget should be spent on marketing and sales.

18.5   RICE

Rice is a good focus for genetic modification because over 3,000,000,000 people in the world use rice as a staple. If one wanted to affect world food supplies, modifying rice is a great way to go about it.

18.5.1   Miracle Rice

In 1960, the International Rice Research Institute (IRRI) was set up in the Philippines. A team of agroscientists worked with and trained other plant breeders from around the world for research into the improvement of rice crops. The IRRI team recognized that rice had the same problems as wheat with "lodging" or falling over when heavy seed-bearing heads were supported by tall, thin, weak stalks. So, they imported the seeds of over 10,000 rice varieties from throughout the world. Many of these were dwarf varieties to be used in crosses with taller varieties. They also used a technique known as *shuttle breeding*, where a crop is grown in one region during the winter and in another region during the summer, effectively cutting the time it took to develop new varieties in half.

In 1962, 38 crosses of various rice plants were made at the IRRI. The eighth cross was between a dwarf variety known as Dee-geo-woo-gen (DGWG), a high-yield, short-statured strain from Taiwan, and Peta, a tall variety from Indonesia known for high vigor, seed dormancy, and resistance to several insects and diseases. This eighth cross turned out to have a great potential. Unfortunately, only 130 seeds were produced by the cross. The way that those 130 seeds ultimately yielded the famous *IR8* variety was a noteworthy endeavor:

- First, the 130 seeds were planted in pots and grown in a screenhouse to produce the first generation (F1) of plants. They were all tall.
- Seeds from the F1 plants were planted in a field to produce about 10,000 second-generation (F2) plants. One-quarter of those plants were dwarf. That meant that dwarfism was a recessive trait controlled by a single gene from the DGWG parent, making the job of producing a commercial variety much easier. Dr. Jennings was so excited that he cabled the good news to Beachell in Texas. "That's when we knew we had it!" Beachell recalled later. It was so exciting that IRRI was able to recruit Beachell to join them in 1963 as Dr. Jennings left to pursue other studies.
- All of the tall plants from the F2 generation were discarded, and seeds from the short, early maturing plants were planted in a nursery exposing plants to the fungus responsible for the plant disease known as Rice Blast. With highly susceptible plants being removed, the remaining plants were kept as the F3 generation.
- From the F3 plants, individual plants were examined and the best ones selected for further cultivation – there were 298 of them in total. Seeds from each of those plants were planted in individual "pedigree rows" in the blast nursery to produce the fourth (F4) generation. Once again, the blast-susceptible progeny were discarded.
- In that F4 generation, in the 288th row, the third plant was selected for further propagation. Using the IRRI's numbering system, the plant was designated as IR8-288-3. This F5 plant became the source for the revolutionary variety that became known simply as IR8.

(From: Ganzel (2007) and Chandler (1982))

The IR8 seed line eventually produced plants that were 100-120 cm tall, with strong stems that supported large heads. The plants needed only 130 days to mature, as opposed to 160-170 days for traditional varieties of rice. "On IRRI's experimental farm in 1965, 23 varieties or lines were placed in yield trials in January and June. Three selections from IRRI's eighth cross led the list in yield. The highest average yields (kg/ha) for the two plantings were 6104 for IR8-246, 6060 for IR8-288-3, and 6047 for IR8-36." "In the 1966 international yield trials, IR8-288-3 performed even more spectacularly than it had done in the more limited 1965 trials. The selection yielded 7034 kg/ha at CRRI in India and 7753 kg/ha at the All India Coordinated Rice Improvement Project at Hyderabad. Per-hectare yields of IR8-288-3 in other countries were 6600 kg in Malaysia, 8000 kg in Mexico, 6710-8200 kg in Bangladesh (at 3 sites), 10,248 kg in Pakistan (Dokri Station), and 6031 kg in Thailand" (Figure 18.7) (*Chandler (1982)*).

IRRI scientists were interested in more than just days to maturity, yield, and degree of lodging. They also observed incidence of disease, seed dormancy, milled rice yields, and the gelatinization temperature and amylose content of the rice starch. It was not enough to produce tons of rice; the product had to have properties that millers could work with and consumers would accept.

Early IR8 was not perfect. It had a chalky grain that detracted from the market appearance of the polished rice. There was substantial breakage during the milling process. In addition, the amylose content of the starch was too high for many Asian consumers. Not that Asian consumers were measuring amylose content in their starch, but rather, a high amylose content would harden after cooking, preventing the nice, soft gel consistency of common dishes such as mochi, puto, and tteok. Moreover, IR8 was susceptible to bacterial blight and to some types of the rice blast disease. However, much effort was later put into improving the quality of the IR8 grain through the removal of the chalkiness of the grain and high amylose content of the starch. The IRRI achieved success on these fronts.

In 1966, a large portion of Asia was going through a period of drought, which could have led to widespread famine. The IRRI responded by making IR8 freely available to the world. Over 2300 farmers came to the IRRI by any means at

FIGURE 18.7 Two rice fields of everyday farmers in the Philippines (1966). The one on the left is a crop of IR8, while the field on the right was planted with the (then) more traditional Intan. The taller crop, which is lodging somewhat in the picture, yielded only half as much rice as the IR8. In 1967, both farmers planted IR8. This scenario demonstrates how IR8 spread so quickly across the country. *From Chandler (1982).*

their disposal to get 2 kg bags of seed (Figure 18.8). The people took buses, rode bicycles, or even walked many miles to get starters for new crops that would ultimately save the lives of many thousands of people throughout Southeast Asia.

IR8, the product from the 1960s, was eventually known as the Miracle Rice. The name came from a newspaper headline in Manila that read "MARCOS GETS MIRACLE RICE," referring to the Philippine President Ferdinand Marcos meeting with IRRI officials and receiving IR8-288-3. The product was very successful, producing several times the yield of traditional rice with a one-month-shorter growing time. The strain was so successful that it was planted throughout Southeast Asia. The problem with the crop's overwhelming success was that a *monoculture* was being established—all of the plants were of the same strain. This lack of genetic diversity made the planet vulnerable to some kind of peril, be it an attack from a virus, microbe, or insect or exposure to a particular adverse weather condition. For example, an insect known as the brown plant hopper developed quite a taste for the plants. When the pests set up residence, farmers had to either combat the infestation or go hungry. As a result, an entire area of the world became dependent upon pesticides. People eventually drifted away from planting the Miracle Rice.

18.5.2 Golden Rice

In 2013, the World Health Organization estimated that 250,000,000 pre-school children suffer from vitamin A (retinoic acid) deficiency. Beta-carotene is a precursor of vitamin A (Figure 18.9) and can be found in foods such as

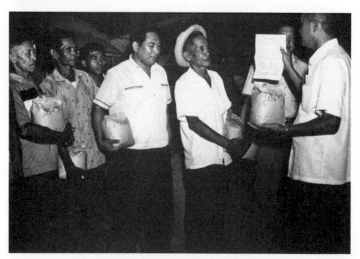

FIGURE 18.8   IR8 spread rapidly in the Philippines because IRRI gave away 2 kg of seed to any farmer who would come and pick it up. In this photo, farmers having received the seed are told of the planting instructions that were included with each bag of seed. *From Chandler (1982).*

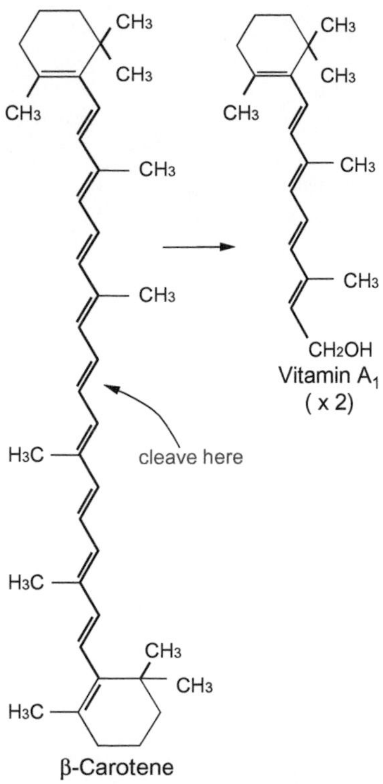

FIGURE 18.9   The cleavage of β-carotene yields two molecules of vitamin A.

carrots, pumpkins, and sweet potatoes and nonorange foods like spinach and kale. Vitamin A deficiency can be much more serious than simply making one have trouble seeing at night—severe cases can be fatal. There are an estimated 250,000-500,000 children who lose their sight each year from vitamin A deficiency, and half of them will die within one year of going blind. Since so many people consume rice, especially in the countries that are afflicted with rampant vitamin A deficiency, rice has been chosen to combat the problem via genetic modification.

To create the first generation of golden rice, gene delivery was used to introduce an entire biosynthetic pathway into the plant cells using promoters that are specific to the endosperm (Figure 18.10). The endosperm was chosen because it is the part of the rice grain that is eaten, and the endosperm plastids are the sites where geranylgeranyl diphosphate is formed. The genes inserted into the rice genome, originating from daffodil and the bacterium *Erwinia uredovora*, caused the cells to produce enzymes that convert geranylgeranyl diphosphate into lycopene. The plant's native enzymes are able to convert lycopene into beta-carotene, and the beta-carotene is converted to retinol (vitamin A) in the animal gut (Figure 18.11).

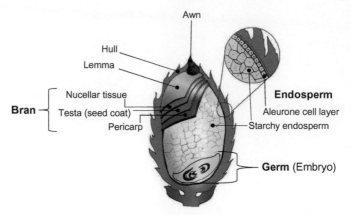

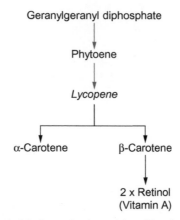

FIGURE 18.11   A biochemical pathway that is used in golden rice to get from geranylgeranyl diphosphate to β-carotene. Bioengineered steps utilizing genes inserted into the rice genome in the pathway are indicated with red arrows.

A second generation of golden rice replaced the daffodil gene (*psy*, encoding phytoene synthase) with the *psy* gene from maize to yield a plant that produced up to 23 times more carotenoids (which include α- and β-carotenes, β-cryptoxanthin, zeaxanthin, and lutein) and preferentially produced α- and β-carotenes over the other carotenoids.

Golden rice received regulatory approval for field trials in the Philippines in 2012. It is expected to receive approval in Bangladesh in 2015, according to Ingo Potrykus, a retired geneticist at the Institute of Plant Sciences in Zurich, Switzerland, and one of the rice's inventors.

The aim of golden rice is to attack the vitamin A deficiency problem via having the rice produce a precursor to vitamin A. One problem is related to patent issues. Once again, it's not the science that has restricted the progress of this technology to the consumer—it's one of many societal issues. With the Safeway tomato puree, it was public perception; with the Flavr Savr tomato, it was legal issues; and with golden rice, the problem has been patent issues again.

A company that creates a new product needs to recoup its development costs. It takes a lot of money to develop a new product, whether it is a drug, a device, or a GM food. For a GM food such as rice, while the company must recoup its investment, the product is nevertheless intended for developing countries that can't afford to pay exorbitant sums for the product. Ethical issues thus arise: how can poor or developing countries be given access to the technology that cost so much to develop?

If you go into biotech with the idea that you're going to save a large number of people of the world, keep in mind whom you are you trying to save and whether they would have the means to pay for whatever you're developing. Why don't we develop a solution and just give it to the consumer? Again, keep in mind that you must recoup enough money to keep your company running so it can continue to do good work, and money is also needed to pay back the investors who gave you the funds you used to develop your product in the first place. A serious issue can arise when the people you're servicing do not have the money to pay for your product. Using golden rice as an example, if the residents of the target countries had enough money to pay for the development of the product, they could have bought vitamin A pills or had a better diet in the first place.

18.6   TERMINATORS AND TRAITORS

There are ways that companies try to get around the issue of getting payment for a newly developed product. Suppose that Suki grows a special strain of tomato that is exceedingly good and easily shipped. Now, suppose that she sells a couple of them to Rachel. Rachel eats one and says, "Wow, these really are great!" So, she plants seeds from the other tomato and grows her own vines, eventually starting her own company to sell these tomatoes that are really the product of Suki and her company. One might argue that there is a problem if Rachel forms a company to sell tomatoes developed by Suki. However, what if Rachel simply continues to grow the tomatoes for her own use without selling them? What if she grows enough to give to all of her neighbors? After all, she purchased the tomatoes and everything in them, including the seeds, so why not grow them? Agriculture companies are aware of this potential situation and have developed a couple of ways to deal with the potential loss of profits.

18.6.1   Terminators

Terminator technology is the term used for engineered plants that produce seeds that cannot be replicated beyond the F1 generation. Suppose that you are a farmer who wants to grow some of these great new GM plants that are resistant to all pests because you would not have to spray your crops. The marketing potential is great because people like the idea of eating food that has never been exposed to pesticides. So, you purchase seeds from the biotech company, plant them, and later harvest a great crop. You sell the crop and make a good profit, so you (of course) want to produce the same crop the next year. However, the company that produced the original seeds used terminator technology, so after you get your harvest, you cannot then go and plant some of the seeds from the plants you just grew—you have to go back to the company and buy more seeds and must do so every single year.

Terminator technology is great for the inventing company because the farmer must repurchase seeds every year, but it is not good for the farmer. Keep in mind that many farmers around the planet are very poor and cannot bear the added expense of repurchasing the seeds every single year. They grow crops for their own survival, and hopefully, they can take enough to market to help pay for farm upkeep plus all of the other expenses they will incur over the coming year. These people often cannot afford to go back to the company and buy seeds every year.

One of the fears associated with the terminator technology is "what if the molecular principle you used to render the seeds sterile spreads to other crops, perhaps by a virus or a pest?" This could potentially render all other crops in the country sterile, so that after this season, we will not be able to grow anymore food because we have no more viable seeds. Even if we have seeds that were stored in a seed bank, it would just be a matter of time until plants from those seeds would also be infected, so the agriculture industry would be completely shut down. While this is not a likely scenario, it is still a concern that is associated with the technology.

18.6.2   Traitors

Traitor technology, which is also called *GURT* (an acronym for genetic use restriction technology), involves a genetic switch that is turned on and off by the use of a chemical additive. For instance, a company might sell you seeds for a crop that will express a gene that kills caterpillars, but it only expresses the gene if you spray the crop with the company's magic potion. The result is that you can grow your crops and you won't have to buy seeds every year, but you do have to keep coming back to the company for the spray.

There are two problems with the traitor technology. One is the same problem as already stated for terminator technology: poor farmers will be tied to the company in that they will have to pay the company every year to be able to grow these crops. The other problem, this time a problem for the company, is that the spray can be analyzed and reverse-engineered by a third-party who can then manufacture and sell the spray at a greatly reduced price. Suppose that you own the biotech company and that you sell the spray for $10 per unit to try to recoup your research and development costs. The third-party company might be able to sell the same spray for $5 per unit because they do not have the development costs that you put into your initial product. Eventually, people start to figure out that this second spray works just as well for half the price and will stop purchasing your spray. Who wins in this situation? Is it the farmer, because he has found out a way to pay half as much for the spray? Does the third-party company win because it is stealing money from the original company that developed the technology? Perhaps, no one wins because the poor farmer is being forced to pay either $10 here or $5 there, when he used to pay nothing; at the same time, the original company is unable to recoup its original investment costs, so it goes bankrupt and no longer is able to invent great products or help humankind.

QUESTIONS

1. Why are the words "gene" and "exon" in quotation marks in Box GM.1 "As an Aside, Let Us Briefly Look at Antisense Technology"?
2. To what does the "8" in "IR8" refer?
3. Give three advantages of the Miracle Rice over traditional rice, such as Peta.
4. Why would a biotechnologist care about the amylose content in the rice he studies?
5. If the Miracle Rice was such a superior crop plant, why is it not being used today?
6. The Safeway Double-Concentrated Tomato Puree (circle all that are true)
   a. was based on the concept that a tomato with less water content would be easier to ship when ripe and also that a tomato with less water content would lower production costs because less reduction (water removal) had to take place to produce a puree.
   b. never caught on in England because the Flavr Savr tomatoes in the United States tasted so poorly.
   c. was a great hit in the United Kingdom until people found out that they were genetically modified.
   d. was marketed as a marvel of genetic engineering and as an example of how science could yield great values for the consumer.
7. Circle all of the following that are true for Bt plants:
   a. Bt plants are Roundup Ready.
   b. Bt is an acronym for "Better than," meaning they have been genetically modified to produce larger, tastier produce than their nongenetically modified counterparts.
   c. Bt plants such as soybeans are resistant to the actions of glyphosate because several copies of the gene for EPSP synthase have been transferred into the plant genome.
   d. Bt plants contain a specific gene from a bacterium that prevents the formation of ice crystals on the plant during light frosts.

8. "Golden rice" gets its name from
   a. β-carotene production.
   b. the high profit gleaned from the sale of this engineered food.
   c. the fact that it uses urea, a nitrogen-containing molecule found in urine (pee), as a nitrogen source.
   d. engineered production of B vitamins, which turn the pee bright yellow.

RELATED READING

Alibhai, M.F., Stallings, W.C., 2001. Closing down on glyphosate inhibition—with a new structure for drug discovery. Proc. Natl. Acad. Sci. U. S. A. 98, 2944–2946.

Amrhein, N., Deus, B., Gehrke, P., Steinrücken, H.C., 1980. The site of the inhibition of the shikimate pathway by glyphosate: II. Interference of glyphosate with chorismate formation in vivo and in vitro. Plant Physiol. 66, 830–834.

"Annex IV": approval registry for field testing of regulated articles. Bureau of Plant Industry, Department of Agriculture, Republic of Philippines. http://biotech.da.gov.ph/upload/annexIV.pdf.

Chandler Jr., R.F., 1982. An Adventure in Applied Science. A History of the International Rice Research Institute, Los Baños/Laguna/Philippines.

Galli-Taliadoros, L.A., Sedgwick, J.D., Wood, S.A., Körner, H., 1995. Gene knock-out technology: a methodological overview for the interested novice. J. Immunol. Methods 181, 1–15.

Ganzel, B., 2007. Farming in the 1950s and 1960s: the green revolution: "Miracle Rice". Wessels Living History Farm, York, Nebraska, U.S.A. http://www.livinghistoryfarm.org/farminginthe50s/crops_17.html (accessed 02.2014.)

Gruys, K.J., Sikorski, J.A., 1999. Inhibitors of Tryptophan, Phenylalanine and Tyrosine Biosynthesis as Herbicides. Dekker, New York.

Haslam, E., 1993. Shikimic Acid: Metabolism and Metabolites. Wiley, New York.

Nayar, A., 2011. Grants aim to fight malnutrition. Nature http://dx.doi.org/10.1038/news.2011.233. http://www.nature.com/news/2011/110414/full/news.2011.233.html.

Paine, J.A., Shipton, C.A., Chaggar, S., Howells, R.M., Kennedy, M.J., Vernon, G., Wright, S.Y., Hinchliffe, E., Adams, J.L., Silverstone, A.L., Drake, R., 2005. Improving the nutritional value of Golden Rice through increased pro-vitamin A content. Nat. Biotechnol. 23, 482–487.

Suszkiw, J., 1999. Tifton, Georgia: a peanut pest showdown. Agric. Res. Magaz. 47, 9.

Ye, X., Al-Babili, S., Klöti, A., Zhang, J., Lucca, P., Beyer, P., Potrykus, I., 2000. Engineering the provitamin A (beta-carotene) biosynthetic pathway into (carotenoid-free) rice endosperm. Science 287, 303–305.

Zhao, J.Z., Cao, J., Li, Y., Collins, H.L., Roush, R.T., Earle, E.D., Shelton, A.M., 2003. Transgenic plants expressing two Bacillus thuringiensis toxins delay insect resistance evolution. Nat. Biotechnol. 21, 1493–1497.

Chapter 19

Patents and Licenses

(Special thanks to John Christie, Executive Director of Technology Transfer, Tulane University)

19.1 TYPES OF PATENTS

Simply holding a patent is not enough to make you rich.

Getting a technology to the market fits with the Edison equation: the process is 2% inspiration and 98% perspiration. A patent application is not limited to the 2%, but it's not the entirety of the 98%, either. Getting to the point of filing a patent application is a good start, but it's not what actually gets a technology to the market (Figure 19.1). A patent is the legal cover for something that you have created of value. It is the deed to your intellectual property, but it's up to you to improve upon it. Just as you may buy some property in the form of land and you may decide to build a house, a parking lot, or a race track on the land, it is also up to you to decide what to do with some intellectual property you have acquired the deed to through patent protection.

The protections that we most commonly deal with in the biotech community are *composition of matter* patents. A composition of matter is basically a chemistry patent on a unique molecule. It describes the molecule, including the structure that is going to be used in your product (e.g., $C_9H_{15}N_5O$ for Rogaine or $C_{33}H_{34}FN_2O_5$ for Lipitor—Figure 19.2). The patent will state that you have the molecule; what the molecule looks like; that the molecule is original; that you either discovered, found, or created it; and that you make certain claims regarding that molecule. You can also have method claims, which include methods of using the molecule to do something (such as grow hair or lower blood cholesterol levels). These are *method of use* claims. Such claims often originally appear in the same patent application as a composition of matter patent because, even with a novel composition of matter, your invention has to be good for something. It is not enough just to say, "I have this original molecule ..."; you must state, "I have this molecule, and it does *A*, *B*, and *C*." So you claim some uses in the original patent application, but some uses of a molecule—and Rogaine is a good example—will be claimed later. The Rogaine patent was initially a composition of matter with claims for use as an antihypertensive. Some time down

An Introduction to Biotechnology. http://dx.doi.org/10.1016/B978-1-907568-28-2.00019-8
Copyright © W T Godbey, 2014.

FIGURE 19.1   Innovation is not unique to your country. Places that one could file a patent application include (top) the Intellectual Property Office of Bloomsbury, England; (middle) the United States Patent and Trademark Office in Alexandria, Virginia; and (bottom) offices of Intellectual Property, India.

the road, somebody figured out that one of these antihypertensive pills could be crumbled up and rubbed on some bald guy's head and he might grow some hair back. So then, somebody else, probably at Lilly (the company at which this work was performed), filed a new patent application claiming a method of using the old composition of matter for an entirely new use. If the new use had been

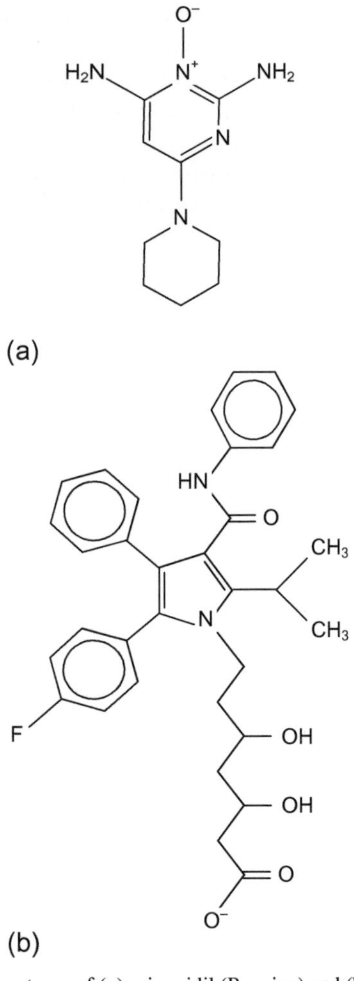

(a)

(b)

FIGURE 19.2 Chemical structures of (a) minoxidil (Rogaine) and (b) atorvastatin (Lipitor).

something that was closely related to hypertension, the company probably could not have patented it because it would not have been novel: it would have been *obvious*. (Patent claims must be "nonobvious.") However, going from usage as an antihypertensive to the realm of hair restoration is far enough away that the new use was freshly patentable.

Another broad category of patents is *medical devices*, which is also a big part of biotechnology, especially as other forms of technology advance. Think of imaging and being able to steer probes and catheters through the body in real time. Luminescent chemicals and such can be thought of more as devices than as treatments. They could serve as diagnostics by providing a new way of getting inside the body for imaging. Alternatively, they might allow a surgeon to find a location for excision. Notice the fine distinction that these examples are not medical treatments. These devices also have broadly different rules for getting a claim approved versus a composition of matter claim. Medical device claims are far more challenging: with a composition of matter claim, you must somehow ultimately prove to a patent examiner that your material is different from the materials that have come before, but devices are almost always built on existing technologies. Devices have much more of a history of "it's like this existing device, only different," so device claims are difficult to prepare and defend. A good legal counsel is needed to write the patent application to distinguish the device from all others. As with a composition of matter application, medical device patent applications also require method of use claims. It is not enough to say that you have an original device; the device must be good in something before patent protection can be awarded.

For example, at one time, Tulane University had a certain medical device that was licensed and was in development. It was an oxygen-sensing catheter that allowed the physician to go through the body to try to find a spot where the oxygen level was at a certain point. It was a combination of two known technologies: an oximeter and a known steerable catheter, to which the oximeter was attached. The patent examiner agreed that Tulane University had come up with a new use for the technologies, but he did not agree that Tulane University had created a new device. Think in terms that anyone can take two existing products off the shelf and tape them together. Tulane University was awarded protection on the method of use claims, but the patent office would not allow protection of the device itself because it was not sufficiently novel.

Aside

Were any secrets revealed in the previous example? The answer is "No," by definition. "Patent" means "to make clear." The philosophy behind a patent is that the inventor announces a new compound, device, or use to the world and explains it so that somebody else in the field could reasonably expect to replicate the invention or use. In exchange for making the technology available for benefit of the rest of society, the inventor(s) is given legal protection to ensure that nobody else can make money off of the discovery without the inventor's approval, for a period of time. The fact that a patent was awarded means that the details are publicly available. One could look up this or any other patent and read the history if so desired.

19.2   LICENSES

After one gets patent protection, the next step, if you want to gain some kind of monetary recognition, is to license the technology. That's where the money is: when a company licenses the rights to use your technology that is protected by a patent.

The above idea is correct. (We're confining this discussion to the biotechnology industry.) Once you have been awarded patent protection, you can find a company that is already in the space and approach them about taking the license to your intellectual property. The patent doesn't necessarily even have to be granted yet. You can get license deals through pending patent applications. Usually, the licensee—the person taking the license—will want to put safeguards in the agreement stating that, if the patent fails to be issued, certain things will happen, because the patent process can take years to get through these days. Preaward licensure is becoming more common because the process is so long and the life of patent protection begins on the day you file for it (not the day it is granted). Your patent life starts to dwindle the day you do the right thing and submit it to the United States Patent and Trademark Office. (Office names vary by country.) Filing starts your 20-year clock. Because of this, the earlier you can find a licensee, the better (if you are actually interested in licensing the technology to an outside party). Potential licensees understand this. Agreements have to be made regarding what will happen if the inventor never gets a valid claim issued on the patent application. Suppose that you have a medical device and it's going to do all sorts of things. There is more patentability in this case, and you will have a market edge just by having a functioning prototype. However, if you're a company, you're basically licensing the technology based on vapor because nobody in the patent office has taken a look at the application and probably will not do so for another 6 months.

Right now, the United States and other patent authorities in the world are aligned in terms of patent life, which is 20 years from the date of filing. There are a couple of ways to extend patent life, but they are usually minimal, perhaps being granted to make up for the extra time that was taken in issuing the original patent, if you can show that the process took longer than normal.

The time will not be significant enough to make a material difference. For instance, the life will not be extended for an additional 10 years. As a result, it is important to find licensing opportunities as early as you possibly can.

Start talking to companies not only because they might license your technology but also because they will give you feedback. The way that you talk to these companies almost always entails *confidentiality agreements* so that you can show them everything you've got and the two parties can talk about it (Figure 19.3). A good confidentiality agreement is usually not very onerous to either party, and such agreements are used as a routine manner. But they are very important, especially when dealing with patentable matter, and you want the two parties to be able to talk freely. Many companies will not talk to you at all without having such an agreement in place. There are different levels of going about it, but these are pretty simple agreements and are very valuable, especially for moving forward with some kind of collaboration, partnership, or transaction.

After you get a confidentiality agreement in place, which is generally a simple matter, you talk to the company and hear its feedback. Even if they're not interested in the technology, it's good to hear why. You can learn a lot from an opportunity that did not develop. Why didn't it? How can we make it better for the next pitch? Consider that you might be pitching to someone who has been in the industry for over 20 years. If they're willing to sit down with you or even just have a phone call or respond to your e-mails, pay attention to what they say.

NONDISCLOSURE AGREEMENT

This Nondisclosure Agreement (the "**Agreement**") is entered into as of the date set forth below by and between the individual or entity identified by signature below ("**Recipient**") and Company Name, Inc. ("**Company**"), having a place of business at 123 Somewhere Street, City, State.

WHEREAS Company has made discoveries or improvements relating to Technology X and/or its use for drug delivery or targeting, or its use as a medical therapy (collectively; "**The Developments**");

WHEREAS Company possesses proprietary and confidential information and materials related to The Developments;

WHEREAS Recipient desires to receive information or materials related to The Developments in order to evaluate a potential business relationship including, but not limited to, consulting, collaboration or provision of services to Company;

WHEREAS Recipient appreciates that Company has expended and continues to expend money and effort to establish a proprietary position with respect to The Developments it has made and that Company considers The Developments and information or materials pertaining thereto to be its confidential property; and

WHEREAS Company is willing to reveal or provide to Recipient information or materials relating to The Developments on a confidential basis.

NOW, THEREFORE, in consideration of the foregoing, and of the mutual covenants, terms and conditions hereinafter expressed, Recipient and Company agree as follows:

1. Confidential Information. "**Confidential Information**" means any information disclosed to Recipient by Company, either directly or indirectly in writing, orally or by inspection of tangible objects, ...

2. Non-use and Nondisclosure. Recipient agrees not to use any Confidential Information for any purpose except ... Recipient agrees not to disclose any Confidential Information to any third parties ...

3. Maintenance of Confidentiality. Recipient agrees that it shall take all reasonable measures to protect the secrecy of and avoid disclosure and unauthorized use of the Confidential Information. ...

FIGURE 19.3   A sample from a confidentiality agreement, also known as a nondisclosure agreement.

It's tremendously useful to hear the insights of experienced people who are already established in the trade. Then, hopefully, during the next conversation, you will be better informed and will be able to address the issues that were raised during the first set of conversations.

So now, you have found a licensee and if it looks serious, if it looks like there may be a license there, you can write up an agreement. The Internet is fabulous for finding things like template agreements or what royalty rates should be used in a given sector (e.g., rates in pharmaceuticals are different from those in diagnostics and rates in vaccines are different from those in devices). You can further break down devices into subclasses: is it an orthopedic device, a surgical device, etc.? You can perform a great deal of research on the web to get an idea of what a ballpark deal should look like.

19.3   AFTER A LICENSE IS GRANTED

19.3.1   The Inventor Will Work with the Licensee After the Patent Has Been Licensed

From the point of view of a university technology transfer office, it would be impossible to license technologies if the inventor were not a participant in a future work with the technology at some level. The inventor knows more about the invention than anybody else on earth. Keep in mind that inventors do not come up with patentable inventions every 5 min; they spent some huge proportion of their lives thinking about it every waking moment and even some sleeping and dreaming moments. They pour their hearts and souls into the invention because it's something they are passionate about. They're not just looking for some new flavor of toothpaste. They might have a medical condition that affects themselves or somebody they know. You've got to have the inventors involved in these things because they know the area very well. Almost every licensee will want to have the involvement of the inventor, as opposed to a clean hand off. As an inventor, you work hard to get the patent, you work hard to get the license, but if the license is granted on January 1, it does not mean you are finished on January 2. The license agreement will include access to know-how (stuff you can't really quantify but it's tremendously important) or a consulting arrangement with the inventor. The inventor has done a ton of work and the company will want access to him. That will simply be part of the deal. Getting a good licensee or collaborator with experience—somebody who is going to work hard to get the invention to the market—is a tremendously valuable part in the education of the young inventor.

19.3.2   Remuneration

Once a licensure agreement has been made, things are contingent upon the eventual success of the product in the market. It's not that today, you were able *to license your invention so you get a check for $1,000,000. How is remuneration typically structured?*

There are multiple levels in a licensing agreement. You can ask for an *upfront fee*, meaning on the day that everybody signs the agreement, the company pays the inventor a lump sum of cash. You can (and should) have a *point of development plan* that identifies milestones that the licensee must meet to ensure that the invention is being aggressively developed. Recall the limited window of protection afforded by a patent. If the inventor licenses the patent to a company that does not pursue bringing the invention to the market, then the inventor would have lost out on a potentially large amount of revenue. In the case of a potential drug, milestones might include when a leading formulation candidate is identified and when toxicology studies will be performed. With clinical trials, a target milestone might be having the first patient enrolled in a phase I trial by a certain date. It can be written into the license agreement that if the company has not met these milestones, then the license comes back to the inventor, and you can put a monetary figure on it. So the company has to make a decision at each point. Will it decide to enroll that first patient into the phase I trial tomorrow and pay the inventor $100,000, or will it instead decide to terminate the license? That's the decision the company has to make: commit more resources to the project or give it back to the licensor (the inventor). If the company chooses to progress through each of the developmental points and the product goes to the market, then a steady royalty stream will be established for the inventor.

Of the licensing-related income that Tulane University has collected in the past 20 years, about 90% has come from royalties from products that were actually sold on the market. All of the other items mentioned above—upfront fees, milestone fees, etc., all of which are six- or seven-figure fees—comprise only a tiny bit of the total package. If you're looking for a payoff, it is in getting a product to the market. There are rare cases where a large upfront fee is paid, but if these products never make it to the market (and the odds dictate that they will not), then the majority of the potential revenue for the inventor is lost. A small upfront fee, combined with diligent effort between the inventor and the company, can eventually yield a royalty stream of well over $1,000,000 a year after a biotechnology product reaches the market.

The licensing structure should be focused on the possibility of success. For the inventor, getting a large upfront fee would come at the price of a reduction in control of the technology down the line. For instance, if a large upfront fee is paid, the company might not have to give the technology back if they do not develop it, or there might be a low royalty rate attached to the product if it reaches the market. A more effective long-term strategy is to have the following:

- A low upfront fee
- Relatively low (or absent) milestone fees
- Control over the development plan retained by the inventor

- A larger royalty rate
- The ability for the inventor to take the technology back if the company does not continue to pursue it

With this type of agreement, the ultimate success is measured in terms of the number of products that have made it to the market. It is also a good strategy because it motivates the inventor to remain highly involved in the development of the product. If the technology is licensed to a small company, which does not have a lot of cash, asking for large upfront fees would end up taking a large portion of their cash that they would otherwise use for the development of the product. Since the end game is to get products to the market, large upfront fees are self-defeating for both the inventor and the licensee.

19.3.3   If the License Is Released Back to the Inventor

If a product is not developed after it is licensed, the licensor (inventor) will want to get it back promptly because the patent clock is still ticking. If the licensee cannot develop the product, the inventor will want to be able to get it into the hands of somebody who can. You want to have a clean break with the licensee, which is achieved through termination language in the license agreement that spells out that the licensee will have no further rights. If you can get the licensor to turn over their data, which is another negotiation point, this will help to make the product more licensable to the next company. You want to keep a positive relationship with the licensing company throughout the development because you need to know what is going on. You will want to know why the company is not moving forward—what failed and at what point did it happen?

If a subsequent company chooses to license the technology, they will be aware that the first license went sour. The subsequent license deal will be less lucrative for the inventor, in part because (1) there will be less patent life left on the technology and (2), now, there is some doubt/risk associated with the product. It's tough to relicense something once it has failed, but if you do not have the rights, you cannot even try. However, there are some things you can do that can help your position. Perhaps, the science in related areas has improved so that the problems can be dealt with. The problems may be in formulation or how to make the drug more bioavailable. The solutions that did not exist 2-3 years ago when the first company gave the technology back may now be readily available.

QUESTIONS

1. What should be the main goal of patenting a technology? Is it to get rich?
2. What should be the main goal of licensing a technology? Is it to get rich?
3. True or false: When you license your technology to a company for a lot of money, you can reasonably consider retiring to live a stress-free life. Explain your answer.

4. Suppose that you have found a process by which you can attach a certain molecule to a constituent of butter to render the compound able to eradicate the AIDS virus (without altering the taste or texture of the butter). What type of patent would you file?

5. Suppose that you have invented a medical device that takes an IV catheter and an ultrasound device and uses them in conjunction to monitor individual cells in the bloodstream and remove cell types that fit within a programmable set of parameters, such as sickle cells or microbes. Would this be patentable as a medical device? Since ultrasound is already used to visualize blood and IV catheters are used to gain access to the blood, would the technology be patentable as a method of use?

Index

Note: Page numbers followed by *f* indicate figures, *b* indicate boxes and *t* indicate tables.